# Applied Regression Analysis in Econometrics

## STATISTICS: Textbooks and Monographs



**Vol. 1:** The Generalized Jackknife Statistic, *H. L. Gray and W. R. Schucany*
**Vol. 2:** Multivariate Analysis, *Anant M. Kshirsagar*
**Vol. 3:** Statistics and Society, *Walter T. Federer*
**Vol. 4:** Multivariate Analysis: A Selected and Abstracted Bibliography, 1957-1972, *Kocherlakota Subrahmaniam and Kathleen Subrahmaniam* (out of print)
**Vol. 5:** Design of Experiments: A Realistic Approach, *Virgil L. Anderson and Robert A. McLean*
**Vol. 6:** Statistical and Mathematical Aspects of Pollution Problems, *John W. Pratt*
**Vol. 7:** Introduction to Probability and Statistics (in two parts), Part I: Probability; Part II: Statistics, *Narayan C. Giri*
**Vol. 8:** Statistical Theory of the Analysis of Experimental Designs, *J. Ogawa*
**Vol. 9:** Statistical Techniques in Simulation (in two parts), *Jack P. C. Kleijnen*
**Vol. 10:** Data Quality Control and Editing, *Joseph I. Naus* (out of print)
**Vol. 11:** Cost of Living Index Numbers: Practice, Precision, and Theory, *Kali S. Banerjee*
**Vol. 12:** Weighing Designs: For Chemistry, Medicine, Economics, Operations Research, Statistics, *Kali S. Banerjee*
**Vol. 13:** The Search for Oil: Some Statistical Methods and Techniques, *edited by D. B. Owen*
**Vol. 14:** Sample Size Choice: Charts for Experiments with Linear Models, *Robert E. Odeh and Martin Fox*
**Vol. 15:** Statistical Methods for Engineers and Scientists, *Robert M. Bethea, Benjamin S. Duran, and Thomas L. Boullion*
**Vol. 16:** Statistical Quality Control Methods, *Irving W. Burr*
**Vol. 17:** On the History of Statistics and Probability, *edited by D. B. Owen*
**Vol. 18:** Econometrics, *Peter Schmidt*
**Vol. 19:** Sufficient Statistics: Selected Contributions, *Vasant S. Huzurbazar (edited by Anant M. Kshirsagar)*
**Vol. 20:** Handbook of Statistical Distributions, *Jagdish K. Patel, C. H. Kapadia, and D. B. Owen*
**Vol. 21:** Case Studies in Sample Design, *A. C. Rosander*
**Vol. 22:** Pocket Book of Statistical Tables, *compiled by R. E. Odeh, D. B. Owen, Z. W. Birnbaum, and L. Fisher*

Vol. 23: The Information in Contingency Tables, *D. V. Gokhale and Solomon Kullback*
Vol. 24: Statistical Analysis of Reliability and Life-Testing Models: Theory and Methods, *Lee J. Bain*
Vol. 25: Elementary Statistical Quality Control, *Irving W. Burr*
Vol. 26: An Introduction to Probability and Statistics Using BASIC, *Richard A. Groeneveld*
Vol. 27: Basic Applied Statistics, *B. L. Raktoe and J. J. Hubert*
Vol. 28: A Primer in Probability, *Kathleen Subrahmaniam*
Vol. 29: Random Processes: A First Look, *R. Syski*
Vol. 30: Regression Methods: A Tool for Data Analysis, *Rudolf J. Freund and Paul D. Minton*
Vol. 31: Randomization Tests, *Eugene S. Edgington*
Vol. 32: Tables for Normal Tolerance Limits, Sampling Plans, and Screening, *Robert E. Odeh and D. B. Owen*
Vol. 33: Statistical Computing, *William J. Kennedy, Jr. and James E. Gentle*
Vol. 34: Regression Analysis and Its Application: A Data-Oriented Approach, *Richard F. Gunst and Robert L. Mason*
Vol. 35: Scientific Strategies to Save Your Life, *I. D. J. Bross*
Vol. 36: Statistics in the Pharmaceutical Industry, *edited by C. Ralph Buncher and Jia-Yeong Tsay*
Vol. 37: Sampling from a Finite Population, *J. Hajek*
Vol. 38: Statistical Modeling Techniques, *S. S. Shapiro*
Vol. 39: Statistical Theory and Inference in Research, *T. A. Bancroft and C.-P. Han*
Vol. 40: Handbook of the Normal Distribution, *Jagdish K. Patel and Campbell B. Read*
Vol. 41: Recent Advances in Regression Methods, *Hrishikesh D. Vinod and Aman Ullah*
Vol. 42: Acceptance Sampling in Quality Control, *Edward G. Schilling*
Vol. 43: The Randomized Clinical Trial and Therapeutic Decisions, *edited by Niels Tygstrup, John M. Lachin, and Erik Juhl*
Vol. 44: Regression Analysis of Survival Data in Cancer Chemotherapy, *Walter H. Carter, Jr., Galen L. Wampler, and Donald M. Stablein*
Vol. 45: A Course in Linear Models, *Anant M. Kshirsagar*
Vol. 46: Clinical Trials: Issues and Approaches, *edited by Stanley H. Shapiro and Thomas H. Louis*
Vol. 47: Statistical Analysis of DNA Sequence Data, *edited by B. S. Weir*
Vol. 48: Nonlinear Regression Modeling: A Unified Practical Approach, *David A. Ratkowsky*
Vol. 49: Attribute Sampling Plans, Tables of Tests and Confidence Limits for Proportions, *Robert E. Odeh and D. B. Owen*
Vol. 50: Experimental Design, Statistical Models, and Genetic Statistics, *edited by Klaus Hinkelmann*
Vol. 51: Statistical Methods for Cancer Studies, *edited by Richard G. Cornell*
Vol. 52: Practical Statistical Sampling for Auditors, *Arthur J. Wilburn*
Vol. 53: Statistical Signal Processing, *edited by Edward J. Wegman and James G. Smith*
Vol. 54: Self-Organizing Methods in Modeling: GMDH Type Algorithms, *edited by Stanley J. Farlow*
Vol. 55: Applied Factorial and Fractional Designs, *Robert A. McLean and Virgil L. Anderson*
Vol. 56: Design of Experiments: Ranking and Selection, *edited by Thomas J. Santner and Ajit C. Tamhane*
Vol. 57: Statistical Methods for Engineers and Scientists. Second Edition, Revised and Expanded, *Robert M. Bethea, Benjamin S. Duran, and Thomas L. Boullion*
Vol. 58: Ensemble Modeling: Inference from Small-Scale Properties to Large-Scale Systems, *Alan E. Gelfand and Crayton C. Walker*

**Vol. 59:** Computer Modeling for Business and Industry, *Bruce L. Bowerman and Richard T. O'Connell*

**Vol. 60:** Bayesian Analysis of Linear Models, *Lyle D. Broemeling*

**Vol. 61:** Methodological Issues for Health Care Surveys, *Brenda Cox and Steven Cohen*

**Vol. 62:** Applied Regression Analysis and Experimental Design, *Richard J. Brook and Gregory C. Arnold*

**Vol. 63:** Statpal: A Statistical Package for Microcomputers – PC-DOS Version for the IBM PC and Compatibles, *Bruce J. Chalmer and David G. Whitmore*

**Vol. 64:** Statpal: A Statistical Package for Microcomputers – Apple Version for the II, II+, and IIe, *David G. Whitmore and Bruce J. Chalmer*

**Vol. 65:** Nonparametric Statistical Inference, Second Edition, Revised and Expanded, *Jean Dickinson Gibbons*

**Vol. 66:** Design and Analysis of Experiments, *Roger G. Petersen*

**Vol. 67:** Statistical Methods for Pharmaceutical Research Planning, *Sten W. Bergman and John C. Gittins*

**Vol. 68:** Goodness-of-Fit Techniques, *edited by Ralph B. D'Agostino and Michael A. Stephens*

**Vol. 69:** Statistical Methods in Discrimination Litigation, *edited by D. H. Kaye and Mikel Aickin*

**Vol. 70:** Truncated and Censored Samples from Normal Populations, *Helmut Schneider*

**Vol. 71:** Robust Inference, *M. L. Tiku, W. Y. Tan, and N. Balakrishnan*

**Vol. 72:** Statistical Image Processing and Graphics, *edited by Edward J. Wegman and Douglas J. DePriest*

**Vol. 73:** Assignment Methods in Combinatorial Data Analysis, *Lawrence J. Hubert*

**Vol. 74:** Econometrics and Structural Change, *Lyle D. Broemeling and Hiroki Tsurumi*

**Vol. 75:** Multivariate Interpretation of Clinical Laboratory Data, *Adelin Albert and Eugene K. Harris*

**Vol. 76:** Statistical Tools for Simulation Practitioners, *Jack P. C. Kleijnen*

**Vol. 77:** Randomization Tests, Second Edition, *Eugene S. Edgington*

**Vol. 78:** A Folio of Distributions: A Collection of Theoretical Quantile-Quantile Plots, *Edward B. Fowlkes*

**Vol. 79:** Applied Categorical Data Analysis, *Daniel H. Freeman, Jr.*

**Vol. 80:** Seemingly Unrelated Regression Equations Models: Estimation and Inference, *Virendra K. Srivastava and David E. A. Giles*

**Vol. 81:** Response Surfaces: Designs and Analyses, *Andre I. Khuri and John A. Cornell*

**Vol. 82:** Nonlinear Parameter Estimation: An Integrated System in BASIC, *John C. Nash and Mary Walker-Smith*

**Vol. 83:** Cancer Modeling, *edited by James R. Thompson and Barry W. Brown*

**Vol. 84:** Mixture Models: Inference and Applications to Clustering, *Geoffrey J. McLachlan and Kaye E. Basford*

**Vol. 85:** Randomized Response: Theory and Techniques, *Arijit Chaudhuri and Rahul Mukerjee*

**Vol. 86:** Biopharmaceutical Statistics for Drug Development, *edited by Karl E. Peace*

**Vol. 87:** Parts per Million Values for Estimating Quality Levels, *Robert E. Odeh and D. B. Owen*

**Vol. 88:** Lognormal Distributions: Theory and Applications, *edited by Edwin L. Crow and Kunio Shimizu*

**Vol. 89:** Properties of Estimators for the Gamma Distribution, *K. O. Bowman and L. R. Shenton*

**Vol. 90:** Spline Smoothing and Nonparametric Regression, *Randall L. Eubank*

**Vol. 91:** Linear Least Squares Computations, *R. W. Farebrother*

**Vol. 92:** Exploring Statistics, *Damaraju Raghavarao*

Vol. 93: Applied Time Series Analysis for Business and Economic Forecasting, *Sufi M. Nazem*
Vol. 94: Bayesian Analysis of Time Series and Dynamic Models, *edited by James C. Spall*
Vol. 95: The Inverse Gaussian Distribution: Theory, Methodology, and Applications, *Raj S. Chhikara and J. Leroy Folks*
Vol. 96: Parameter Estimation in Reliability and Life Span Models, *A. Clifford Cohen and Betty Jones Whitten*
Vol. 97: Pooled Cross-Sectional and Time Series Data Analysis, *Terry E. Dielman*
Vol. 98: Random Processes: A First Look, Second Edition, Revised and Expanded, *R. Syski*
Vol. 99: Generalized Poisson Distributions: Properties and Applications, *P.C. Consul*
Vol. 100: Nonlinear $L_p$-Norm Estimation, *René Gonin and Arthur H. Money*
Vol. 101: Model Discrimination for Nonlinear Regression Models, *Dale S. Borowiak*
Vol. 102: Applied Regression Analysis in Econometrics, *Howard E. Doran*
Vol. 103: Continued Fractions in Statistical Applications, *K.O. Bowman and L.R. Shenton*
Vol. 104: Statistical Methodology in the Pharmaceutical Sciences, *Donald A. Berry*

**ADDITIONAL VOLUMES IN PREPARATION**

# Applied Regression Analysis in Econometrics

HOWARD E. DORAN
The University of New England
Armidale, New South Wales
Australia



MARCEL DEKKER, INC. New York and Basel

Library of Congress Cataloging-in-Publication Data

Doran, Howard E.
Applied regression analysis in econometrics / Howard E. Doran.
p. cm. -- (Statistics, textbooks and monographs ; vol.102)
Rev. ed. of: Single equation methods in econometrics.
Bibliography: p.
Includes index.
ISBN 0-8247-8049-3 (alk. paper)
1. Econometrics. 2. Regression analysis. I. Doran, Howard E., Single equation methods in econometrics. II. Title.
III. Series: Statistics, textbooks and monographs ; v. 102.
HB139.D67 1989
330'.028--dc19 89-1195
CIP

This book is printed on acid-free paper.



MARCEL DEKKER, INC.
270 Madison Avenue, New York, New York 10016

Current printing (last digit):
10 9 8 7 6 5 4 3 2 1

PRINTED IN THE UNITED STATES OF AMERICA

# Preface

The book originated in a one-semester course that has been taught for a number of years at the University of New England. This course has catered to two types of students. The first of these have been second- and third-year undergraduates majoring in economics, agricultural economics, or financial administration. The second group has consisted of postgraduate students enrolled in Diploma and Master's programs. Almost all of these students were not training to be specialist econometricians, but were seeking to augment their major study area with some empirical skills. Because of this background their preparation in mathematics and statistics was relatively limited.

The book was written with such students in mind. Care has been exercised to avoid the trap of presenting a series of "cookbook recipes." Rather, emphasis is given to the development of an appreciation of the assumptions underlying the linear statistical model. Such an understanding then forms the basis for informed modeling of real-world situations.

It is assumed that readers have a grounding in high school algebra and have completed a course in elementary statistics at a tertiary level. Calculus and matrix algebra are, apart from a few isolated cases, avoided in the first part of the book. In the second, more advanced section, limited use of differential calculus has been made.

Even though a background in basic statistics would normally include a study of simple regression, this topic is covered in considerable detail in Chapter 2, which is one of the most important in the book. This chapter relates the classical assumptions to the properties of least squares estimators and forms an essential basis for all that follows. The multiple regression model is then introduced through the simplest case of two regressor variables.

By presenting the theoretical framework in the context of one and two regressors, all essential features of the general model can be discussed using only high school algebra. While the avoidance of matrix algebra undoubtedly results in some sacrifice of mathematical elegance and generality, such an approach does make the theory of linear regression accessible to students whose mathematical confidence and maturity are not high.

An essential feature of the book is that experience with a computing package is integrated into the general treatment. It is the author's view that a study of applied regression analysis can be made far more stimulating if real problems can be tackled. This enables the student to experiment, and the interaction between theory and practice provides an additional stimulus toward mastery of the subject.

This book is a revision and extension of an earlier version entitled Single Equation Methods in Econometrics: Applied Regression Analysis by H. E. Doran and J. W. B. Guise. It was published by the University of New England in its Teaching Monograph Series. I wish to record my indebtedness to the university for its willingness to release copyright to enable this book to be published.

Many colleagues and friends have contributed to the book. First, the fundamental contribution of the late John Guise must be stressed. Many of the basic ideas of the lecture course are his, and Chapter 10 was entirely written by him shortly before his unexpected death. Second, special thanks are due to Bill Griffiths and Prasada Rao, who have both taught the course; their input and discussions have been very helpful. Tim Coelli and Martin Mantle have willingly offered help and encouragement. Third, I record my thanks to Diane Dancer, who shouldered the administrative tasks of teaching a large class to free me to finish the manuscript. Finally, thanks are due to Sue Nano and Marlene Youman, who cheerfully and competently typed the original text.

Howard E. Doran

# Contents

# Applied Regression Analysis in Econometrics

# Part I
# BASIC REGRESSION ANALYSIS

# 1
# Introduction and Overview

## 1.1 INTRODUCTION

The study of regression analysis in an economic context is concerned with the formulation and uses of models of economic relationships. An essential aspect of this formulation is the use of data.

Let us consider an economist who is trying to explain variations in the demand for some commodity. Economic theory indicates that both the price of the commodity and the income level would probably exert some influence, so a model of the form

$$Q = f(P, Y)$$

is postulated. We have used Q to indicate the quantity demanded, P the price, and Y the income level. In addition, it would probably be assumed that increases in price would lead to decreases in demand, and that those with higher incomes would demand more than those on lower incomes. Readers who are familiar with calculus could set down these conditions concisely as $\partial Q/\partial P < 0$ and $\partial Q/\partial Y > 0$.

Such a model implies that if the function f were specified, and P and Y given, the demand Q would be determined exactly. However, we all know that human behavior is far more complex than this. Any group of individuals having precisely the same income and facing the same price would exhibit variation in quantities demanded. This variation would reflect different tastes, values, priorities, and so on. In fact, the same individual's demand would probably vary from time to time due to changes in mood, weather, or any of innumerable other influences. A much more realistic model would therefore be

$$Q = f(P, Y) + u,$$

where u is a variable included to capture all the variations in demand Q over and above those induced by price fluctuations and differing income levels.

In general, a model will be defined as a single algebraic equation of the form

$$Y = f(X_1, X_2, \ldots, X_K) + u, \tag{1.1.1}$$

where Y is a variable whose movements we are seeking to describe or explain. The variables $X_1, X_2, \ldots, X_K$ are known as regressors or explanatory variables and are assumed to bear a causal relationship to the dependent variable Y.[1] The additional term u (known as the disturbance) is a random variable, which is included to account for the fact that movements in Y are not completely explained by the variables $X_1, X_2, \ldots, X_K$.[2]

Using this definition, model formulation involves choice of the regressors $X_1, X_2, \ldots, X_K$, specification of the function $f(\cdot)$[3] and estimation of any unknown constants (parameters) appearing in $f(\cdot)$ or characterizing the behavior of the disturbance u.

Once a model has been formulated, it may be used to test economic hypotheses, predict values of the dependent variable Y for given values of $X_1, X_2, \ldots, X_K$, or, in some cases, control the values that Y takes.

In the context of our simple demand model, we may wish to test hypotheses about price and income elasticities or predict likely values of demand for preassigned values of price and income.

## 1.2 THE LINEAR REGRESSION MODEL

Two further fundamental assumptions about the model will now be made.

(a) We will assume that the function $f(\cdot)$ is linear in the unknown parameters. Thus we assume that

$$f(X_1, X_2, \ldots, X_K) = \beta_0 + \beta_1 X_1 + \cdots + \beta_K X_K, \tag{1.2.1}$$

where $\beta_0, \beta_1, \ldots, \beta_K$ are unknown parameters. The essential ideas here are that none of the regressor variables $X_1, X_2, \ldots, X_K$ depend on an unknown parameter, and none of the parameters depend on any other parameter.

Returning to the demand example, we see that

$$Q = \beta_0 + \beta_1 P + \beta_2 Y + u$$

and

$$\ln Q = \beta_0 + \beta_1 \ln P + \beta_2 \ln Y + u$$

are both linear models. However, a model of the form

$$Q = \beta_0 + \beta_1 P + \beta_1^2 Y + u$$

would not be a linear model. A formulation such as

$$Q = \beta_0 + \beta_1 P + \beta_2 Y^{\delta} + u$$

may or may not be linear. If $\delta$ is a known constant (e.g., $\delta = 2$), then the model is linear. On the other hand, if $\delta$ were unknown, the regressor $Y^{\delta}$ would involve an unknown parameter, making the model nonlinear.

(b) The values that the regressor variables $X_1$, $X_2$, ..., $X_K$ take are assumed to be fixed numbers. An immediate implication of this assumption is that whereas the X's may cause Y, Y cannot cause the X's. As Y is a random variable (due to the influence of the disturbance u) if Y caused the X's, they also would be random variables and not fixed in repeated samples. We will therefore assume that any causation is in the direction of the X's to Y, but not from Y to the X's. Schematically, we could represent the situation as follows:

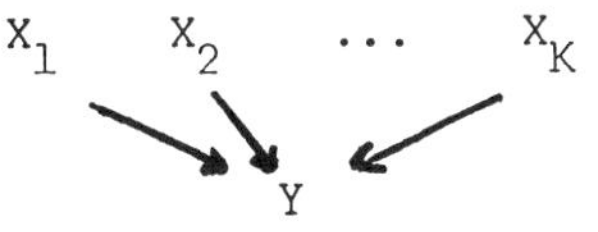


The assumptions so far have referred only to what we could term the "systematic" part of the model, namely, $f(X_1, X_2, \ldots, X_K)$. Clearly, this part of the model will be used for testing economic hypotheses, prediction, and control. However, the disturbance u is an absolutely vital part of the specification of a linear model. In our initial study of regression analysis we will be making some very restrictive assumptions about this random variable.

## 1.3 THE METHOD OF LEAST SQUARES

Having specified the linear function

$$f(X_1, X_2, \ldots, X_K) = \beta_0 + \beta_1 X_1 + \cdots + \beta_K X_K$$

up to the unknown parameters $\beta_0$, $\beta_1$, ..., $\beta_K$ the first task in a regression analysis is to complete the model formulation by using observations on the variables Y, $X_1$, $X_2$, ..., $X_K$ to estimate the parameters. Naturally, we would like to use a method that uses the available data in some optimal way. A method that can be shown to have this property is the method of least squares. We will be defining carefully the exact sense in which this method of estimation is optimal and showing how optimality depends crucially on the assumptions underlying the model. If statistical analysis is not to degenerate into a collection of "cookbook recipes," users must develop an awareness of how the model assumptions determine the method of approach.

## 1.4 TOWARD THE "REAL WORLD": SOME PROBLEM AREAS

The first part of our study involves a thorough examination of the properties of least squares estimators and the mechanical aspects of computation, hypothesis testing, and prediction. However, in an applied analysis, researchers must critically assess the assumptions underlying the linear model and question whether they are a realistic description of their context. If the assumptions are found to be unrealistically simple or inappropriate, the method of least squares can no longer be assumed optimal. In such cases, the method of analysis may have to be varied. An important part of any regression analysis is therefore the use of diagnostic procedures designed to detect ways in which the underlying assumptions may have been violated. We now list some common problems that may be encountered.

### Misspecification

An implicit assumption underlying least squares estimation is that the model has been correctly specified. In particular, it is assumed that both the choice of regressor variables and the functional form relating them is correct. In practice such an assumption would probably never be realistic. It follows that an important aspect of a study of regression should include consideration of the effects of misspecification. In the first part of our study we will consider only the effects of leaving out relevant regressors and/or including irrelevant regressors. An examination of these effects enables us to determine when they are likely to detract significantly from the analysis, and when they are not likely to matter very much. A framework can be developed that may enable us to recognize symptoms of misspecification.

For example, the demand situation discussed earlier may be analyzed using many different linear models. Two such models that have often been used are

$$Q = \beta_0 + \beta_1 P + \beta_2 Y + u$$

and

$$\ln Q = \beta_0 + \beta_1 \ln P + \beta_2 \ln Y + u.$$

Under some circumstances (e.g., the range of observations on the variables is small) both models may provide an adequate description of reality. In other cases, one model may be much better than the other. For the purpose of this discussion, let us assume that the second model is better. We have so far made no attempt to include the effects of price variations in substitutes and complements. If we denote by P* and P** the prices of two substitutes and/or complements, a different model would be

$$\ln Q = \beta_0 + \beta_1 \ln P + \beta_2 \ln Y + \beta_3 \ln P^* + \beta_4 \ln P^{**} + u.$$

Two questions now arise. Given we never know the true model, what will be the effects on estimates and predictions if P* and P** are omitted when they really do affect variations in Q? Again, if they in fact have no relevance in explaining movements in Q, do we pay a penalty for including them "just in case"? Diagnosing possible misspecifications and analyzing their effects is a vital part of analysis.

## Multicollinearity

When we consider the linear model

$$Y = \beta_0 + \beta_1 X_1 + \cdots + \beta_K X_K + u,$$

the parameter $\beta_1$ may be interpreted as the marginal effect on the average value of Y when $X_1$ is increased by one unit, provided that the variables $X_2, \ldots, X_K$ are held fixed. In a controlled environment such as a laboratory, it may be possible to physically control the regressors. However, in the typical situation facing an economist, all the variables $X_1, X_2, \ldots, X_K$ are varying together. Any chance we may have of isolating the effect of $X_1$ will therefore depend on whether observations on $X_1$ have some characteristics which are quite distinct from those of $X_2, \ldots, X_K$. The situation in which observations on some (or all) of the regressors exhibit the same characteristics is known as multicollinearity. This results in poor estimates of the parameters because it is virtually impossible to disentangle the individual effects. If the object of the analysis is only the prediction of Y, this may not matter. However, where an accurate estimate of the $\beta$'s is essential (e.g., hypothesis testing or control) multicollinearity poses a serious problem. We may view multicollinearity as data deficiency, and our efforts to overcome it revolve around attempts to augment the sample with information from other sources.

## Heteroscedasticity and Autocorrelation

As mentioned earlier, the disturbance u is a random variable that is assumed to have very particular properties. These properties are likely to be quite realistic in the context of controlled experiments. However, in the uncontrolled economic environment, it is very unlikely that they will hold—even approximately. The phenomena known as heteroscedasticity and autocorrelation are very common properties of economic data and are just two cases in which the classical disturbance assumptions are quite inappropriate. To ignore these problems and apply least squares in a mechanical way would result in inefficient use of data information and perhaps, invalid statistical inferences. In such cases our method of approach should be modified.

## Stochastic (i.e., Random) Regressors

The optimality of the method of least squares depends crucially on the assumption that the regressor variables are fixed in repeated samples. We have already seen that this implies that causality is from the X's to Y, but not vice versa. Once again, such an assumption is entirely realistic in a laboratory context in which the X's are set by the experimenter.

In economic models, however, this assumption would often not be appropriate. We mention two very common situations as illustrations:

(i) Often there is feedback from Y to some, or all, of the X's. In other words, there is simultaneous causality. In contrast to the situation depicted in Section 1.2, we now have the schematic representation

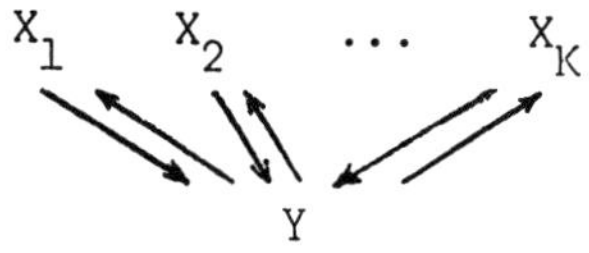


For example, in our simple demand model

$$\ln Q = \beta_0 + \beta_1 \ln P + \beta_2 \ln Y + u,$$

it may be quite realistic to assume that whereas income (Y) influences demand (Q), the reverse is not the case. However, because equilibrium price and quantity arise from the interaction of demand and supply, it is likely that Q influences P as well as P influencing Q. In such a situation, it logically follows that P cannot be fixed and must be a random variable.

(ii) Economic models are often inherently dynamic. By this we mean that the value the dependent variable Y takes in a given time period depends not only on the values of the regressors $X_1$, $X_2$, ..., $X_K$ but also on the value it had in the preceding period. For example, if we denote by $Y_t$, $X_{1t}$, and $u_t$ the values of Y, $X_1$, and u at time t, a model of the form

$$Y_t = \beta_0 + \beta_1 X_{1t} + \beta_2 Y_{t-1} + u_t$$

is dynamic. We know that $Y_t$ is a random variable, for all values of t. Thus the regressor $Y_{t-1}$ must also be random.

In both cases (i) and (ii), the assumption that the regressors are fixed, an assumption underlying the effectiveness of the least squares approach, has been violated. A reappraisal and perhaps some modification may be needed if we are to be able to use the estimated model with any degree of confidence.

### Qualitative Dependent Variables

It sometimes happens that the variable we are seeking to explain is qualitative rather than quantitative. For example, an econometrician may be investigating the factors that influence a consumer to buy a new car. The decision is either "yes" or "no" and this becomes the dependent variable.[4] Models that analyze qualitative dependent variables are somewhat different from the linear regression model (1.1.1) and (1.1.2) and require a different approach. We will be examining two such models, the probit and logit models.

In each of the five cases sketched above—misspecification, multicollinearity, heteroscedasticity and autocorrelation, stochastic regressors, and qualitative dependent variables—the basic approach to estimation has to be critically evaluated and, perhaps, modified. This takes regression analysis a long way from the mechanical, prescriptive procedure that it may appear to be in the introductory phases of the study.

## 1.5 COMPUTING AND REGRESSION

The computational burden involved in regression analysis is very considerable. Once a regression model involves more than two regressors, analysis becomes impracticable without the use of an electronic computer. For this reason, any course that purports to teach applied regression analysis will, of necessity, involve the student with computers. Nowadays, this does not necessitate any programming skill, as every computer establishment has regression packages available. Use of such a package requires knowledge only on how to set the problem up, enter the necessary information into the computer, and interpret the output.

Facility in the use of a computer package can provide a great stimulus in developing the "art" of regression analysis. It is a common experience for students who have completed an academic discipline to feel totally helpless and inept when facing the real situation. One reason for this is that often the only examples they have encountered have been artificially refined and simplified to make them suitable for classroom discussion and analysis. The ready availability of computer facilities means that even the beginner can now try his or her hand at fairly realistic problems. The continual interaction between theory and practice can create a "feel" for the use of this particular technique of analysis which no amount of algebraic manipulation can impart. Not only can students learn to appreciate what a computer can do, they can, more importantly, learn what a computer cannot do for them.

In this book we have treated computer experience as an integral part of the study of applied regression analysis. The particular package we use for most of the book is MINITAB, chosen because of its widespread availability and easy operation. In the later part of our study, we make use of the package SHAZAM, which is a specialized econometric program possessing some features not found in MINITAB. Most sections in the book (apart from the discussion on simple regression) show the sequence of commands needed to accomplish a particular task, and displays showing actual output are common. It should be stressed at this point that all regression packages are essentially the same. Experience with one forms a valuable basis for use of another. The differences lie only in the conventions used for introducing input and displaying output.

## 1.6 ORGANIZATION OF THE BOOK

The book is divided into two parts. The first part should be regarded as giving the basic groundwork for regression analysis. The second part covers essentially three more advanced topics: stochastic regressors (including the briefest of introductions to systems of equations), qualitative dependent variables, and some additional discussion on specification testing. The latter topic has been the subject of intense research over the last five to ten years, but so far very little of the resulting methodology has appeared in books intended for the practitioner. The last chapter is an attempt to make some of these techniques available to readers who are not trained in theoretical econometrics. In the second part, basic calculus has been used where necessary (up to the level of partial differentiation) and less effort has been made to describe in detail the steps needed for the computations with either MINITAB or SHAZAM. By the time the reader reaches these chapters, he or she will find that using the computer has become quite simple.

NOTES

1. When dealing with general theory, the symbol Y will always refer to the dependent variable. In particular contexts, as in the demand example, other symbols may be used and Y may refer to one of the regressor variables. This confusion arises through different conventions between economists and statisticians.

2. This definition of a model is obviously rather narrow. For example, a model does not have to be expressed in the form of an algebraic relationship. Again, a model may consist of many algebraic equations, or the disturbance u need not be additive as in equation (1.1.1).

3. We will sometimes use the shorthand notation $f(\cdot)$ for the function $f(X_1, X_2, \ldots, X_K)$. Such a function will usually involve constants as well as the variables $X_1, X_2, \ldots, X_K$.

4. Such a variable is known as a binary choice variable.

# 2

# Simple Regression

## 2.1 THE MODEL AND ASSUMPTIONS

Simple regression is the analysis of equation (1.1.1) (estimation, hypothesis testing, and prediction) when the function f(·) is linear and there is only one regressor variable in the equation. When there is more than one regressor, the model is known as the multiple regression model. Thus the simple linear regression model is given by

$$Y_i = \beta_0 + \beta_1 X_i + u_i, \tag{2.1.1}$$

where the subscript 1 is dropped from $X_{1i}$ for convenience.

We are going to interpret the model (2.1.1) as specifying the mechanism through which observations on the dependent variable $Y_i$ are generated. In particular, we are assuming that the actual value that $Y_i$ takes when the regressor takes the value $X_i$ consists of the sum of two parts: (a) a systematic component $\beta_0 + \beta_1 X_i$, and (b) a random component $u_i$. This interpretation is illustrated in Figure 2.1.1 for two values, $X_1$ and $X_2$.

The purpose of regression analysis is to use the observation pairs $(X_1, Y_1), (X_2, Y_2), \ldots, (X_N, Y_N)$ to try to reconstruct the systematic component. This means that we will be using the observations to estimate the unknown parameters $\beta_0$ and $\beta_1$. This reconstruction may then be used for statistical inference, prediction, or control. Because the $u_i$, known as the disturbances, are random, they can never be predicted. However, it may be possible to estimate some characteristics of their behavior.

An essential part of the linear model (2.1.1) is the assumptions that are made regarding the nature of the disturbances $u_i$. We assume that $u_i$ is a random variable, drawn from a probability distribution having the following properties:[1]

$$\text{(a)} \quad E(u_i) = 0 \qquad i = 1, 2, \ldots, N. \tag{2.1.2}$$

$$\text{(b)} \quad E(u_i^2) = \sigma^2 \qquad i = 1, 2, \ldots, N. \tag{2.1.3}$$

$$\text{(c)} \quad E(u_i u_j) = 0 \qquad (i \neq j) \tag{2.1.4}$$

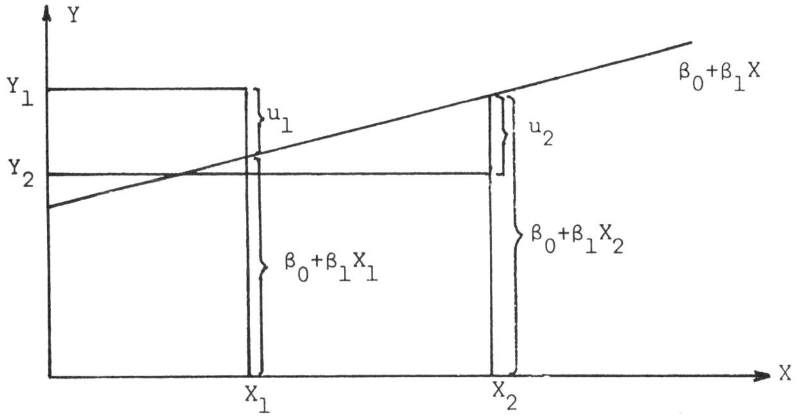


Figure 2.1.1. Generation of observations on the dependent variable Y.

In addition to the above, we will assume that the $X_i$ are not random variables, but fixed numbers. That is, if it were possible to repeat the experiment that generated the $Y_i$, they (the $Y_i$) would be different, but the $X_i$ would be exactly the same.

Assumptions (a) and (b) imply that each $u_i$ is drawn from a distribution which has mean zero and variance $\sigma^2$, because

$$\begin{aligned} \text{var}(u_i) &= E(u_i^2) - [E(u_i)]^2 \\ &= E(u_i^2), \quad \text{as } E(u_i) = 0 \text{ from (a)} \\ &= \sigma^2. \end{aligned} \qquad (2.1.5)$$

This means that the variance of $u_i$ can be anything finite but that it must be constant for all i, that is, for each observation (i = 1, 2, ..., N).

The assumptions (a), (b), and (c) can be used to deduce the properties of the dependent variable $Y_i$. First, because it is generated as the sum of nonrandom and random components, it must be a random variable. Second, taking expectations of (2.1.1), conditional on the value $X_i$, we have

$$E[Y_i \mid X_i] = \beta_0 + \beta_1 X_i. \qquad (2.1.6)$$

This implies that the mean of $Y_i$ depends on the value of $X_i$, and geometrically we can visualize these means (for different values of X) as all lying on the straight line $\beta_0 + \beta_1 X$.

Third, the variance of $Y_i$, for a given value $X_i$ of the regressor variable, is given by

$$\text{var}[Y_i \mid X_i] = \sigma^2. \qquad (2.1.7)$$

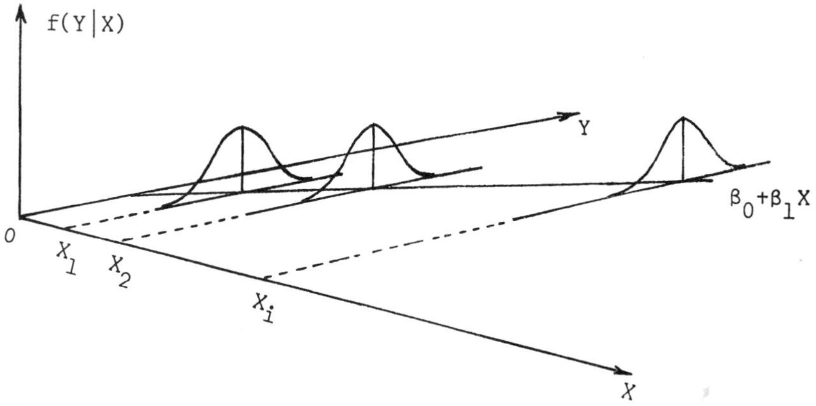


Figure 2.1.2 Conditional probability density functions of Y, for different values of the regressor X.

Thus, although the mean of $Y_i$ depends on the value of the corresponding $X_i$, its variance does not. These relationships are perhaps best appreciated by a diagram, shown in Figure 2.1.2. It can be seen that the distributions of the $Y_i$ corresponding to different $X_i$ differ only in the location of the means.

Assumption (c) implies that the covariance between two disturbances associated with two different observations will be zero.

$$\begin{aligned} \text{cov}(u_i u_j) &= E(u_i u_j) - E(u_i)E(u_j) \\ &= E(u_i u_j) \quad \text{from (a)} \\ &= 0 \quad \text{if } i \neq j \text{ from (c)} \end{aligned} \tag{2.1.8}$$

This assumption means that any two disturbances are uncorrelated. It will always be satisfied if $u_i$ and $u_j$ are independent, that is, if there is no relationship between any two different disturbances. This property of the disturbances in turn implies that

$$\text{cov}(Y_i, Y_j) = 0. \tag{2.1.9}$$

The assumptions and some of the implications have been set out somewhat formally, and we will now examine their meanings in the context of a simple example.

Suppose that in equation (2.1.1) $Y_i$ represents the quantity of tomatoes purchased from Colworths Supermarket in the ith week, and $X_i$ represents the price of tomatoes in Colworths during that week. Two other variables that may influence $Y_i$ and which have not been explicitly specified in equation (2.1.1), and whose influence therefore will be taken into account in the dis-

$$N\hat{\beta}_0 + \hat{\beta}_1 \Sigma X_i = \Sigma Y_i, \tag{2.2.4}$$

$$\hat{\beta}_0 \Sigma X_i + \hat{\beta}_1 \Sigma X_i^2 = \Sigma X_i Y_i. \tag{2.2.5}$$

Equations (2.2.4) and (2.2.5) are known as normal equations. They are linear equations in the unknowns $\hat{\beta}_0$ and $\hat{\beta}_1$ and can be solved provided that[6]

$$N \Sigma X_i^2 - (\Sigma X_i)^2 \neq 0.$$

Assuming then that this condition holds, we may solve for $\hat{\beta}_0$ and $\hat{\beta}_1$ and obtain

$$\hat{\beta}_1 = \frac{\Sigma X_i Y_i - (\Sigma X_i)(\Sigma Y_i)/N}{\Sigma X_i^2 - (\Sigma X_i)^2/N} \tag{2.2.6}$$

and

$$\hat{\beta}_0 = \bar{Y} - \hat{\beta}_1 \bar{X} = \frac{1}{N}(\Sigma Y_i - \hat{\beta}_1 \Sigma X_i) \tag{2.2.7}$$

where $\bar{X} = \Sigma X_i/N$ and $\bar{Y} = \Sigma Y_i/N$. We now define variables $x_i$ and $y_i$ by

$$x_i = X_i - \bar{X} \quad \text{and} \quad y_i = Y_i - \bar{Y}. \tag{2.2.8}$$

It is then easily shown that

$$\hat{\beta}_1 = \frac{\Sigma x_i y_i}{\Sigma x_i^2}. \tag{2.2.9}$$

Equations (2.2.7) and (2.2.9) define the least squares estimates of $\beta_0$ and $\beta_1$, or (geometrically) the line of best fit. The method described is called <u>ordinary least squares</u>,[7] which we will abbreviate to OLS.

The estimated regression line for the Colworths data may now be found. The data are summarized by the five quantities

$$\Sigma X = 179, \qquad \Sigma Y = 1740,$$

$$\Sigma X^2 = 3621, \qquad \Sigma Y^2 = 371{,}000,$$

$$\Sigma XY = 26{,}350.$$

These five numbers are now used to find the three numbers $\Sigma x^2$, $\Sigma y^2$, and $\Sigma xy$.

$$\Sigma x^2 = \Sigma X^2 - \frac{(\Sigma X)^2}{N} = 3621 - \frac{179^2}{10} = 416.9,$$

$$\Sigma y^2 = \Sigma Y^2 - \frac{(\Sigma Y)^2}{N} = 371{,}000 - \frac{1740^2}{10} = 68{,}240,$$

$$\Sigma xy = \Sigma XY - \frac{(\Sigma X)(\Sigma Y)}{N} = 26{,}350 - \frac{(179)(1740)}{10} = -4796$$

From (2.2.9),

$$\hat{\beta}_1 = \frac{\Sigma xy}{\Sigma x^2} = \frac{-4796}{416.9} = -11.504.$$

Substituting this $\hat{\beta}_1$ into (2.2.7) yields

$$\hat{\beta}_0 = \frac{1}{N}(\Sigma Y - \hat{\beta}_1 \Sigma X) = \frac{1}{10}[1740 - (-11.504)(179)]$$

$$= 379.922.$$

The line of best fit is thus given, correct to three figures, by

$$\hat{Y}_i = 380 - 11.5X_i.$$

This means that for a 1-cent increase in the price of tomatoes we would

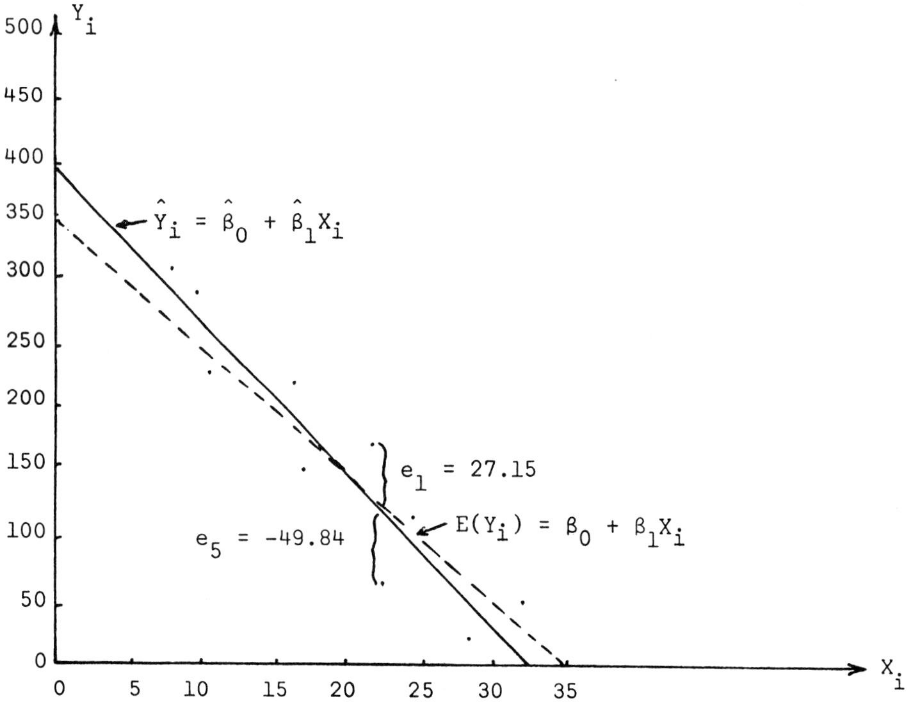


Figure 2.2.1. Real and estimated lines for Colworths data.

estimate the average quantity purchased in one week to fall by 11.5 kg. This line is plotted, along with the observations, in Figure 2.2.1.

The residuals, $e_i$, can be calculated once $\hat{\beta}_0$ and $\hat{\beta}_1$ have been determined.

$$e_1 = Y_1 - \hat{Y}_1 = 200 - [379.922 - 11.504(18)] = 27.15$$

$$\begin{array}{llllll} e_2 = & 38.18 & e_5 = & -49.84 & e_8 = & -45.86 \\ e_3 = & 35.12 & e_6 = & -34.88 & e_9 = & 11.13 \\ e_4 = & -42.32 & e_7 = & 26.17 & e_{10} = & 35.20 \end{array}$$

The real line, showing $E[Y_i | X_i]$, is the dashed line in Figure 2.2.1. As can be seen, it does not coincide with the estimated line, and apart from an extraordinary fluke, never would. As the true line in a real situation would never be known, it is important to be able to assess how close the estimated line is likely to be. This involves consideration of the properties of least squares estimators and their distributions.

## An Interpretation of $\hat{\beta}_1$

The right-hand side of equation (2.2.9) can be rewritten to given an appealing interpretation of the slope estimator $\hat{\beta}_1$. Because $\Sigma x_j^2$ is the same as $\Sigma x_i^2$, we have

$$\hat{\beta}_1 = \frac{\Sigma x_i y_i}{\Sigma x_j^2}$$

$$= \frac{\Sigma x_i^2 (y_i/x_i)}{\Sigma x_j^2}$$

$$= \sum_i \frac{x_i^2}{\Sigma x_j^2} \frac{y_i}{x_i}.$$

That is,

$$\hat{\beta}_1 = \sum_i w_i \frac{y_i}{x_i}, \qquad (2.2.10)$$

where

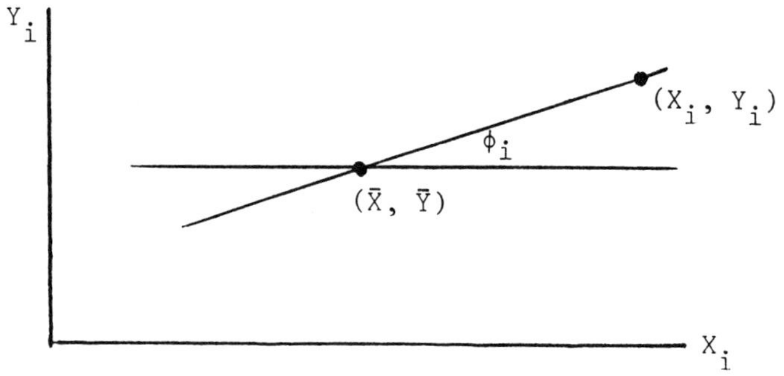


Figure 2.2.2. Slope of line joining the point $(X_i, Y_i)$ to $(\bar{X}, \bar{Y})$.

$$w_i = \frac{x_i^2}{\Sigma x_j^2}. \tag{2.2.11}$$

This expression for $w_i$ implies that $w_i \geq 0$ for all i, and that $\Sigma w_i = 1$. Therefore, equation (2.2.10) shows that $\hat{\beta}_1$ is a weighted average of the ratios $y_i/x_i$. From Figure 2.2.2,

$$\frac{y_i}{x_i} = \frac{Y_i - \bar{Y}}{X_i - \bar{X}} = \tan \phi_i$$

which is the slope of the line joining $(X_i, Y_i)$ to $(\bar{X}, \bar{Y})$. The weighting factor $w_i$ is the proportion of the total sum of squares of X (about the mean $\bar{X}$) attributable to the ith observation. As we shall see, $\Sigma x^2$ can be interpreted as the information content of the sample. Thus $\hat{\beta}_1$ is a weighted average of individual slopes, where the weighting factor reflects the relative importance of each observation.

## 2.3 PROPERTIES OF LEAST SQUARES ESTIMATORS[8]

Having defined the deviation form of a variable $x_i = X_i - \bar{X}$, we will derive some results that will prove useful in later work.

$$\Sigma x = \Sigma(X - \bar{X}) = \Sigma X - N\bar{X} = N\bar{X} - N\bar{X} = 0.$$

Thus, for <u>any</u> variable $X_i$,

$$\Sigma x = 0, \tag{2.3.1}$$

$$\Sigma xy = \Sigma(X - \bar{X})y = \Sigma Xy - \bar{X}\,\Sigma y = \Sigma Xy. \tag{2.3.2}$$

In a similar way, we can show that

$$\Sigma xy = \Sigma xY, \tag{2.3.3}$$

$$\Sigma x^2 = \Sigma xX. \tag{2.3.4}$$

We will now turn to some fundamental properties of OLS estimators. Only the estimator $\hat{\beta}_1$ will be considered, but in each case an exactly analogous result holds for $\hat{\beta}_0$.

## OLS Estimators Are Linear

$$\hat{\beta}_1 = \frac{\Sigma x_i y_i}{\Sigma x_j^2} = \frac{\Sigma x_i Y_i}{\Sigma x_j^2} = \sum_i \frac{x_i}{\Sigma x_j^2} Y_i .$$

Thus

$$\hat{\beta}_1 = \sum_i b_i Y_i, \tag{2.3.5}$$

where

$$b_i = \frac{x_i}{\Sigma x_j^2}. \tag{2.3.6}$$

Because the $x_i$ are fixed numbers, and the $b_i$ are functions of the x's alone, the $b_i$ are also constants. Thus equation (2.3.5) defines $\hat{\beta}_1$ as a linear combination of the observations on the dependent variable. It is in this sense that OLS estimators are said to be linear.

It should be noted that this property depends crucially on the assumption that observations on the regressor variable $X_i$ are fixed in repeated samples. If the $X_i$ were stochastic, the $b_i$ would not be constants and (2.3.5) would not define a linear combination of the $Y_i$. The significance of this property lies in the fact that the distribution of a linear function can often be derived easily.

## OLS Estimators Are Unbiased

Consider any linear estimator $\tilde{\beta}_1$, defined by

$$\tilde{\beta}_1 = \sum_i c_i Y_i,$$

where the $c_i$ are constants. Then

$$E(\tilde{\beta}_1) = E\Big(\sum_i c_i Y_i\Big) = \sum_i c_i E(Y_i) = \sum c_i E(\beta_0 + \beta_1 X_i + u_i)$$

$$= \sum_i c_i(\beta_0 + \beta_1 X_i) \quad \text{[because } X_i \text{ is constant and } E(u_i) = 0]$$

$$= \beta_0 \sum_i c_i + \beta_1 \sum_i c_i X_i.$$

$\tilde{\beta}_1$ is an unbiased estimator if and only if

$$E(\tilde{\beta}_1) = \beta_1,$$

and it follows that a sufficient condition for this is that

$$\Sigma c_i = 0 \quad \text{and} \quad \Sigma c_i X_i = 1.$$

Now from (2.3.6),

$$\Sigma b_i = \frac{\Sigma x_i}{\Sigma x_j^2} = 0 \quad \text{[from (2.3.1)]} \tag{2.3.7}$$

and

$$b_i X_i = \frac{\Sigma x_i X_i}{\Sigma x_j^2} = \frac{\Sigma x_i^2}{\Sigma x_j^2} = 1 \quad \text{[from (2.3.4)]}. \tag{2.3.8}$$

Thus the OLS estimator $\hat{\beta}_1$ satisfies the sufficient condition and is therefore an unbiased linear estimator.

### OLS Estimators Have Minimum Variance Among All Linear, Unbiased Estimators

Consider again any linear estimator $\tilde{\beta}_1$. It can be shown [see, e.g., Johnston (1972, 21-22)] that

$$\mathrm{var}(\hat{\beta}_1) \leq \mathrm{var}(\tilde{\beta}_1). \tag{2.3.9}$$

These properties are summarized by saying that OLS estimators are best linear unbiased estimators (BLUE).

### Properties of the Residuals

Consider the residual $e_i$, given by

$$e_i = Y_i - \hat{Y}_i = Y_i - (\hat{\beta}_0 + \hat{\beta}_1 X_i)$$

$$= Y_i - (\bar{Y} - \hat{\beta}_1 \bar{X} + \hat{\beta}_1 X_i) = (Y_i - \bar{Y}) - \hat{\beta}_1 (X_i - \bar{X}).$$

That is,

$$e_i = y_i - \hat{\beta}_1 x_i. \tag{2.3.10}$$

We can now immediately deduce two properties that characterize the residuals. First,

$$\Sigma e = \Sigma(y - \hat{\beta}_1 x) = \Sigma y - \hat{\beta}_1 \Sigma x.$$

From (2.3.1), $\Sigma y = \Sigma x = 0$, and hence

$$\Sigma e = 0. \tag{2.3.11}$$

Second,

$$\Sigma eX = \Sigma(y - \hat{\beta}_1 x)X = \Sigma yX - \hat{\beta}_1 \Sigma xX.$$

Using (2.3.3) and (2.3.4) gives

$$\Sigma eX = \Sigma xy - \hat{\beta}_1 \Sigma x^2 = \Sigma xy - \Sigma xy \quad \text{[from (2.2.9)]}.$$

Table 2.3.1

| $i$ | $e_i$ | $X_i$ | $e_i X_i$ |
|---|---|---|---|
| 1 | 27.15 | 18 | 488.70 |
| 2 | 38.13 | 12 | 457.56 |
| 3 | 35.12 | 10 | 351.20 |
| 4 | -42.32 | 25 | -1058.00 |
| 5 | -49.84 | 20 | -996.80 |
| 6 | -34.88 | 10 | -348.80 |
| 7 | 26.17 | 24 | 628.08 |
| 8 | -45.86 | 16 | -733.76 |
| 9 | 11.13 | 14 | 155.82 |
| 10 | 35.20 | 30 | 1056.00 |
| | $\Sigma e_i = 0.00$ | | $\Sigma e_i X_i = 0.00$ |

Thus

$$\Sigma eX = 0. \qquad (2.3.12)$$

We demonstrate these properties in Table 2.3.1 using the residuals generated by the Colworths example. These two properties are so fundamental that they are sometimes taken as defining the line of best fit [see Rao and Miller (1971, 8)]. That is, we choose as the line of best fit the particular line which generates residuals satisfying (2.3.11) and (2.3.12). This line will be identical to that obtained using the least squares principle.

## 2.4 WHEN WILL OLS ESTIMATORS BE "GOOD"?

We have seen that provided the assumptions about the disturbances $u_i$ and the regressor observations $X_i$ hold, the OLS estimators are BLUE. However, this does not mean that they are "good" in the sense that there is a high probability that the estimates $\hat{\beta}_0$ and $\hat{\beta}_1$ are close to the unknown parameters $\beta_0$ and $\beta_1$. Even though OLS estimators are best (when compared with all other linear, unbiased estimators) they may still be very poor. Because OLS estimators are unbiased, the criterion of "goodness" reduces to a consideration of their variance. Estimators that are good have a small variance and those which are not have a large variance. It becomes natural, then, to investigate the situations in which the variances of OLS estimators will be small.

We have already seen that $\hat{\beta}_1$ is a linear estimator. In particular, $\hat{\beta}_1 = \Sigma_i b_i Y_i$, where $b_i$ is given by (2.3.6). Thus

$$\hat{\beta}_1 = \sum_i b_i(\beta_0 + \beta_1 X_i + u_i) = \beta_0 \sum_i b_i + \beta_1 \sum_i b_i X_i + \sum b_i u_i.$$

Using the fact that $\Sigma b_i = 0$ and $\Sigma b_i X_i = 1$, we have

$$\hat{\beta}_1 - \beta_1 = \sum_i b_i u_i. \qquad (2.4.1)$$

The variance of $\hat{\beta}_1 - \beta_1$ is now easily derived:

$$\text{var}(\hat{\beta}_1 - \beta_1) = \text{var}\Big(\sum_i b_i u_i\Big)$$

$$= \sum_i b_i^2 \text{ var}(u_i) \qquad \text{(because the } u_i \text{ are uncorrelated)}$$

$$= \sigma^2\Big(\sum_i b_i^2\Big).$$

Now[9]

$$\sum_i b_i^2 = \sum_i \left(\frac{x_i}{\Sigma_j x_j^2}\right)^2 = \sum_i \frac{x_i^2}{(\Sigma_j x_j^2)^2} = \frac{1}{\Sigma_i x_i^2}.$$

Therefore,

$$\text{var}(\hat{\beta}_1 - \beta_1) = \frac{\sigma^2}{\Sigma x^2}.$$

Subtracting a constant (here $\beta_1$) from a random variable (here $\hat{\beta}_1$) does not change its variance. Thus $\text{var}(\hat{\beta}_1 - \beta_1) = \text{var}(\hat{\beta}_1)$, implying that

$$\text{var}(\hat{\beta}_1) = \frac{\sigma^2}{\Sigma x^2}. \tag{2.4.2}$$

It can be shown in a similar way that

$$\text{var}(\hat{\beta}_0) = \frac{\sigma^2}{\Sigma x^2} \frac{\Sigma X^2}{N}. \tag{2.4.3}$$

We note in passing that the covariance between $\hat{\beta}_0$ and $\hat{\beta}_1$ can be shown to be

$$\text{cov}(\hat{\beta}_0, \hat{\beta}_1) = \frac{-\bar{X}\sigma^2}{\Sigma x^2}. \tag{2.4.4}$$

We will now consider the expression for $\text{var}(\hat{\beta}_1)$ and show that it can be given an interpretation which greatly helps our intuitive perception of when the estimate $\hat{\beta}_1$ is likely to be close to $\beta_1$. We have obtained the estimated line as

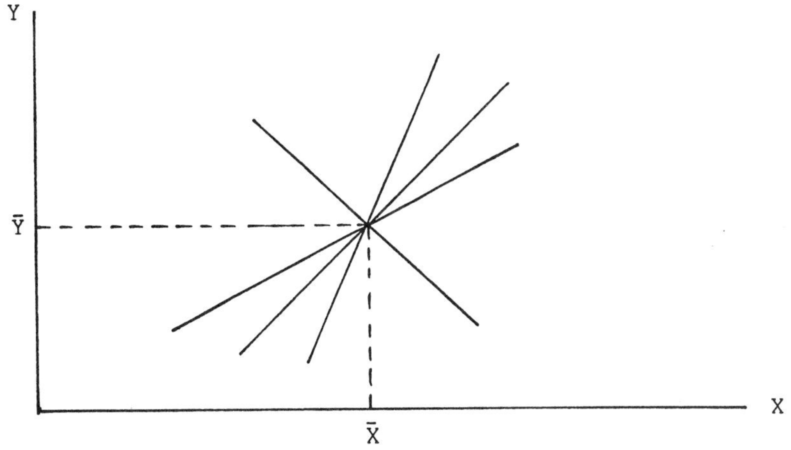


Figure 2.4.1. Possible lines of best fit.

$$\hat{Y}_i = \hat{\beta}_0 + \hat{\beta}_1 X_i,$$

where

$$\hat{\beta}_0 = \bar{Y} - \hat{\beta}_1 \bar{X}.$$

If we rearrange the latter equation, we have

$$\bar{Y} = \hat{\beta}_0 + \hat{\beta}_1 \bar{X},$$

which shows that the estimated line always passes through the point $(\bar{X}, \bar{Y})$. We can thus visualize the line of best fit as swiveling around the point $(\bar{X}, \bar{Y})$, as shown in Figure 2.4.1.

Consider two samples of the same size but one in which $\Sigma x^2$ is large and the other where $\Sigma x^2$ is small. The two possibilities are illustrated in Figure 2.4.2. In Figure 2.4.2(a), where all observations are fairly close to $\bar{X}$ and so $\Sigma x^2$ is small, it is difficult to predict the position of the estimated least squares line. Any additional observation that lay much farther from $\bar{X}$ would almost certainly have a big influence on the value of $\hat{\beta}_1$. Hence $\hat{\beta}_1$ would be unreliable and would have a large variance. On the other hand, in Figure 2.4.2(b), where the observations are not clustered around the mean, it is fairly obvious where the least squares line will be, and one would also feel confident that this would be quite close to the real line. In this case $\Sigma x^2$ would be relatively large and $\text{var}(\hat{\beta}_1)$ relatively small.

We may therefore interpret $\Sigma x^2$ as a measure of the information content that the sample has about the parameter $\beta_1$. If $\Sigma x^2$ is large (the sample con-

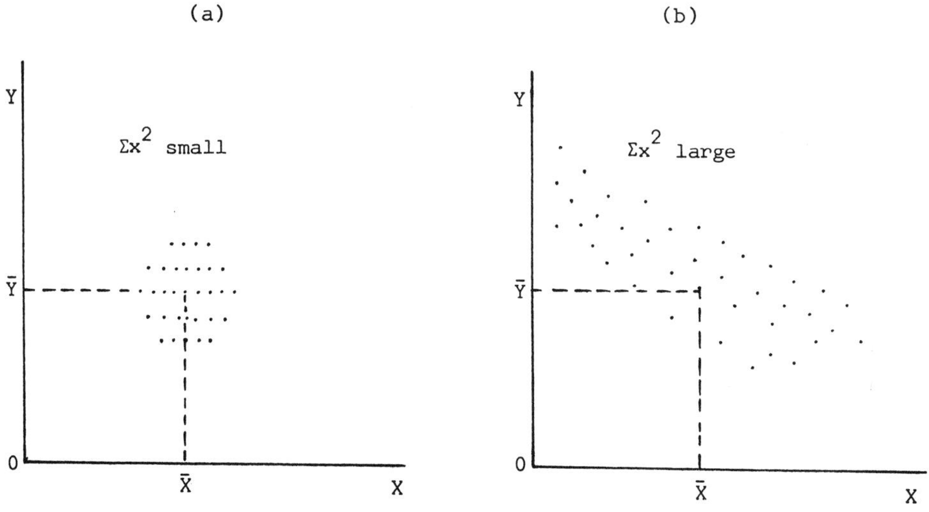


Figure 2.4.2. Data characterized by small and large values of $\Sigma x^2$.

tains a lot of information about $\beta_1$), $\hat{\beta}_1$ will be a reliable estimator of $\beta_1$. It is important to notice from equation (2.4.2) that when we say that $\Sigma x^2$ is large, we mean large relative to $\sigma^2$. To conclude this discussion, the two factors that lead to a large $\Sigma x^2$ are, first, a large sample size, and second, a wide spread of $X_i$ values. These then are the conditions that will give rise to $\hat{\beta}_1$ being a good estimate of $\beta_1$.

## 2.5 DISTRIBUTION OF LEAST SQUARES ESTIMATORS

Finding the line of best fit can be regarded as merely a descriptive analysis of the data—a way of capturing the main features of the observation pairs $(X_1, Y_1), \ldots, (X_N, Y_N)$. In most situations, however, it is desirable or necessary to use this description as a basis for statistical inference. This necessitates a knowledge of the probability distributions of the random variables $\hat{\beta}_0$ and $\hat{\beta}_1$. To be able to derive these distributions it is necessary to make an assumption about the distribution of the disturbances $u_i$. Accordingly, we add the following assumption to those given in Section 2.1:

ASSUMPTION The disturbances $u_i$ are drawn from a normal distribution. Then because

$$\hat{\beta}_1 = \beta_1 + \sum_i b_i u_i$$

is a linear combination of normal random variables, it will itself be normally distributed.

When we combine this result with earlier expressions for the means and variances of $\hat{\beta}_1$ and $\hat{\beta}_0$, we have the following:

THEOREM 2.5.1 If the disturbances $u_i$ are normally distributed,

$$\hat{\beta}_1 \sim N\left(\beta_1, \frac{\sigma^2}{\Sigma x^2}\right) \tag{2.5.1}$$

and

$$\hat{\beta}_0 \sim N\left(\beta_0, \frac{\sigma^2}{\Sigma x^2} \frac{\Sigma X^2}{N}\right), \tag{2.5.2}$$

where the symbol "~" means "is distributed as." In an alternative form, we may write

$$\frac{\hat{\beta}_1 - \beta_1}{\sigma/(\Sigma x^2)^{\frac{1}{2}}} \sim N(0, 1) \tag{2.5.3}$$

and

$$\frac{\hat{\beta}_0 - \beta_0}{\sigma(\Sigma X^2/N\ \Sigma x^2)^{\frac{1}{2}}} \sim N(0,1). \qquad (2.5.4)$$

In practice, $\sigma^2$ is an unknown parameter of the model. This means that it is never possible to calculate the variances of $\hat{\beta}_1$ and $\hat{\beta}_0$ or to use (2.5.3) and (2.5.4) for hypothesis testing or the construction of confidence intervals. However, it is possible to estimate $\sigma^2$ and hence obtain estimates of the variances of $\hat{\beta}_1$ and $\hat{\beta}_0$.

We define an estimator of $\sigma^2$ by

$$\hat{\sigma}^2 = \frac{1}{N-2}\sum_{i=1}^{N} e_i^2, \qquad (2.5.5)$$

and it can be shown that $\hat{\sigma}^2$ is an unbiased estimator of $\sigma^2$. This definition of $\hat{\sigma}^2$ is in fact not very convenient for computations because it involves finding all the residuals $e_i$ ($i = 1, 2, \ldots, N$). Two entirely equivalent expressions, which are much better suited to the computation of $\hat{\sigma}^2$, are easily derived using equation (2.3.10). We have

$$\begin{aligned}\Sigma e^2 &= \Sigma(y - \hat{\beta}_1 x)\\ &= \Sigma y^2 - 2\hat{\beta}_1\ \Sigma xy + \hat{\beta}_1^2\ \Sigma x^2.\end{aligned}$$

Now by (2.2.9),

$$\hat{\beta}_1 = \frac{\Sigma xy}{\Sigma x^2},$$

which immediately implies that

$$\hat{\sigma}^2 = \frac{1}{N-2}(\Sigma y^2 - \hat{\beta}_1\ \Sigma xy) \qquad (2.5.6)$$

or

$$\hat{\sigma}^2 = \frac{1}{N-2}(\Sigma y^2 - \hat{\beta}_1^2\ \Sigma x^2). \qquad (2.5.7)$$

EXAMPLE Consider again the Colworths data. We derived the residuals on page 21. Using the definition (2.5.5) directly gives us

$$\begin{aligned}\hat{\sigma}^2 &= \frac{1}{8}(27.15^2 + 38.13^2 + \cdots + 35.20^2)\\ &= \frac{1}{8} \times 13{,}067 = 1633\end{aligned}$$

Using (2.5.6), with $\Sigma y^2 = 68{,}240$, $\Sigma xy = -4796$, and $\hat{\beta}_1 = -11.504$, we find that

$$\hat{\sigma}^2 = \frac{1}{8}[68,240 - (-11.504)(-4796)]$$

$$= \frac{1}{8} \times 13,067 = 1633$$

It can be shown that when $\sigma$ is replaced by $\hat{\sigma}$ in (2.5.3) and (2.5.4) we have a slight modification, which can be stated in the following important result:

THEOREM 2.5.2

$$\frac{\hat{\beta}_1 - \beta_1}{S_{\hat{\beta}_1}} \sim t_{N-2} \tag{2.5.8}$$

and

$$\frac{\hat{\beta}_0 - \beta_0}{S_{\hat{\beta}_0}} \sim t_{N-2}, \tag{2.5.9}$$

where

$$S^2_{\hat{\beta}_1} = \frac{\hat{\sigma}^2}{\Sigma x^2} \tag{2.5.10}$$

$$S^2_{\hat{\beta}_0} = \frac{\hat{\sigma}^2}{\Sigma x^2} \frac{\Sigma X^2}{N} \tag{2.5.11}$$

and $t_{N-2}$ stands for the t-distribution with N - 2 degrees of freedom. $S_{\hat{\beta}_1}$ and $S_{\hat{\beta}_0}$ are known as the standard errors of $\hat{\beta}_1$ and $\hat{\beta}_0$, respectively, and are estimates of the standard deviations.

Theorem 2.5.2 is the basis for all inference about $\beta_1$ and $\beta_0$.

## 2.6 INFERENCE WITH THE REGRESSION MODEL

Once the line of best fit has been found, that is, $\beta_0$, $\beta_1$, and $\sigma^2$ have been estimated, we are in a position to move beyond a mere description of the data and to make inferences about the underlying model. The areas of inference fall into the three categories of confidence interval construction, hypothesis testing, and prediction.[10]

Confidence Intervals

Considering the estimator $\hat{\beta}_1$, equation (2.5.8) enables us to write

$$P\left(-t_{\epsilon/2} < \frac{\hat{\beta}_1 - \beta_1}{S_{\hat{\beta}_1}} < t_{\epsilon/2}\right) = 1 - \epsilon, \tag{2.6.1}$$

where $0 < \epsilon < 1$ and $t_{\epsilon/2}$ is defined implicitly by

$$P(|t_{N-2}| > t_{\epsilon/2}) = \epsilon. \tag{2.6.2}$$

Rearranging (2.6.1) gives

$$P[\hat{\beta}_1 - (t_{\epsilon/2})(S_{\hat{\beta}_1}) < \beta_1 < \hat{\beta}_1 + (t_{\epsilon/2})(S_{\hat{\beta}_1})] = 1 - \epsilon,$$

where the interval from $\hat{\beta}_1 - (t_{\epsilon/2})(S_{\hat{\beta}_1})$ to $\hat{\beta}_1 + (t_{\epsilon/2})(S_{\hat{\beta}_1})$, with numerical quantities from a given sample substituted for $\hat{\beta}_1$ and $S_{\hat{\beta}_1}$, is known as a $100(1 - \epsilon)$ percent confidence interval for $\beta_1$. An equivalent expression for a confidence interval for $\beta_0$ can be constructed in exactly the same manner. This means that the probability of obtaining a sample such that the interval $\hat{\beta}_1 - (t_{\epsilon/2})(S_{\hat{\beta}_1})$ to $\hat{\beta}_1 + (t_{\epsilon/2})(S_{\hat{\beta}_1})$ contains $\beta_1$ is $1 - \epsilon$. Or, if a large number of samples were taken and a $\hat{\beta}_1$ and $S_{\hat{\beta}_1}$ were calculated for each sample, in $100(1 - \epsilon)$ percent of the samples the interval would include $\beta_1$ and in $100\epsilon$ percent of the cases it would not. It is in this sense that a confidence interval provides additional information about $\beta_1$.

Clearly, if the interval is small, we have probably estimated $\beta_1$ with considerable precision and conversely, if the interval is large, we really have very little idea of the value of $\beta_1$. The width of the confidence interval is $2(t_{\epsilon/2})(S_{\hat{\beta}_1})$, so the standard error $S_{\hat{\beta}_1}$ is an indicator of the precision of estimation—the smaller $S_{\hat{\beta}_1}$, the more information we have about $\beta_1$.

Returning to the tomato example, we have

$$S_{\hat{\beta}_1} = \left(\frac{\hat{\sigma}^2}{\Sigma x^2}\right)^{\frac{1}{2}} = \left(\frac{1633}{416.9}\right)^{\frac{1}{2}} = 1.979,$$

$$S_{\hat{\beta}_0} = \left[\frac{\hat{\sigma}^2}{\Sigma x^2}\left(\frac{\Sigma X^2}{N}\right)\right]^{\frac{1}{2}} = \left[\left(\frac{1633}{416.9}\right)\left(\frac{3621}{10}\right)\right]^{\frac{1}{2}} = 37.66.$$

Taking $\epsilon = 0.05$, and given $N - 2 = 8$, $t_{\epsilon/2} = 2.306$ and the 95 percent confidence interval for $\beta_1$ is

$$-11.504 \pm 2.306 \times 1.979,$$

that is,

$-16.068$ to $-6.940$.

Similarly, the 95 percent confidence interval for $\beta_0$ is

$380 \pm 2.306 \times 37.66$

or

$293.2$ to $466.8$.

In the case of a 99 percent confidence interval for $\beta_1$ ($\epsilon = 0.01$, $t_{\epsilon/2} = 3.355$), the interval is

$-11.504 \pm 3.355 \times 1.979$

or

$-18.144$ to $-4.864$.

## Hypothesis Testing

Often an important reason for regression analysis is to test hypotheses about the parameters $\beta_1$ and/or $\beta_0$. It may be assumed, for example, that $\beta_1$ has the known value $\beta_1^*$. If this (null) hypothesis is true, then by (2.5.8) the t-ratio, given by

$$t = \frac{\hat{\beta}_1 - \beta_1^*}{S_{\hat{\beta}_1}} \tag{2.6.3}$$

has a $t_{N-2}$ distribution.

The null hypothesis

$$H_0: \beta_1 = \beta_1^* \tag{2.6.4}$$

can then be tested by computing (2.6.3), that is, evaluating t and determining whether the value of the ratio is consistent with $t_{N-2}$. Statisticians define this rather vague notion in terms of the probability of incorrectly rejecting the null hypothesis $H_0$ in favor of the alternative hypothesis, that is, the probability of commiting a type I error.

If we wish to control the probability of a type I error to at most $\epsilon$ (for example, $\epsilon = 0.05$), decision rules will be determined by the alternative hypothesis $H_1$. We distinguish three cases:

(i) $H_1: \beta_1 \neq \beta_1^*$. This is a two-tail test and the decision rule would be:
Reject $H_0$ (in favor of $H_1$) if

$$|t| > t_{\epsilon/2}, \tag{2.6.5}$$

where $t_{\epsilon/2}$ is defined by (2.6.2).

(ii) $H_1: \beta_1 > \beta_1^*$ (one-tail test). The decision rule is:
Reject $H_0$ if

$$t > t_\epsilon, \tag{2.6.6}$$

where $t_\epsilon$ is such that

$$P(t_{N-2} > t_\epsilon) = \epsilon. \tag{2.6.7}$$

(iii) $H_1: \beta_1 < \beta_1^*$ (one-tail test). The decision rule is:
Reject $H_0$ if

$$t < -t_\epsilon, \tag{2.6.8}$$

where $t_\epsilon$ is defined in (2.6.7).

These three situations are depicted in Figure 2.6.1. Exactly the same procedure is followed for tests about $\beta_0$, except that the t-ratio is defined in this case by (2.5.9).

A common null hypothesis which is tested is that of $\beta_1 = 0$. If $\beta_1 = 0$, the true model is

$$Y_i = \beta_0 + u_i,$$

implying that the variable $X_i$ is irrelevant in explaining variation in $Y_i$. As it is always important to know whether the regressor variable is relevant, this hypothesis test is carried out routinely. In this case ($\beta_1^* = 0$), the t-ratio becomes

$$t = \frac{\hat{\beta}_1}{S_{\hat{\beta}_1}}. \tag{2.6.9}$$

It is useful to note that if $\epsilon = 0.05$, and N - 2 is not too small (say, N - 2 > 12),

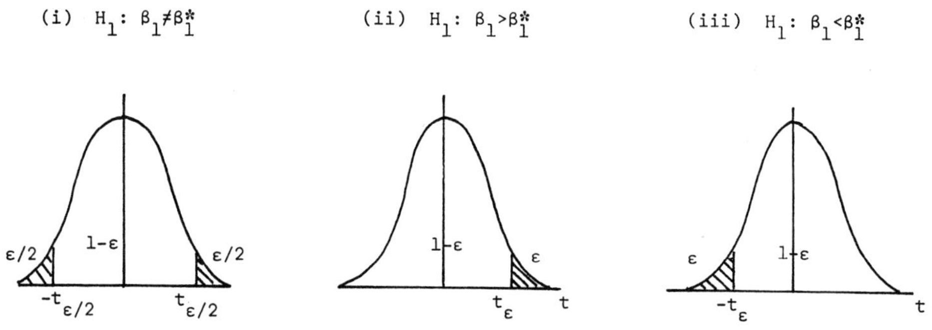


Figure 2.6.1. The regions of rejection of $H_0$ are shaded.

then $t_{\epsilon/2}$ is approximately equal to 2. This gives rise to the very useful rule of thumb that if the estimate $\hat{\beta}_1$ is more than twice its standard error $S_{\hat{\beta}_1}$, then $X_i$ is relevant in explaining variations in $Y_i$. As with all such rules, this must be treated with caution, and we will see subsequently that many more factors need to be taken into consideration when specifying the model.

In economic analysis we frequently have a priori information on the parameters that may enable us to specify a one-tail test, which is more sensitive than a two-tail test. For example in the tomato consumption model it would be quite certain that if price ($X_i$) is relevant at all in explaining variations in consumption, it would affect consumption inversely. The appropriate hypotheses would therefore be

$$H_0: \beta_1 = 0$$

$$H_1: \beta_1 < 0.$$

Taking $\epsilon = 0.05$ (that is, testing at the 5 percent level of significance) we would reject $H_0$ if $t < -1.860$. If we did not use our economic knowledge of the effect of price, we would specify the alternative hypothesis as

$$H_1: \beta_1 \neq 0,$$

and consequently would reject $H_0$ only if $|t| > 2.306$. This is clearly not as sensitive a test. Using our computed values $\hat{\beta}_1 = -11.504$ and $S_{\hat{\beta}_1} = 1.979$, we have that if $H_0$ is true,

$$t = \frac{-11.504}{1.979} = -5.813$$

and thus $H_0$ is rejected, implying that price has an inverse effect on the consumption of tomatoes.

## Prediction

Another important use of an estimated regression model is prediction. We distinguish two types and illustrate them using the Colworths tomato data.

<u>Prediction of the (Conditional) Mean</u>. Let us assume that $X_i$ takes some given, fixed value $X_0$. We know that

$$E[Y_i | X_0] = \beta_0 + \beta_1 X_0$$

and there are many situations in which we may want to predict this mean. Graphically, we are predicting the point on the true line corresponding to $X_0$. It seems natural to predict this by the point corresponding to $X_0$ on the <u>estimated</u> line. We therefore define the predictor $\hat{Y}_0$ by

$$\hat{Y}_0 = \hat{\beta}_0 + \hat{\beta}_1 X_0. \tag{2.6.10}$$

$\hat{Y}_0$ is a linear combination of the random variables $\hat{\beta}_0$ and $\hat{\beta}_1$, and is therefore a random variable having a mean and variance. Taking expectations of (2.6.10), we have

$$E(\hat{Y}_0) = E(\hat{\beta}_0) + E(\hat{\beta}_1)X_0$$
$$= \beta_0 + \beta_1 X_0$$

because $\hat{\beta}_0$ and $\hat{\beta}_1$ are unbiased estimators of $\beta_0$ and $\beta_1$, respectively. Thus

$$E(\hat{Y}_0) = E[Y_i | X_0], \tag{2.6.11}$$

and $\hat{Y}_0$ is seen to be an unbiased predictor of $E[Y_i | X_0]$.

From the properties of linear combinations of random variables,

$$\mathrm{var}(\hat{Y}_0) = \mathrm{var}(\hat{\beta}_0) + \mathrm{var}(\hat{\beta}_1)X_0^2 + 2\,\mathrm{cov}(\hat{\beta}_0, \hat{\beta}_1)X_0.$$

Substituting expressions (2.4.2), (2.4.3), and (2.4.4) we see (after a little simplification) that

$$\mathrm{var}(\hat{Y}_0) = \sigma^2\left[\frac{1}{N} + \frac{(X_0 - \bar{X})^2}{\Sigma x^2}\right]. \tag{2.6.12}$$

It we further assume that the disturbances $u_i$ are normally distributed, it can be shown, analogously to Theorem 2.5.2, that

$$t = \frac{\hat{Y}_0 - E[Y_i | X_0]}{\hat{\sigma}[1/N + (X_0 - \bar{X})^2/\Sigma x^2]^{\frac{1}{2}}} \sim t_{N-2}. \tag{2.6.13}$$

This result forms the basis for inference about the conditional mean $E[Y_i | X_0]$.

EXAMPLE Suppose it has been claimed that when the price of tomatoes is 15 cents/kg the average consumption is 220 kg/week. Test this claim using the Colworths data.

Here

$$H_0: E[Y | X_0 = 15] = 220,$$

$$H_1: E[Y | X_0 = 15] \neq 220.$$

From the estimated regression model

$$\hat{Y}_0 = \hat{\beta}_0 + \hat{\beta}_1 X_0$$
$$= 380 - (11.5)(15).$$

That is,

$$\hat{Y}_0 = 207.5.$$

We have calculated

$$\bar{X} = \frac{179}{10} = 17.9,$$

$$\hat{\sigma} = (1633)^{\frac{1}{2}} = 40.41,$$

$$\Sigma x^2 = 416.9.$$

Thus, by (2.6.13)

$$t = \frac{207.5 - 220}{40.41[1/10 + (15 - 17.9)^2/416.9]^{\frac{1}{2}}} = -\frac{12.5}{14.01} = -0.892.$$

At the 5 percent level of significance, we will reject $H_0$ only if $|t| > 2.306$, and thus failing to reject $H_0$, the claim about average consumption is accepted.

Alternatively, we may wish to construct a 95 percent confidence interval for the true average consumption when $X_0 = 15$. The $100(1 - \epsilon)$ percent confidence interval would be given in general by

$$\hat{Y}_0 \pm t_{\epsilon/2}\hat{\sigma}\left[\frac{1}{N} + \frac{(X_0 - \bar{X})^2}{\Sigma x^2}\right]^{\frac{1}{2}}$$

and in this case ($\epsilon = 0.05$, $t_{\epsilon/2} = 2.306$) by

$$207.5 \pm 2.306 \times 14.01$$

That is, the confidence interval is

175.2 to 239.8.

Prediction of a Particular Value of $Y_i$. To distinguish this type of prediction from the former, consider the following situation: The manager has decided to maintain the price of tomatoes next week at 15 cents/kg. He wants to find a 95 percent confidence interval for the quantity of tomatoes that will be demanded. Notice that he is not interested in this case in average demand when $X_0 = 15$, but demand next week.

We will denote by $Y_0$ the particular value of Y we are trying to predict. Then

$$Y_0 = \beta_0 + \beta_1 X_0 + u_0,$$

where $u_0$ is the value of the disturbance. As we are still able to predict only on the basis of our estimated model, we will define the predictor of $Y_0$ by $\hat{Y}_0$, given by (2.6.10). In other words, whether we are predicting a mean or a particular value we will use exactly the same predictor.

Consider now the error in prediction $\hat{Y}_0 - Y_0$.

$$\begin{aligned}\hat{Y}_0 - Y_0 &= \hat{\beta}_0 + \hat{\beta}_1 X_0 - (\beta_0 + \beta_1 X_0 + u_0) \\ &= (\hat{\beta}_0 - \beta_0) + (\hat{\beta}_1 - \beta_1)X_0 - u_0.\end{aligned}$$

Once again, using the result from linear combinations of random variables, we are able to show that

$$\text{var}(\hat{Y}_0 - Y_0) = \sigma^2\left[1 + \frac{1}{N} + \frac{(X_0 - \bar{X})^2}{\Sigma x^2}\right]. \tag{2.6.14}$$

Comparing (2.6.12) and (2.6.14), we have that

$$\text{var}(\hat{Y}_0 - Y_0) = \text{var}(\hat{Y}_0) + \sigma^2 \geq \text{var}(\hat{Y}_0). \tag{2.6.15}$$

The appropriate distributional result for prediction of a particular value of the dependent variable is

$$t = \frac{\hat{Y}_0 - Y_0}{\hat{\sigma}[1 + 1/N + (X_0 - \bar{X})^2/\Sigma x^2]^{\frac{1}{2}}} \sim t_{N-2}. \tag{2.6.16}$$

We may now use this result to obtain the 95 percent confidence interval for next week's demand, given $X_0 = 15$.

$$\hat{\sigma}\left[1 + \frac{1}{N} + \frac{(X_0 - \bar{X})^2}{\Sigma x^2}\right]^{\frac{1}{2}} = 42.77$$

Thus the confidence interval is

$$207.5 \pm 2.306 \times 42.77$$

or

$$108.9 \text{ to } 306.1$$

We notice that the confidence interval for the prediction of the mean is narrower than that for the particular value of the dependent variable, even though exactly the same predictor is used. This will always be the case, and is a direct result of equation (2.6.15).

Finally, we note that in both (2.6.12) and (2.6.14) the variances depend on $(X_0 - \bar{X})^2$. The variances will be a minimum when $X_0 = \bar{X}$ and increase the farther $X_0$ is away from $\bar{X}$. This means that prediction will be most precise when $X_0 = \bar{X}$, with precision decreasing for values of $X_0$ away from $\bar{X}$.

## 2.7 A MEASURE OF GOODNESS OF FIT

By drawing the line of best fit on the scatter diagram (see, for example, Figure 2.2.1) it is possible to assess visually whether the data have been effectively described, or explained, by the fitted line. Obviously, the better the fit, the more confidently we will use the model for inference and prediction. Although a graphical assessment may sometimes be adequate, it would be useful to have a single number whose value could be used as an index of goodness of fit. When we consider multiple regression, such an index becomes even more important because simple graphical methods can no longer be used.

Let us consider the total variation of the dependent variable about its mean $\bar{Y}$, measured by $\Sigma y^2$. Now

$$\Sigma y_i^2 = \Sigma (Y_i - \bar{Y})^2$$

$$= \Sigma [(\hat{Y}_i - \bar{Y}) + (Y_i - \hat{Y}_i)]^2$$

$$= \Sigma [(\hat{\beta}_0 + \hat{\beta}_1 X_i - \bar{Y}) + e_i]^2.$$

Because $\hat{\beta}_0 = \bar{Y} - \hat{\beta}_1 \bar{X}$, it follows that

$$\hat{\beta}_0 + \hat{\beta}_1 X_i - \bar{Y} = \hat{\beta}_1 (X_i - \bar{X}) = \hat{\beta}_1 x_i.$$

Thus

$$\Sigma y_i^2 = \Sigma (\hat{\beta}_1 x_i + e_i)^2$$

$$= \hat{\beta}_1^2 \Sigma x_i^2 + \Sigma e_i^2 + 2\hat{\beta}_1 \Sigma x_i e_i.$$

Also,

$$\Sigma x_i e_i = \Sigma (X_i - \bar{X}) e_i = \Sigma X_i e_i - \bar{X} \Sigma e_i.$$

Now, by the fundamental properties of OLS residuals given in (2.3.11) and (2.3.12),

$$\Sigma X_i e_i = \Sigma e_i = 0,$$

implying that the cross-product term $\Sigma x_i e_i$ is zero. Thus

$$\Sigma y_i^2 = \hat{\beta}_1^2 \Sigma x_i^2 + \Sigma e_i^2$$

or

$$\Sigma y_i^2 = \Sigma \hat{y}_i^2 + \Sigma e_i^2. \tag{2.7.1}$$

$\Sigma y_i^2$ is called the <u>total sum of squares</u> (TSS), $\Sigma \hat{y}_i^2$ is called the <u>explained sum of squares</u> (ESS) and measures the variation that is explained by the model. Finally, $\Sigma e_i^2$ is called the <u>residual sum of squares</u> (RSS) and is a measure of the amount of variation not explained by the model. Using this notation, we have the important result

$$\text{TSS} = \text{ESS} + \text{RSS}. \tag{2.7.2}$$

Clearly, if ESS represents a high proportion of TSS the model has explained most of the variation in $Y_i$. That is, the line fits the data well. This leads us to define a measure of goodness of fit, denoted by $r^2$, as

$$r^2 = \frac{\text{ESS}}{\text{TSS}} = \frac{\Sigma \hat{y}^2}{\Sigma y^2}. \tag{2.7.3}$$

Because ESS and RSS are sums of squares they can never be negative, and this implies immediately that

$$0 \leq \text{ESS} \leq \text{TSS}.$$

Dividing by TSS, we obtain

$$0 \leq r^2 \leq 1. \tag{2.7.4}$$

An $r^2$ value close to 1 represents a very good fit to the data, whereas a low $r^2$ means a poor fit. The basic interpretation of $r^2$ is as a measure of the proportion of variation in the dependent variable which is explained by the regression.

We now present two further interpretations of $r^2$:

(a) $$\Sigma \hat{y}^2 = \hat{\beta}_1^2 \Sigma x^2 = \hat{\beta}_1 \Sigma xy = \Sigma(\hat{\beta}_1 x)y = \Sigma y\hat{y}$$

Thus

$$r^2 = \frac{\Sigma \hat{y}^2}{\Sigma y^2} = \frac{(\Sigma \hat{y}^2)^2}{\Sigma y^2 \, \Sigma \hat{y}^2} = \frac{(\Sigma y\hat{y})^2}{\Sigma y^2 \, \Sigma \hat{y}^2}.$$

But $(\Sigma y\hat{y})^2/\Sigma y^2 \, \Sigma \hat{y}^2$ is simply the square of the sample correlation between $Y_i$ and its predicted value $\hat{Y}_i$. This interpretation holds for multiple as well as simple regression and enables us to use $r^2$ as an index of the predictive performance of the estimated model. If it has predicted well with the sample observations (as measured by high $r^2$) it is likely to predict nonsample values of $Y_i$ with some accuracy.

(b) $$\begin{aligned} r^2 &= \frac{\Sigma \hat{y}^2}{\Sigma y^2} = \frac{\Sigma \hat{y}^2 \, \Sigma x^2}{\Sigma y^2 \, \Sigma x^2} = \frac{\hat{\beta}_1^2 \Sigma x^2 \, \Sigma x^2}{\Sigma y^2 \, \Sigma x^2} \\ &= \frac{(\hat{\beta}_1 \Sigma x^2)^2}{\Sigma y^2 \, \Sigma x^2} = \frac{(\Sigma xy)^2}{\Sigma y^2 \Sigma x^2} \quad \text{[by (2.2.9)]}. \end{aligned} \tag{2.7.5}$$

But $(\Sigma xy)^2/\Sigma y^2 \, \Sigma x^2$ is the square of the sample correlation between $Y_i$ and $X_i$. This interpretation of the measure of goodness of fit has no relevance in the context of multiple regression.

Finally, computation of $r^2$ is best achieved by use of one of the following three formulas:

$$r^2 = \frac{\Sigma \hat{y}^2}{\Sigma y^2} = \frac{\hat{\beta}_1^2 \Sigma x^2}{\Sigma y^2} \tag{2.7.6}$$

or

$$r^2 = \frac{\hat{\beta}_1(\hat{\beta}_1 \Sigma x^2)}{\Sigma y^2} = \frac{\hat{\beta}_1 \Sigma xy}{\Sigma y^2} \tag{2.7.7}$$

or

$$r^2 = \frac{ESS}{TSS} = 1 - \frac{RSS}{TSS}$$

$$= 1 - \frac{\Sigma e^2}{\Sigma y^2} \tag{2.7.8}$$

when $\Sigma e^2$ has been found in order to estimate $\sigma^2$ [see (2.5.6)].

EXAMPLE Using the Colworths tomato data, we have

$$\hat{\beta}_1 = -11.504,$$

$$\Sigma xy = -4796,$$

$$\Sigma y^2 = 68,240.$$

Thus, from (2.7.7),

$$r^2 = \frac{(-11.504)(-4796)}{68,240} = 0.809,$$

and we see that the estimated regression model has explained about 81 percent of the variation in demand for tomatoes.

## 2.8 MODELS THAT HAVE NO INTERCEPT

It sometimes happens that a model is specified without an intercept. That is, $\beta_0 = 0$. Consider the following two examples:

(a) The Keynesian consumption function may be written in the form

$$E[C_i | Y_i] = \beta_0 + \beta_1 Y_i$$

where $C_i$ is consumption and $Y_i$ income. Here $\beta_1$ is the marginal propensity to consume (MPC). The average propensity to consume (APC) is defined by

$$APC = \frac{E[C_i | Y_i]}{Y_i}$$

$$= \frac{\beta_0}{Y_i} + \beta_1.$$

If it is believed that marginal and average propensities to consume are equal, then

$$\beta_1 = \frac{\beta_0}{Y_i} + \beta_1,$$

implying that $\beta_0 = 0$. Thus the consumption function would take the form

$$E[C_i \mid Y_i] = \beta_1 Y_i.$$

The corresponding regression model

$$C_i = \beta_1 Y_i + u_i$$

has no intercept.

(b) Suppose that we have the model

$$Y_i = \beta_0 + \beta_1 X_i + u_i$$

and divide throughout by $\sqrt{X_i}$. Then

$$\frac{Y_i}{\sqrt{X_i}} = \beta_0 \frac{1}{\sqrt{X_i}} + \beta_1 \sqrt{X_i} + v_i,$$

where $v_i = u_i/\sqrt{X_i}$. This is one example of a transformation of the model. As we shall see later, there are many situations in which transformation of the model is necessary prior to estimation. It is important to realize that though $\beta_0$ is the intercept in the original model, the transformed model does not have an intercept. This is seen clearly when we define the intercept as the parameter associated with the "variable" which only ever takes the value 1.

We will now consider the model

$$Y_i = \beta_1 X_i + u_i. \tag{2.8.1}$$

If we apply the least squares principle in the same way as in Section 2.2, we have

$$\Sigma e_i^2 = \Sigma (Y_i - \hat{\beta}_1 X_i)^2,$$

which can be minimized by setting

$$\frac{\partial \Sigma e_i^2}{\partial \hat{\beta}_1} = 0.$$

That is,

$$-2 \Sigma (Y_i - \hat{\beta}_1 X_i) X_i = 0,$$

yielding

$$\hat{\beta}_1 = \frac{\Sigma X_i Y_i}{\Sigma X_i^2} . \tag{2.8.2}$$

This equation gives the least squares estimator of $\beta_1$ analogously to (2.2.9), with one important difference—the variables $X_i$ and $Y_i$ are not corrected for the mean.

The residuals $e_i$ are given by

$$e_i = Y_i - \hat{Y}_i = Y_i - \hat{\beta}_1 X_i,$$

where $\hat{\beta}_1$ is now given by (2.8.2). It can be shown that (2.3.12), namely,

$$\Sigma e_i X_i = 0, \tag{2.8.3}$$

still holds but (2.3.11) does not. That is,

$$\Sigma e_i \neq 0. \tag{2.8.4}$$

The fact that the residuals satisfy only one constraint (2.8.3) instead of the two constraints (2.3.11) and (2.3.12) has important implications:

(a) The number of degrees of freedom is N - 1 when there is no intercept, rather than N - 2. Thus an unbiased estimator of $\sigma^2$ is in this case $\Sigma e^2/(N - 1)$ and not $\Sigma e^2/(N - 2)$ as given in (2.5.5). Furthermore, the distributional results concerning $\hat{\beta}_1$ and the predictor $\hat{Y}_0$ apply only if $t_{N-2}$ is replaced by $t_{N-1}$.

(b) The partitioning (2.7.2) of sums of squares holds only when $\Sigma e = 0$. When there is no intercept,

$$TSS \neq ESS + RSS.$$

Thus the foundation for the definition of $r^2$ as a measure of goodness of fit collapses, and the three equivalent expressions (2.7.6), (2.7.7), and (2.7.8) usually give different answers. In fact, it is possible to obtain $r^2$ values which are greater than 1 or less than zero—depending on the particular definition that is used. A measure of goodness of fit that does satisfy (2.7.4) is

$$r^2 = \frac{(\Sigma XY)^2}{\Sigma X^2 \, \Sigma Y^2} \tag{2.8.5}$$

$$= \frac{\hat{\beta}_1^2 \, \Sigma X^2}{\Sigma Y^2} \tag{2.8.6}$$

$$= \hat{\beta}_1 \frac{\Sigma XY}{\Sigma Y^2} \tag{2.8.7}$$

but here quantities are not mean corrected. For a full discussion of these issues, the reader is referred to Aigner (1971, 85-90).

## 2.9 MODELS THAT ARE NONLINEAR IN THE VARIABLES

An essential feature of the model (2.1.1) is that it is linear in the parameters $\beta_0$ and $\beta_1$. There are often situations in which a model is linear in the parameters but the regressor variable is nonlinear. Such models can be expressed in the standard form by a transformation of the variables. We will cite two examples to illustrate the procedure.

(a) The Phillips Curve.[11] The relationship between changes in the wage rate and the level of unemployment is known as the Phillips curve. If we define $W_t$ to be the wage rate at time t, and

$$\dot{W}_t = \frac{W_t - W_{t-1}}{W_{t-1}}$$

to be the rate of change in the wage rate, it may be postulated that $\dot{W}_t$ is directly proportional to the rate of excess demand for labor $D_t^*$. Thus

$$\dot{W}_t = \alpha D_t^*.$$

Furthermore, $D_t^*$ would be related inversely to the unemployment rate $R_t$. A model that expresses the way a small decrease in $R_t$ would have a much larger effect on $D_t^*$ when $R_t$ is small than would the same decrease when $R_t$ is large is

$$D_t^* = c + d\frac{1}{R_t}.$$

Thus

$$\dot{W}_t = \alpha\left(c + d\frac{1}{R_t}\right)$$

and $\dot{W}_t$ is related linearly to the (nonlinear) variable $1/R_t$.

An appropriate regression model would be

$$Y_t = \beta_0 + \beta_1 X_t + u_t,$$

where

$$Y_t = \frac{W_t - W_{t-1}}{W_{t-1}}$$

and

$$X_t = \frac{1}{R_t}.$$

(b) Let us suppose we are investigating a demand function for some commodity. We will assume that the quantity demanded $Q_i$ is determined by the price $P_i$. Also suppose that it is believed that the elasticity of demand does not depend on the price, but remains constant. Elasticity is defined as $(dQ/dP)(P/Q)$. Thus we are assuming that

$$\frac{P}{Q}\frac{dQ}{dP} = \beta_1,$$

where $\beta_1$ is some (unknown) constant. If we integrate this relationship with respect to P, we obtain

$$\ln Q = \beta_0 + \beta_1 \ln P,$$

where $\beta_0$ is an arbitrary constant. The appropriate regression model under the assumption of constant elasticity would be

$$Y_i = \beta_0 + \beta_1 X_i + u_i,$$

where $Y_i = \ln Q_i$ and $X_i = \ln P_i$. The parameter estimate $\hat{\beta}_1$ would be the estimate of the elasticity of demand.

## 2.10 AN EXAMPLE

We close this chapter by presenting an example of a simple regression analysis which is a little more realistic than the Colworths example that has been used so far. This example has a double purpose. First, it provides another illustration of the computations that are involved; second and more important, we will often refer back to this example to illustrate later concepts.

Consider the data shown in Table 2.10.1. The dependent variable is private consumption (denoted by $C_i$) and the regressor variable is gross national expenditure (denoted by $Y_i$).[12] We will estimate the model

$$C_i = \beta_0 + \beta_1 Y_i + u_i, \qquad (2.10.1)$$

which is the simplest representation of the Keynesian absolute income hypothesis. Here $\beta_1$ is the marginal propensity to consume (MPC) and we assume that $0 < \beta_1 < 1$. The parameter $\beta_0$ measures autonomous consumption at zero income level. We assume that $\beta_0 \geq 0$. We note that average propensity to consume (APC), defined by $C_i/Y_i$, is given by

$$\frac{C_i}{Y_i} = \frac{\beta_0}{Y_i} + \beta_1.$$

Thus the model implies that

Table 2.10.1[a]

| Year | Private consumption [billions ($10^9$) of dollars] | Gross national expenditure [billions ($10^9$) of dollars] |
|---|---|---|
| 1948-49 | 7.403 | 10.390 |
| 1949-50 | 7.793 | 11.379 |
| 1950-51 | 8.299 | 12.618 |
| 1951-52 | 8.258 | 13.306 |
| 1952-53 | 8.089 | 11.588 |
| 1953-54 | 8.565 | 12.746 |
| 1954-55 | 9.098 | 13.874 |
| 1955-56 | 9.370 | 14.230 |
| 1956-57 | 9.450 | 13.895 |
| 1957-58 | 9.748 | 14.629 |
| 1958-59 | 10.035 | 15.449 |
| 1959-60 | 10.680 | 16.474 |
| 1960-61 | 10.840 | 17.222 |
| 1961-62 | 11.149 | 16.720 |
| 1962-63 | 11.840 | 18.359 |
| 1963-64 | 12.572 | 19.511 |
| 1964-65 | 13.148 | 21.497 |
| 1965-66 | 13.539 | 21.987 |
| 1966-67 | 14.196 | 23.095 |
| 1967-68 | 14.992 | 24.236 |
| 1968-69 | 15.810 | 26.248 |
| 1969-70 | 16.804 | 27.517 |
| 1970-71 | 17.515 | 28.517 |
| 1971-72 | 18.322 | 29.323 |
| 1972-73 | 19.420 | 30.697 |
| 1973-74 | 20.591 | 34.118 |
| 1974-75 | 21.420 | 34.367 |
| 1975-76 | 22.068 | 34.717 |
| 1976-77 | 22.661 | 35.906 |
| 1977-78 | 22.953 | 35.698 |
| 1978-79 | 23.743 | 37.776 |
| 1979-80 | 24.340 | 37.560 |
| 1980-81 | 25.088 | 40.139 |
| 1981-82 | 26.073 | 42.421 |

[a]All figures are at average 1966-67 prices.

Source: Australian National Accounts 1980-81.

$$\text{APC} > \text{MPC}, \tag{2.10.2}$$

except if $\beta_0 = 0$, in which case APC = MPC.

The first step in estimating the regression relationship is to find the five quantities $\Sigma Y$, $\Sigma Y^2$, $\Sigma C$, $\Sigma C^2$, and $\Sigma CY$:

$$N = 34,$$

$$\Sigma Y = 798.209, \qquad \Sigma C = 505.904,$$

$$\Sigma Y^2 = 21{,}966.716157, \qquad \Sigma C^2 = 8{,}711.732776,$$

$$\Sigma CY = 13{,}827.641483.$$

These numbers are now used to compute the three mean-adjusted quantities $\Sigma y^2$, $\Sigma c^2$, and $\Sigma cy$.

$$\Sigma y^2 = 21{,}966.716157 - \frac{798.209^2}{34} = 3227.3747,$$

$$\Sigma c^2 = 8711.732776 - \frac{505.904^2}{34} = 1184.1193,$$

$$\Sigma cy = 13{,}827.641483 - \frac{(505.904)(798.209)}{34} = 1950.6672.$$

These eight numbers provide a complete summary of the sample information with respect to a simple regression analysis.

$$\hat{\beta}_1 = \frac{\Sigma cy}{\Sigma y^2} = \frac{1950.6672}{3227.3747} = 0.60441$$

$$\hat{\beta}_0 = \frac{1}{N}(\Sigma C - \hat{\beta}_1 \Sigma Y) = 0.68996$$

$$\hat{\sigma}^2 = \frac{1}{N-2}(\Sigma c^2 - \hat{\beta}_1 \Sigma cy)$$

$$= \frac{1}{32}(1184.1193 - 0.60441 \times 1950.6672).$$

Thus

$$\hat{\sigma}^2 = 0.15989,$$

$$S_{\hat{\beta}_1} = \left(\frac{\hat{\sigma}^2}{\Sigma y^2}\right)^{\frac{1}{2}} = \left(\frac{0.15989}{3227.3747}\right)^{\frac{1}{2}} = 0.00704,$$

$$S_{\hat{\beta}_0} = \left(\frac{\hat{\sigma}^2}{\Sigma y^2}\,\frac{\Sigma Y^2}{N}\right)^{\frac{1}{2}} = \left(\frac{0.15989 \times 21{,}966.716}{3227.3747 \times 34}\right)^{\frac{1}{2}} = 0.1789,$$

$$r^2 = \frac{\hat{\beta}_1 \Sigma cy}{\Sigma c^2} = \frac{0.60441 \times 1.9506672}{1184.1193} = 0.996.$$

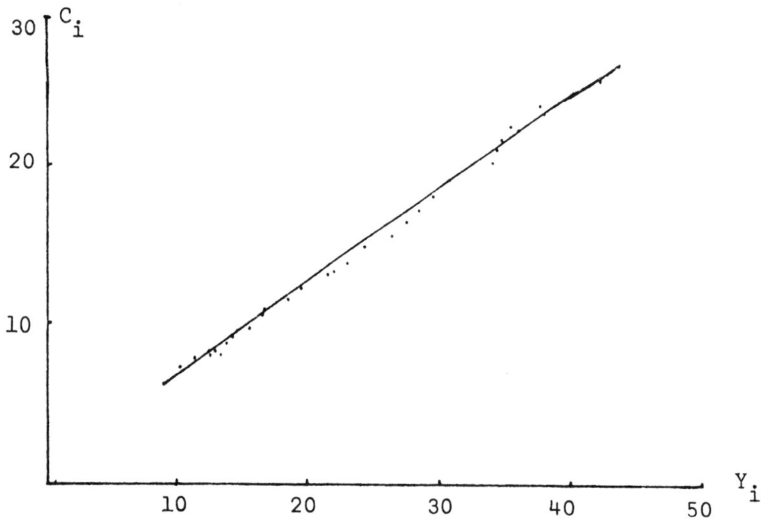


Figure 2.10.1. Scatter diagram and line of best fit for consumption-income data.

We are thus able to report the results[13] of the estimation as

$$\hat{C}_i = \underset{(0.179)}{0.690} + \underset{(0.007)}{0.604}Y_i; \quad r^2 = 0.996, \quad \hat{\sigma}^2 = 0.160.$$

See Figure 2.10.1 for the line of best fit.

Let us now use the estimates to test two hypotheses:

(a) Test the hypothesis that the marginal propensity to consume is 0.7, with $\epsilon = 0.01$.

$$H_0: \beta_1 = 0.7,$$

$$H_1: \beta_1 \neq 0.7.$$

If $H_0$ is true,

$$t = \frac{\hat{\beta}_1 - 0.7}{S_{\hat{\beta}_1}} \sim t_{32}.$$

Here,

$$t = \frac{0.60441 - 0.7}{0.00704} = -13.58.$$

The critical value $t_{0.005} \approx 2.75$. Thus, as $|t| > t_{0.005}$, $H_0$ is rejected and the null hypothesis is not consistent with the data.

(b) Test the hypothesis that the marginal and average propensities to consume are equal ($\epsilon = 0.01$):[14]

$$H_0 : \beta_0 = 0$$

$$H_1 : \beta_0 > 0.$$

If $H_0$ is true,

$$t = \frac{\hat{\beta}_0 - 0}{S_{\hat{\beta}_0}} \sim t_{32}.$$

Here

$$t = \frac{0.690 - 0}{0.1789} = 3.86.$$

The critical value $t_{0.01} = 2.46$, so $H_0$ is rejected and we conclude that APC > MPC.

Let us suppose now that we had assumed that MPC = APC, implying that $\beta_0 = 0$. The appropriate model would then be

$$C_i = \beta_1 Y_i + u_i$$

and by (2.8.2),

$$\hat{\beta}_1 = \frac{\Sigma CY}{\Sigma Y^2}$$

$$= \frac{13,827.6414}{21,966.7162} = 0.6295.$$

Also, by (2.8.7),

$$r^2 = \frac{\hat{\beta}_1 \Sigma CY}{\Sigma C^2} = \frac{0.6295 \times 13,827.641}{8711.733} = 0.999.$$

Thus the goodness of fit, as measured by $r^2$, is higher for the model which assumes that MPC = APC, even though we have rejected this hypothesis. This apparent inconsistency is resolved when we recall that for a model with no intercept, $r^2$ measures the proportion of variation about the origin, whereas for a model with an intercept, $r^2$ measures the proportion of variation about the mean which is explained by the regression.

## NOTES

1. The significance of assumption (b) is that $E(u_i^2)$ does <u>not</u> depend on i. That is, it is the same for all disturbances.

2. In practice, because the values of the parameters $\beta_0$ and $\beta_1$ are never known, the values of the disturbances $u_i$ can never be known.

3. Although mathematically $\beta_0$ always will have this interpretation, there are many practical instances where the regression model is considered valid (a good approximation) only for a certain range of the independent variable. If this range does not include $X_i = 0$, one is not forced to follow this interpretation.

4. This derivation of the least squares estimators may be omitted by the reader who is unfamiliar with calculus.

5. We will frequently abbreviate $\Sigma_{i=1}^{N} X_i$ to $\Sigma X_i$ or even $\Sigma X$ when there is no risk of confusion.

6. This condition will be violated only when all the values of $X_i$ are identical. In this case it is impossible to estimate $\beta_0$ and $\beta_1$ using least squares.

7. The term ordinary is used to distinguish it from other least squares methods, for example, weighted least squares, two-stage least squares, and three-stage least squares.

8. It is important to distinguish between an estimator, which is a random variable, and an estimate, which is the value the estimator takes from a given sample. Some writers use different symbols to make this distinction. We do not make any notational distinction, believing that the context always makes clear which we are considering. For example, we will use $\hat{\beta}_1$ to denote both the estimator of $\beta_1$ and the estimate of $\beta_1$ found from a particular sample. The properties of least squares estimators are very important. However, the nontechnical reader may omit the proofs.

9. Here we have used the fact that i and j are dummy suffixes and thus $\Sigma_i x_i^2 = \Sigma_j x_j^2$.

10. Much of the inference is quite similar to confidence intervals and hypothesis testing for population means. This material is covered in all elementary statistics courses and our treatment will therefore be quite brief.

11. Reference is made to Kelejian and Oates (1981, 98-101) for a full discussion. We use their notation.

12. Note that here the regressor variable is denoted by $Y_i$. This is not in accordance with our previous usage but is conventional among economists.

13. This is the conventional way to report the results of a regression analysis. Some authors give t-ratios defined by $\hat{\beta}_0/S_{\hat{\beta}_0}$ and $\hat{\beta}_1/S_{\hat{\beta}_1}$ in parentheses under $\hat{\beta}_0$ and $\hat{\beta}_1$, respectively.

14. We have incorporated our prior knowledge that $\beta_0 \geq 0$ to set up a one-tail test—this is a more sensitive test.

# 3

# The MINITAB Computer Package

## 3.1 INTRODUCTION

In Chapter 2 we studied simple regression, and it was apparent that the computations required were quite considerable. In subsequent chapters, we will be considering the multiple regression model

$$Y_i = \beta_0 + \beta_1 X_{1i} + \cdots + \beta_K X_{Ki} + u_i.$$

We will soon recognize that even for $K = 2$, numerical calculation becomes exceedingly laborious and once $K > 2$ regression analysis on a desk calculator becomes impracticable. It follows that if we are to gain practical experience, it will be necessary to learn how to use a computer.

There are many very good regression packages available which enable users to carry out the necessary computations without having to learn how to write programs. Use of a package involves following conventions in data input and program commands, and finally, correctly interpreting the resulting output. It is worth pointing out right at the beginning that the computer can only carry out the instructions which are given to it. I can <u>never</u> determine what problem has to be solved. The computer is a valuable and necessary complement to a sound theoretical basis—never a substitute.

The particular package we will be using for almost all our work is called MINITAB, a general statistical package that has been chosen because of its wide availability and simplicity of use. In what follows we will be giving the reader enough information about MINITAB to start using it in a regression context. This is not intended to be comprehensive, and the reader is strongly urged to obtain the handbook by Ryan, Joiner, and Ryan (1985), or the MINITAB Reference Card, available from most computer facilities.

## 3.2 GENERAL INFORMATION ON MINITAB

The MINITAB package consists of commands that operate basically on columns of data and constants. A particular column is identified by the letter C followed by a number. For example, C1, C15 represent columns 1 and 15, respectively. A constant is identified by K followed by a number (e.g., K3). There is a dictionary of commands which enable operations to be carried out on the columns of data and constants.

EXAMPLE

```
DELETE ROWS 4 AND 5 FROM COLUMN C3
```

A MINITAB command is identified by at most four letters. In the example above, the only essential elements are

```
DELE 4 5 C3
```

Any text beyond this is purely for the convenience of the user. We will follow the convention of using lowercase letters to denote the nonessential parts of a command. Thus the example above could be written

```
DELEte rows 4 and 5 from column C3
```

## 3.3 DATA MANIPULATION

### Input/Output

A necessary part of a MINITAB program is to define the contents of the columns. The two main commands available for this are READ and SET. For example,

```
READ the following numbers into C1 and C2
    0.1    5.2
    3.7    6.1
    1.8    4.6
    END
SET the following numbers into C5
    0.1  5.2  3.7  6.1  1.8  4.6
    END
```

We recall that these commands may simply be written as

READ C1 C2
SET C5

In the examples above, the data are entered directly into the columns. A far more usual situation is when the data have already been stored in an external file. The commands READ and SET can then be used to transfer these data directly into the nominated columns.

Let us suppose that the file WORK.DTA contains five columns of data. Then

READ 'WORK.DTA' C1-C5

places these data into columns C1 to C5. (Where column numbers are consecutive, it is only necessary to write the first and last, joined by -.) The command

SET 'WORK.DTA' C10

places the numbers in the five columns into the single column C10 by rows. That is, the first row observations are followed by the second row observations, and so on.

Sometimes a user wants to save for future use some or all of the columns of data generated in a session. The WRITe command can be used for this purpose. Suppose that the columns 1 to 20 have been used and it is desired to save columns 5, 6, 10, 15, 16, 17, and 18. These columns can be stored in a file (say, 'SAVE.DTA') as follows:

WRITe 'SAVE.DTA' C5 C6 C10 C15-18

The PRINt command can be used to display the data contained in some or all of the columns. For example,

PRINt C1-4 C6

would cause the observations in these columns to be displayed on the terminal.

## The LET Command

This is the most useful command for defining columns from existing columns or constant. It has the general form

LET C (or K) = mathematical expression.

The arithmetic operators

+ addition
- subtraction
* multiplication
/ division
** exponentiation [e.g., C2**3 = $(C2)^3$]

can be used in the right-hand side, together with functions. For example,

```
LET C7  = 2*C3 + 500
LET K6  = SUM(C2) + K5
LET C15 = LOGE(SQRT(C6)/(C1 + C4))
```

The command

```
LET C2(5) = 6*K1 - 82
```

gives a value to the fifth row of C2.

## Subcommands

Some commands have subcommands. These are used for special options or to convey additional information. To use a subcommand, put a semicolon at the end of the main command line and type each subcommand on a new line ending the line with a semicolon. The final subcommand line is terminated with a period. As far as we are concerned, at least initially, the only commands that have subcommands are COPY and REGRESSION.

## Data Correction and Manipulation

It is often necessary to be able to correct errors and manipulate columns. MINITAB offers many possibilities but we will consider here only six commands: INSErt, DELEte, ERASe, COPY, STACk, and LAG.

INSErt. Rows may be inserted into columns by INSErt between rows K, and K of C ··· C. For example,

```
INSE 3 4 C4 C5
6.2 7.9
END
```

would place a new row between rows 3 and 4 of columns C4 and C5. The new row consists of the numbers 6.2 and 7.9.

<u>DELEte</u>. This command deletes rows from columns and takes the form

DELEte rows K, ..., K of columns C ··· C

for example,

DELE 4, 6, 7, 8, 10  C1-C6.

Where rows are consecutive, the command can be abbreviated by a colon. Thus the example may be written

DELE 4 6:8 10 C1-C6

<u>ERASe</u>. Columns and constants that are no longer needed can be erased. This prevents the worksheet from becoming unnecessarily cluttered and saves storage space. The general form is

ERASe E, ..., E

where E stands for either columns or constants.

<u>COPY</u>. Along with the LET command, COPY is the most versatile tool for data manipulation. It has two subcommands, USE and OMIT. The following examples illustrate its use.

COPY C2 into K1-K6

This copies the column C2, assumed to have six rows, into the constants K1-K6. If C2 has more than six rows, we could use

COPY C2 K1-K6;
USE 1:6.

COPY K1-K6 C2 would construct the column C2 from the constants K1, K2, ..., K6.

COPY C4-C8 into C12-C16;
USE 26:40.

would copy rows 26 to 40 of columns C4 to C8 into the new columns C12 to C16.

COPY C4-C8 into C12-C16;
OMIT 26:40.

would copy C4 to C8 into C12 to C16, omitting rows 26 to 40.

STACk. This command enables columns to be constructed by joining other columns and constants together. The general form is STACk (E, ..., E) on (E, ..., E) and store in (C, ..., C). For example,

```
STACk (C3, C4) (C1, C2) into (C5, C6)
```

This forms C5 by joining C1 onto the end of C3, and C6 by joining C2 onto the end of C4. If parentheses contain only one argument, they may be omitted. For example,

```
STACk C3 C1 C5
```

would create C5, as in the preceding example.

LAG. Econometric models with time-series data often contain lagged variables. The form of the LAG command is

```
LAG K C1 C2
```

which lags C1 by K periods and stores it in C2. The first K rows would not contain any numbers. An alternative form is

```
LAG C1 C2
```

and this lags C1 by one period and stores it in C2. Here the first observation of C2 is missing. LAG may be used as a function with the LET command. For example,

```
LET C6 = LOGE(C4/LAG(C2)).
```

## 3.4 PLOTS AND HISTOGRAMS

A very useful aspect of MINITAB is the ease with which graphs of data can be produced. Of the many available options, we will mention only three. The command

```
PLOT Y in C against X in C
```

will produce a scatter diagram of Y against X. For example, if Y is in C9 and X in C2,

PLOT C9 C2

will produce the required scatter plot.

The command MPLOt (multiple plot) enables the user to graph several variables on the same axes. Thus

MPLOt C3 C1 C5 C1 C6 C1

would plot the three variables C3, C5, and C6 against C1. The first pair of columns are plotted using the symbol A, the second pair the symbol B, and so on. If more than one point falls on the same spot, a count is given.

Finally, the HISTogram command constructs a histogram from the data stored in a particular column. This command takes the form

HIST C5

## 3.5 REGRESSION

The REGRession command takes the form

REGR Y in C against the K regressors in C, ..., C storing the standardized residuals in C, predicted Y in C.

For example,

REGR C1 4 C6 C7 C8 C9 C10 C11

would regress the dependent variable in C1 against the four regressors C6 through C9, storing the standardized residuals in C10 and the predictions in C11. The parameters C10 and C11 are optional in the example above. Thus

REGR C1 4 C6-C9

would perform this regression, except that the standardized residuals and predictions would not be stored for possible future analysis.

The REGR command contains many subcommands, the most important being:

| | |
|---|---|
| NOCOnstant | the model has no intercept ($\beta_0 = 0$) |
| COEFficients C | places $\hat{\beta}$'s in C |
| RESIduals C | stores the residuals in C |
| DW | prints the Durbin-Watson statistic |

For example,

```
REGR C2 4 C1 C3-C5;
NOCO;
RESI C10.
```

would regress C2 (dependent variable) against the regressors C1, C3 to C5. The intercept is omitted and the residuals are stored in C10 for future use.

## Controlling and Improving the Output

There are a number of ways of improving the output.

(a) Columns (variables) can be named. The general form of the NAME command is

NAME C 'NAME' ... C 'NAME'.

Example: Suppose that we wish to use the data of Table 2.10.1 and have stored C in C1 and Y in C2. We may name these columns by

NAME C1 'CONS' C2 'INCOME'

These names will then appear in the output and will help in interpretation.

(b) Comments can be introduced. The command NOTE allows the user to devote that line to a comment which will appear in the output.

(c) Paging can be controlled. The command NEW PAGE causes the printer to skip to a new page before continuing the output.

(d) The amount of detail in the output can be controlled. The command BRIEf takes the general form

BRIEf K,

where K = 1, 2, or 3. This command can be used either preceding the REGR command or as a subcommand. If the former, it stays in effect until it is canceled by a subsequent BRIEf command. If it is used as a subcommand, it applies only to that regression. It is usually preferable to use BRIEf as a subcommand. The values of K have the following effects:

K = 1 output contains only most essential information
K = 2 more detail
K = 3 maximum amount of detail

The default value of K is K = 1. That is, if BRIEf is not used, the value of K is automatically set at 1.

<u>Example:</u>

```
REGR  C2  1  C3;
BRIE  3.
```

Rather than attempting to describe the levels of output, we encourage the reader to experiment.

## 3.6 A THREE-FILE SYSTEM

Files can be used to great advantage in conjunction with MINITAB. We particularly recommend the use of a three-file system, such that data, MINITAB instructions, and output are stored in three separate files. This enables the data and commands to be edited separately and means that the program is not lost when the MINITAB session is terminated. A MINITAB instruction file has the form

```
 OUTFILE 'FILE.OUT'
[MINITAB COMMANDS
|          .
|          .
|          .
|          .
|          .
L STOP
```

If this file is called, for example, 'FILE.INS', it is activated by the command

```
EXEC 'FILE.INS'
```

## 3.7 AN EXAMPLE

Using the consumption and income data of Table 2.10.1, we will now use MINITAB to perform the analysis of Section 2.10. In particular, we will

(a) Regress consumption against income;
(b) Superimpose the line of best fit on the scatter diagram;
(c) Construct a histogram of the residuals; and
(d) Regress consumption against income, omitting the intercept.

If we assume that the data of Table 2.10.1 are stored in a file called 'CONS.DTA', then (a), (b), (c), and (d) can be achieved by the following program, which we will store as 'CONS. INS'.

```
OUTFILE 'CONS.OUT'
READ 'CONS.DTA' C1 C2
NAME C1 'CONS' C2 'INCOME'
REGR C1 1 C2-C4;
RESI C5.
NOTE THE FITTED OBSERVATIONS IN C4
NOTE THE RESIDUALS IN C5
NAME C4 'FIT' C5 'RESID'
MPLOT C1 C2 C4 C2
HIST C5
REGR C1 1 C2;
NOCO.
STOP
```

This program is now performed by the command

```
EXEC 'CONS.INS'
```

as soon as the MINITAB prompt is received.

DISPLAY 3.7.1

```
MTB > EXEC 'CONS.INS'
MTB > OUTFILE 'CONS.OUT'
MTB > READ 'CONS.DTA' C1 C2
      34 ROWS READ
 ROW          C1          C2

   1       7.403     10.390
   2       7.793     11.379
   3       8.299     12.618
   4       8.258     13.306
  .    .    .

MTB > NAME C1 'CONS' C2 'INCOME'
MTB > REGR C1 1 C2-C4;
SUBC> RESI C5.

The regression equation is
CONS = 0.687 + 0.604 INCOME

Predictor          Coef          Stdev       t-ratio
Constant         0.6875         0.1791          3.84
INCOME         0.604475       0.007045         85.80

s = 0.4002        R-sq = 99.6%      R-sq(adj) = 99.6%
Analysis of Variance

SOURCE        DF           SS            MS
Regression     1       1179.3        1179.3
Error         32          5.1           0.2
Total         33       1184.4
```

```
MTB > NOTE THE FITTED OBSERVATIONS IN C4
MTB > NOTE RESIDUALS IN C5
MTB > NAME C4 'FIT' C5 'RESID'
MTB > MPLOT C1 C2 C4 C2

         -                                                                 2
         -                                                             2
    24.0+                                                          3
         -                                                     22   B
         -                                                   32
         -                                                   A
         -                                             2
    18.0+                                           22
         -                                      B 2
         -                                  B   A
         -                               B2 A
         -                              2A
    12.0+                         2 2
         -                    222
         -                44 2
         -            A 42
         -          2 3
     6.0+
         -
          ----+---------+---------+---------+---------+---------+--
            12.0      18.0      24.0      30.0      36.0      42.0
        A = CONS vs. INCOME        B = FIT vs. INCOME
MTB > HIST C5

Histogram of RESID   N = 34

Midpoint   Count
    -0.8       2  **
    -0.6       2  **
    -0.4       5  *****
    -0.2       2  **
     0.0       9  *********
     0.2       7  *******
     0.4       5  *****
     0.6       1  *
     0.8       0
     1.0       1  *

MTB > REGR C1 1 C2;
SUBC> NOCO.

The regression equation is
CONS = 0.629 INCOME

Predictor         Coef          Stdev     t-ratio
Noconstant
INCOME        0.629456       0.003214      195.86

s = 0.4763
Analysis of Variance

SOURCE          DF          SS          MS
Regression       1      8703.6      8703.6
Error           33         7.5         0.2
Total           34      8711.0
```

The following points could be noted:

(a) The REGR commands regress consumption in C1 against the single regressor income in C2. In the first of these regressions the residuals are stored in C5 and the predictions $\hat{C}$ in C4.

(b) The MPLOT command plots the two graphs C1 against C2 and C4 against C2 on the same graph. That is, in this case C against Y (scatter diagram—symbol A) and $\hat{C}$ against Y (line of best fit—symbol B) are plotted on the same graph.

(c) The NOCO subcommand results in C being regressed on Y with no intercept.

(d) All MINITAB sessions are terminated by the command

STOP

The results of the analysis are shown in Display 3.7.1. The reader can verify that apart from slight differences due to rounding, the results are identical to those of Section 2.10.

# 4

# Multiple Regression

## 4.1 THE MODEL AND ASSUMPTIONS

In Chapter 2 we studied the simple regression model and used it to demonstrate important principles that carry over into the more complex multiple regression situation. For example, we discussed the interpretation of the assumptions which are made about the disturbance $u_i$, and using relatively easy algebra, were able to see how these assumptions are crucial in establishing the BLUE property of OLS estimators.

The multiple regression model, with K explanatory (or regressor) variables, is defined by

$$Y_i = \beta_0 + \beta_1 X_{1i} + \cdots + \beta_K X_{Ki} + u_i. \tag{4.1.1}$$

Although the simple regression model is widely used, equation (4.1.1) is the more common model because in most instances $Y_i$ is influenced by more than one variable and we can usually obtain data on at least some of these additional explanatory variables. For example, in the estimation of demand curves, the demand for some commodity will depend not only on the price of that commodity but also on the prices of complements and substitutes and on income. In consumption function estimation, aggregate consumption might depend not only on income but also on consumption in the preceding period and some measure of wealth. Thus it is necessary to extend the analysis of Chapter 2 to cover models such as (4.1.1).

It is crucially important that the implications of the model (4.1.1) be understood. As in the simple regression model, we interpret (4.1.1) as specifying the mechanism by which observations $Y_i$ on the dependent variable are generated. In particular, by writing

$$Y_i = E[Y_i | X_1, \ldots, X_K] + u_i,$$

where

$$E[Y_i | X_1, \ldots, X_K] = \beta_0 + \beta_1 X_{1i} + \cdots + \beta_K X_{Ki},$$

we are emphasizing the fact that the mean of Y depends on the values of the variables $X_1, X_2, \ldots, X_K$, and that the observed value of Y involves a deviation $u_i$ from that mean. If we abbreviate the notation of the conditional mean to $E[Y|X]$, it is clear that the parameter $\beta_1$ is given by[1]

$$\beta_1 = \frac{\partial}{\partial X_1} E[Y|X]. \tag{4.1.2}$$

Because $\beta_1$ is defined as a partial derivative, it should be interpreted as the marginal effect of a unit change in $X_1$ on the conditional mean of Y when the other variables $X_2, \ldots, X_K$ are held constant. This feature is the main difference between simple and multiple regression. When we are estimating $\beta_1$ we are trying to disentangle the effect of $X_1$ on Y from all other effects. Let us consider two examples to highlight the implied problem.

EXAMPLE 1 Suppose that an agricultural research station is conducting experiments on the effects of fertilizer and irrigation on a particular crop. If Y is crop yield, $X_1$ amount of fertilizer, and $X_2$ amount of water applied, a suitable model may be

$$Y_i = \beta_0 + \beta_1 X_{1i} + \beta_2 X_{2i} + u_i.$$

Here $\beta_1$ is the marginal effect of a unit change in fertilizer application on average yield when the amount of water applied is held fixed. In such a situation, in which the regressor variables are controllable, $\beta_1$ could be estimated by holding $X_2$ at some fixed level $X_2^*$, say. Then

$$\begin{aligned} Y_i &= (\beta_0 + \beta_2 X_2^*) + \beta_1 X_{1i} + u_i \\ &= \beta_0^* + \beta_1 X_{1i} + u_i. \end{aligned}$$

As $X_1$ is varied, estimation of $\beta_1$ becomes a problem of simple regression.

EXAMPLE 2 Consider the estimation of a demand curve, mentioned above. If Y is demand, $X_1$ price of the commodity, $X_2$ price of a substitute, and $X_3$ income, we may postulate that

$$Y_i = \beta_0 + \beta_1 X_{1i} + \beta_2 X_{2i} + \beta_3 X_{3i} + u_i.$$

We recall that $\beta_1$ is the marginal effect of own-price on demand, holding the price of the substitute and the income level fixed. The problem here is that substitute price and income cannot be held fixed—they covary with the price of the product. Somehow the effect of own-price has to be disentangled from the effects of the other variables.

It should be clear that if we are to be successful in isolating the effect of $X_1$ in the (usual) situation in which all the regressor variables are covarying, we must develop methods to hold the other variables constant artificially.

The same assumptions are made about the disturbance term, namely,

(a) $E(u_i) = 0$, (b) $E(u_i^2) = \sigma^2$, (c) $E(u_i u_j) = 0$ for $i \neq j$

and, of course, these assumptions have the same interpretation as in the simple regression model. In addition, it is assumed that the X's are not random variables. Under these assumptions the least squares estimators will be best, linear, unbiased estimators. We will further assume that the number of observations is greater than the number of parameters to be estimated (i.e., $N > K + 1$), and that no exact linear relationships exist between regressor variables. We will return to this assumption shortly.

In the following sections least squares estimators and their distributions will be discussed. All the relevant expressions (formulas) are relatively simple provided that one is prepared to use matrix algebra. However, as it is desirable to keep this material accessible to those readers who are not competent with matrix algebra, matrices will be introduced as sparingly as possible. This avoidance of matrices prevents, to some extent, the general specification of expressions such as least squares estimates and their standard errors for the case of $K + 1$ coefficients. Instead, expressions for the case $K = 2$ will be given to illustrate the differences between simple and multiple regression. When $K > 2$ these expressions become too complicated to make their presentation worthwhile in anything other than matrix algebra.

In simple regression least squares estimates and associated quantities are readily found using a desk calculator. With multiple regression it is much easier to use a computer program, so all multiple regression problems will be solved using the MINITAB computer program. However, to illustrate "what the computer does," an example for the case $K = 2$ will be solved in this chapter. Because many general expressions are not given and because the computer will carry out almost all the calculations, there is a danger of losing sight of the interpretation of many statistical quantities and regarding them simply as something that occurs in a given place on the computer output. The interpretation for most concepts is the same in the multiple regression model. Thus, unless an additional interpretation of some concept is given in this chapter, the reader should always return to Chapter 2 to ensure that he or she has the correct interpretation of the concept.

## 4.2 A METHOD OF ESTIMATION

To gain insights into multiple regression it is only necessary to consider the case of two regressor variables. For the moment, then, we will restrict our attention to the model

$$Y_i = \beta_0 + \beta_1 X_{1i} + \beta_2 X_{2i} + u_i. \tag{4.2.1}$$

We have already seen that $\beta_1$ is the marginal effect that a unit change in $X_1$ has on the average value of Y when $X_2$ remains constant. If $X_2$ is constant, it will have no effect on the changes occurring in $X_1$ and Y, and it is only these changes that have any bearing on the estimation of $\beta_1$.

Thus an equivalent interpretation of $\beta_1$ is the marginal effect on average Y of a unit change in $X_1$ once the effect of $X_2$ has been removed from both $X_1$ and Y. This suggests that a procedure we could use to estimate $\beta_1$ is the following: Remove the effect of $X_2$ from both $X_1$ and Y and then carry out a simple regression of "purged" Y on "purged" $X_1$. More formally,

(a) Regress Y on $X_2$ and save the residuals, which we will denote by $e_{y2}$;
(b) Regress $X_1$ on $X_2$ and save the residuals $e_{12}$;
(c) Regress $e_{y2}$ on $e_{12}$.

The slope coefficient from this regression would be an estimate of $\beta_1$. Two points should be noted here. First, the residual $e_{y2}$ is simply the variable the variable Y after the effect of $X_2$ has been removed. A similar remark applies to $e_{12}$. Second, a fundamental property of OLS residuals is that they sum to zero [see (2.3.11)]. Thus $e_{y2}$ and $e_{12}$ are in effect already mean corrected, and we can write down the estimate $\hat{\beta}_1$ immediately from (2.2.9) as

$$\hat{\beta}_1 = \frac{\Sigma e_{y2} e_{12}}{\Sigma e_{12}^2}. \tag{4.2.2}$$

Consider, first, the numerator of (4.2.2):

$$\Sigma e_{y2} e_{12} = \Sigma (y - \hat{\beta}_{y2} x_2)(x_1 - \hat{\beta}_{12} x_2),$$

where $\hat{\beta}_{y2}$ and $\hat{\beta}_{12}$ are the slope coefficients obtained by regressing Y on $X_2$ and $X_1$ on $X_2$, respectively. After a little algebra we obtain

$$\Sigma e_{y2} e_{12} = \hat{\beta}_{y1} \Sigma x_1^2 - \hat{\beta}_{y2} \hat{\beta}_{12} \Sigma x_2^2.$$

The denominator $\Sigma e_{12}^2$ of (4.2.2) is seen, by (2.7.8), to be $(1 - r_{12}^2)\Sigma x_1^2$, where $r_{12}$ is the sample correlation coefficient between $X_1$ and $X_2$. Thus

$$\hat{\beta}_1 = \frac{\hat{\beta}_{y1} \Sigma x_1^2 - \hat{\beta}_{y2} \hat{\beta}_{12} \Sigma x_2^2}{(1 - r_{12}^2)\Sigma x_1^2}$$

$$= \frac{\hat{\beta}_{y1} - \hat{\beta}_{y2}(\hat{\beta}_{12} \Sigma x_2^2 / \Sigma x_1^2)}{1 - r_{12}^2}$$

Also, by (2.2.9),

$$\hat{\beta}_{12}\frac{\Sigma x_2^2}{\Sigma x_1^2} = \frac{\Sigma x_1 x_2}{\Sigma x_2^2}\frac{\Sigma x_2^2}{\Sigma x_1^2} = \frac{\Sigma x_1 x_2}{\Sigma x_1^2} = \hat{\beta}_{21}.$$

where $\hat{\beta}_{21}$ is the slope estimate obtained when $X_2$ is regressed on $X_1$. Thus we finally obtain an expression for $\hat{\beta}_1$ in the form

$$\hat{\beta}_1 = \frac{\hat{\beta}_{y1} - \hat{\beta}_{y2}\hat{\beta}_{21}}{1 - r_{12}^2}. \tag{4.2.3}$$

This expression for $\hat{\beta}_1$ is interesting because it can be used to deduce important principles that apply in the general multiple regression case.

(a) Suppose that $X_1$ and $X_2$ are uncorrelated.[2] Then $r_{12} = 0$ and it follows that $\hat{\beta}_{21} = 0$. Thus, from (4.2.3),

$$\hat{\beta}_1 = \hat{\beta}_{y1}.$$

Thus, if $X_1$ and $X_2$ are orthogonal, the estimate of the marginal effect of $X_1$ will be the same whether or not $X_2$ is included in the model. Conversely, if $r_{12} \neq 0$ (and this is almost invariably the case in an economic context), then

$$\hat{\beta}_1 \neq \hat{\beta}_{y1},$$

and inclusion of $X_2$ will alter the estimate of $\beta_1$. We can reasonably infer that if $|r_{12}|$ is very small, the effect on the estimate of $\beta_1$ from the inclusion or exclusion of $X_2$ will be very slight, whereas if $|r_{12}|$ is large, the effect will be correspondingly greater.

(b) Let us suppose that there is an exact linear relationship between $X_1$ and $X_2$. That is, it is possible to find two constants $k_0$ and $k$, say, such that

$$X_2 = k_0 + kX_1.$$

Then

$$\bar{X}_2 = k_0 + k\bar{X}_1,$$

and subtracting these two equations gives us

$$x_2 = kx_1. \tag{4.2.4}$$

By following through the algebra used to derive (4.2.3), under the condition $x_2 = kx_1$, it is easily shown that

$$\hat{\beta}_{y1} = \hat{\beta}_{y2}\hat{\beta}_{21}$$

and

$$r_{12}^2 = 1.$$

Thus, in this case, both the numerator and denominator of (4.2.3) are zero, and $\hat{\beta}_1$ is consequently indeterminate. This is the reason for the earlier assumption that no exact linear relationship exists between any of the regressor variables.

(c) Suppose now that $X_1$ and $X_2$ are highly correlated but do not have an exact linear relationship. In this case

$$\hat{\beta}_{y1} \approx \hat{\beta}_{y2}\hat{\beta}_{21}$$

$$r_{12}^2 \approx 1.$$

Thus $\hat{\beta}_1$ will be determined by the ratio of two very small numbers, and a very slight change in the values of any (or all) of $\hat{\beta}_{y1}$, $\hat{\beta}_{y2}$, $\hat{\beta}_{21}$, $r_{12}^2$ could give rise to a very large change in the value of $\hat{\beta}_1$. Thus $\hat{\beta}_1$ is likely to be a very unreliable estimator of $\beta_1$.

This situation of high correlation between regressor variables is very common with economic data and poses serious problems in the accurate estimation of the parameters of a model. It is known as the problem of <u>multicollinearity</u>.

EXAMPLE To illustrate the problems that can arise when $X_1$ and $X_2$ are highly correlated, a simulation experiment was conducted. Two samples of size 20 were generated from the model

$$Y_i = 1 + 3X_{1i} + 2X_{2i} + u_i,$$

where

$$X_{1i} \sim N(0, 1),$$

$$X_{2i} = X_{1i} + \epsilon_i \quad (\epsilon_i \text{ independent of } X_{1i}),$$

$$\epsilon_i \sim N(0, 0.0448^2),$$

$$u_i \sim N(0, 0.1^2).$$

This choice for the variance of $\epsilon_i$ means that the theoretical correlation between $X_1$ and $X_2$ is 0.999.

The sample results from the simple regressions of Y on $X_1$, Y on $X_2$, and $X_2$ on $X_1$ were as follows:

| Sample 1 | Sample 2 |
|---|---|
| $\hat{\beta}_{y1} = 4.98068$ | $\hat{\beta}_{y1} = 4.95402$ |
| $\hat{\beta}_{y2} = 5.00213$ | $\hat{\beta}_{y2} = 5.01599$ |
| $\hat{\beta}_{21} = 0.995089$ | $\hat{\beta}_{21} = 0.98623$ |
| $r_{12}^2 = 0.999$ | $r_{12}^2 = 0.998$ |

We observe that each of the sample estimates $\hat{\beta}_{y1}$, $\hat{\beta}_{y2}$, $\hat{\beta}_{21}$, and $r_{12}^2$ is very stable. In fact, in every case the difference between the sample results is less than 1 percent. However, when we compute the numerator and denominator of (4.2.3), we obtain

| | Sample 1 | Sample 2 |
|---|---|---|
| $\hat{\beta}_{y1} - \hat{\beta}_{y2}\hat{\beta}_{21}$ | 0.003115 | 0.007100 |
| $1 - r_{12}^2$ | 0.001 | 0.002 |

Thus, the numerators have changed by 128 percent, and the denominators by 100 percent, leading to very unstable estimates of $\hat{\beta}_1$. An estimator can be regarded as reliable only if it varies very little from sample to sample.

## 4.3 LEAST SQUARES ESTIMATORS AND THEIR DISTRIBUTIONS

In Section 4.2 we developed an estimator for $\beta_1$ in the two-variable model (4.2.1), based on the idea of removing the effects of $X_2$ from the other variables $X_2$ and Y. We obtained

$$\hat{\beta}_1 = \frac{\hat{\beta}_{y1} - \hat{\beta}_{y2}\hat{\beta}_{21}}{1 - r_{12}^2}. \tag{4.3.1}$$

The expression (4.3.1), although useful for conveying some principles of estimation in a multiple regression context, is not convenient for actually obtaining $\hat{\beta}_1$ from given observations of Y, $X_1$, and $X_2$. We know from Chapter 2 that

$$\hat{\beta}_{y1} = \frac{\Sigma yx_1}{\Sigma x_1^2},$$

$$\hat{\beta}_{y2} = \frac{\Sigma yx_2}{\Sigma x_2^2},$$

$$\hat{\beta}_{21} = \frac{\Sigma x_1 x_2}{\Sigma x_1^2},$$

$$r_{12}^2 = \frac{(\Sigma x_1 x_2)^2}{\Sigma x_1^2 \Sigma x_2^2}.$$

Substituting these formulas into (4.3.1), we obtain

$$\hat{\beta}_1 = \frac{\Sigma x_1 y \Sigma x_2^2 - \Sigma x_2 y \Sigma x_1 x_2}{\Sigma x_1^2 \Sigma x_2^2 - (\Sigma x_1 x_2)^2}. \tag{4.3.2}$$

In an exactly similar way, we have

$$\hat{\beta}_2 = \frac{\Sigma x_2 y \Sigma x_1^2 - \Sigma x_1 y \Sigma x_1 x_2}{\Sigma x_1^2 \Sigma x_2^2 - (\Sigma x_1 x_2)^2}. \tag{4.3.3}$$

Formulas (4.3.2) and (4.3.3) are convenient expressions for estimating $\hat{\beta}_1$ and $\hat{\beta}_2$, but are they in fact least squares estimators? If we follow the least squares principle, $\hat{\beta}_1$ and $\hat{\beta}_2$ (as well as $\hat{\beta}_0$) would be obtained by minimizing the residual sum of squares. That is,

$$\sum_i e_i^2 = \sum_i [Y_i - (\hat{\beta}_0 + \hat{\beta}_1 X_{1i} + \hat{\beta}_2 X_{2i})]^2$$

is a minimum, implying that

$$\frac{\partial \Sigma e_i^2}{\partial \hat{\beta}_0} = \frac{\partial \Sigma e_i^2}{\partial \hat{\beta}_1} = \frac{\partial \Sigma e_i^2}{\partial \hat{\beta}_2} = 0.$$

The normal equations in this case are

$$N\hat{\beta}_0 + \hat{\beta}_1 \Sigma X_{1i} + \hat{\beta}_2 \Sigma X_{2i} = \Sigma Y_i$$

$$\hat{\beta}_0 \Sigma X_{1i} + \hat{\beta}_1 \Sigma X_{1i}^2 + \hat{\beta}_2 \Sigma X_{1i} X_{2i} = \Sigma X_{1i} Y_i$$

$$\hat{\beta}_0 \Sigma X_{2i} + \hat{\beta}_1 \Sigma X_{1i} X_{2i} + \hat{\beta}_2 \Sigma X_{2i}^2 = \Sigma X_{2i} Y_i.$$

Solving these equations for $\hat{\beta}_1$, $\hat{\beta}_2$, and $\hat{\beta}_0$ we find $\hat{\beta}_1$ and $\hat{\beta}_2$ given by (4.3.2) and (4.3.3), respectively, and

$$\hat{\beta}_0 = \bar{Y} - \hat{\beta}_1 \bar{X}_1 - \hat{\beta}_2 \bar{X}_2. \tag{4.3.4}$$

Thus the expressions (4.3.2) and (4.3.3) for $\hat{\beta}_1$ and $\hat{\beta}_2$, derived using some intuitive concepts of removing the effects of variables, turn out to be precisely the least squares estimators. As in the case of simple regression, these estimators are best, linear, unbiased estimators provided that the assumptions in Section 4.1 hold.

Derivation of the variances of the estimators $\hat{\beta}_1$ and $\hat{\beta}_2$ is somewhat complicatd when matrix algebra is not used. However, the final expressions are[3]

$$\text{var}(\hat{\beta}_1) = \frac{\sigma^2}{\Sigma x_1^2 (1 - r_{12}^2)} \tag{4.3.5}$$

and

$$\text{var}(\hat{\beta}_2) = \frac{\sigma^2}{\Sigma x_2^2 (1 - r_{12}^2)}. \tag{4.3.6}$$

It is instructive to consider the interpretation of (4.3.5). In Chapter 2 we defined the information content which the sample contains about $\beta_1$ as $\Sigma x^2$. We saw that $\text{var}(\hat{\beta}_1)$ is the inverse of the size of the information content relative to $\sigma^2$.

When we consider a model with two regressors, $\text{var}(\hat{\beta}_1)$ will be small only if the variation in $X_1$ is large when $X_2$ is held fixed. That is, $\text{var}(\hat{\beta}_1)$ is determined by the information content of $X_1$, which is not also contained in $X_2$. Let us illustrate by a simple example. Suppose that $\Sigma x_1^2 = 100$. If $X_1$ is regressed on $X_2$ and $r_{12}^2 = 0.95$, this means that 95 percent of the variation in $X_1$ is explained by $X_2$. Thus only 5 percent of the variation in $X_1$ contributes to the estimation of $\beta_1$. We may measure the information content in this case as $0.05 \Sigma x_1^2 = 0.05 \times 100 = 5$. If this is large relative to $\sigma^2$, $\hat{\beta}_1$ will be precisely estimated, but otherwise $\hat{\beta}_1$ will be an unreliable estimator.

In general, then, $1 - r_{12}^2$ is the proportion of variation in $X_1$ that is not common to $X_2$, and $(1 - r_{12}^2) \Sigma x_1^2$ is the autonomous variation in $X_1$. Thus (4.3.5) tells us that it is the autonomous information content of $X_1$ relative to $\sigma^2$ which determines the variance of $\hat{\beta}_1$ and hence the precision with which $\beta_1$ is estimated.

Having introduced some basic concepts of multiple regression in the framework of a two-variable model, we will now generalize many of the results from the simple regression model to the case of the K regressor-variables model (4.1.1).

Residuals

The residuals from a least squares regression satisfy the K + 1 restrictions

$$\Sigma e = 0; \quad \Sigma eX_1 = \Sigma eX_2 = \cdots = \Sigma eX_K = 0. \tag{4.3.7}$$

Thus there are N - (K + 1) = N - K - 1 degrees of freedom. If the model has no intercept, then in general $\Sigma e \neq 0$, but the constraints involving the regressors still hold. This implies that there are K constraints and hence N - K degrees of freedom in this case.

Estimating $\sigma^2$

$\sigma^2 = E(u_i^2)$ and is the average value of $u_i^2$. An estimator of $\sigma^2$ is therefore the average value of $e_i^2$. As there are N - K - 1 independent pieces of information in the residuals $e_1, e_2, \ldots, e_N$, an unbiased estimator of $\sigma^2$ is $\hat{\sigma}^2$, given by

$$\hat{\sigma}^2 = \frac{\Sigma e^2}{N - K - 1}. \tag{4.3.8}$$

If there is no intercept, an unbiased estimator is

$$\hat{\sigma}^2 = \frac{\Sigma e^2}{N - K}. \tag{4.3.9}$$

The generalization of (2.5.6), giving a formula for computing $\hat{\sigma}^2$, is

$$\hat{\sigma}^2 = \frac{1}{N - K - 1}(\Sigma y^2 - \hat{\beta}_1 \Sigma x_1 y - \hat{\beta}_2 \Sigma x_2 y \cdots \hat{\beta}_K \Sigma x_K y)$$

$$= \frac{1}{N - K - 1}\left(\Sigma y^2 - \sum_{k=1}^{K} \hat{\beta}_k \Sigma x_k y\right) \tag{4.3.10}$$

or

$$\hat{\sigma}^2 = \frac{1}{N - K}\left(\Sigma Y^2 - \sum_{k=1}^{K} \hat{\beta}_k \Sigma X_k Y\right) \tag{4.3.11}$$

when there is no intercept.

A Measure of Goodness of Fit

In Chapter 2 we defined a measure of goodness of fit $r^2$ by

$$r^2 = \frac{ESS}{TSS} = 1 - \frac{RSS}{TSS},$$

and this measure could be interpreted in three ways:

(i) The proportion of variation of the dependent variable Y explained by the regression;
(ii) The square of the sample correlation between Y and $\hat{Y}$; and
(iii) The square of the sample correlation between Y and X.

In the case of multiple regression, we denote the measure by $R^2$, still defined by

$$R^2 = \frac{ESS}{TSS} = 1 - \frac{RSS}{TSS}$$

$$= 1 - \frac{\Sigma y^2 - \Sigma_{k=1}^{K} \hat{\beta}_k \Sigma x_k y}{\Sigma y^2}$$

That is,

$$R^2 = \frac{\Sigma_{k=1}^{K} \hat{\beta}_k \Sigma x_k y}{\Sigma y^2}. \tag{4.3.12}$$

With respect to $R^2$, interpretations (i) and (ii) still apply, but (iii) is no longer relevant. When there is no intercept, we define $R^2$ by

$$R^2 = \frac{\Sigma_{k=1}^{K} \hat{\beta}_k \Sigma X_k Y}{\Sigma Y^2} \tag{4.3.13}$$

and the interpretation of $R^2$ is the proportion of variation of Y <u>about the origin</u> which is explained by the regression.

## The Distribution of $\hat{\beta}_1$

Provided that $u_i$ is normally distributed, Theorem 2.5.2 generalizes in a perfectly straightforward way, and we have

$$\frac{\hat{\beta}_1 - \beta_1}{S_{\hat{\beta}_1}} \sim t_{N-K-1}, \tag{4.3.14}$$

where now

$$S^2_{\hat{\beta}_1} = \frac{\hat{\sigma}^2}{\Sigma x_1^2 (1 - R_1^2)} \tag{4.3.15}$$

and $t_{N-K-1}$ stands for the t distribution with parameter N - K - 1. $R_1^2$ is the $R^2$ that would be obtained if we regressed $X_1$ on the remaining regressors $X_2, \ldots, X_K$, and is the generalization of $r_{12}^2$ in (4.3.5). This result on the distribution of $\hat{\beta}_1$ (and the equivalent results for $\hat{\beta}_2, \ldots, \hat{\beta}_K$) are used to construct confidence intervals and test hypotheses about individual $\beta$'s in exactly the same way as in Chapter 2.

## 4.4 COVARIANCE MATRIX AND FURTHER HYPOTHESIS TESTS

Each of the estimators $\hat{\beta}_0, \hat{\beta}_1, \ldots, \hat{\beta}_K$ has, of course, a variance, which is a measure of the variability the estimator would display when applied to different samples. Also, for every pair of estimators there is a covariance. Both the variances and covariances are functions of the unknown parameter $\sigma^2$. By estimating $\sigma^2$, we can also estimate the variances and covariances. We define

$$s^2_{\hat{\beta}_i} = \text{estimated variance of } \hat{\beta}_i \quad (i = 0, 1, \ldots, K);$$

$$s_{\hat{\beta}_i\hat{\beta}_j} = \text{estimated covariance between } \hat{\beta}_i \text{ and } \hat{\beta}_j \quad (i \neq j;\ i, j = 0, 1, \ldots, K).$$

These numbers can be very conveniently displayed in a (K + 1) × (K + 1) matrix $\hat{V}$, known as the estimated covariance matrix. Thus

$$\hat{V} = \begin{bmatrix} s^2_{\hat{\beta}_0} & s_{\hat{\beta}_0\hat{\beta}_1} & s_{\hat{\beta}_0\hat{\beta}_2} & \cdots & s_{\hat{\beta}_0\hat{\beta}_K} \\ s_{\hat{\beta}_0\hat{\beta}_1} & s^2_{\hat{\beta}_1} & s_{\hat{\beta}_1\hat{\beta}_2} & \cdots & s_{\hat{\beta}_1\hat{\beta}_K} \\ s_{\hat{\beta}_0\hat{\beta}_2} & s_{\hat{\beta}_1\hat{\beta}_2} & s^2_{\hat{\beta}_2} & \cdots & s_{\hat{\beta}_2\hat{\beta}_K} \\ \vdots & \vdots & \vdots & & \vdots \\ s_{\hat{\beta}_0\hat{\beta}_K} & s_{\hat{\beta}_1\hat{\beta}_K} & s_{\hat{\beta}_2\hat{\beta}_K} & \cdots & s^2_{\hat{\beta}_K} \end{bmatrix}. \qquad (4.4.1)$$

The estimated covariance matrix $\hat{V}$ has two main uses, apart from displaying the variances and covariances. First, it is used in prediction, and we will consider this in the next section. Second, it provides the necessary information that enables us to test hypotheses about a linear combination of the parameters $\beta_i$.

There are often situations in which we wish to test hypotheses of the following form:

$$H_0: \alpha_0\beta_0 + \alpha_1\beta_1 + \cdots + \alpha_k\beta_k = h$$

$$H_1: \alpha_0\beta_0 + \alpha_1\beta_1 + \cdots + \alpha_k\beta_k \neq h,$$

where $\alpha_0, \alpha_1, \ldots, \alpha_k$ and h are known constants.

Using summation notation, we have

$$H_0: \sum_{k=0}^{K} \alpha_k\beta_k = h$$

$$H_1: \sum_{k=0}^{K} \alpha_k\beta_k \neq h. \tag{4.4.2}$$

To illustrate why such hypotheses are important, we will consider two examples:

EXAMPLE 1 Suppose that the demand for a commodity can be expressed in the form

$$\ln Q_i = \beta_0 + \beta_1 \ln P_i + \beta_2 \ln P_i^* + \beta_3 \ln Y_i + u_i,$$

where $Q_i$ is demand, $P_i$ the price of the commodity, $P_i^*$ the price of a substitute or complement, and $Y_i$ the level of income. It is often assumed that there is "no money illusion," that is, if each of P, P*, and Y are increased by the same proportion, the demand Q remains unchanged. Let $\lambda$ be the proportion. Then, if there is no money illusion,

$$\begin{aligned}\ln Q_i &= \beta_0 + \beta_1 \ln(\lambda P_i) + \beta_2 \ln(\lambda P_i^*) + \beta_3 \ln(\lambda Y_i) + u_i \\ &= \beta_0 + \beta_1 \ln P_i + \beta_2 \ln P_i^* + \beta_3 \ln Y_i + (\beta_1 + \beta_2 + \beta_3) \ln \lambda + u_i.\end{aligned}$$

Therefore, if there is no money illusion, $\beta_1 + \beta_2 + \beta = 0$.

If we want to test this assumption, we would have

$$H\ : \beta_1 + \beta_2 + \beta_3 = 0$$

$$H_1: \beta_1 + \beta_2 + \beta_3 \neq 0.$$

This is of the same form as (4.4.2), where $K = 3$, $\alpha_0 = 0$, $\alpha_1 = \alpha_2 = \alpha_3 = 1$, and $h = 0$. Notice that the $\alpha$'s and h are known constants.

EXAMPLE 2 Suppose in Example 1 that there are two substitutes or complements, having prices $P_1^*$ and $P_2^*$. The model would then be

$$\ln Q_i = \beta_0 + \beta_1 \ln P_i + \beta_2 \ln P^*_{1i} + \beta_3 \ln P^*_{2i} + \beta_4 \ln Y_i + u_i.$$

In this context it may be important to test whether or not price variations in the two substitutes/complements exert the same influence. We would set up the test as

$$H_0 : \beta_2 = \beta_3$$

$$H_1 : \beta_2 \neq \beta_3$$

This again is a special case of (4.4.2). Writing $\beta_2 = \beta_3$ as $\beta_2 - \beta_3 = 0$, we have here $K = 4$, $\alpha_0 = \alpha_1 = \alpha_4 = 0$, $\alpha_2 = 1$, $\alpha_3 = -1$, and $h = 0$. Notice again that the hypotheses completely determine the constants $\alpha_0, \alpha_1, \ldots, \alpha_K$ and h.

How then do we test hypotheses about a linear combination of the $\beta$'s? An intuitively reasonable approach would be to run the regression and obtain the estimates $\hat{\beta}_0, \hat{\beta}_1, \ldots, \hat{\beta}_K$. Then, given that $\alpha_0, \alpha_1, \ldots, \alpha_k$ are always known, we could construct the linear combination $\Sigma\, \alpha_k \hat{\beta}_k$, which we will denote by $\hat{h}$. We could now compare $\hat{h}$ with the known value h, and $H_0$ would be accepted if $\hat{h}$ was close to h. This procedure is identical to earlier hypothesis tests except that now the statistic on which we focus our attention is $\hat{h}$ instead of, say, $\hat{\beta}_1$. It stands to reason that we are going to need the standard error of $\hat{h}$, which will be denoted by $S_{\hat{h}}$.

Formulas for $S^2_{\hat{h}}$ are rather untidy if a matrix formulation is not used. We will therefore relax our practice of not using matrix algebra for this particular section and the section on prediction. To introduce the use of matrices in this context, we will consider a simple example. Let us suppose that we want to find the variance of the linear combination $\alpha_1\hat{\beta}_1 + \alpha_2\hat{\beta}_2$, where $\alpha_1$ and $\alpha_2$ are known constants. From elementary statistics it is well known that

$$\text{var}(\alpha_1\hat{\beta}_1 + \alpha_2\hat{\beta}_2) = \alpha_1^2\,\text{var}(\hat{\beta}_1) + \alpha_2^2\,\text{var}(\hat{\beta}_2) + 2\alpha_1\alpha_2\,\text{cov}(\hat{\beta}_1, \hat{\beta}_2).$$

This expression may be written in matrix notation as

$$\text{var}(\alpha_1\hat{\beta}_1 + \alpha_2\hat{\beta}_2) = [\alpha_1 \quad \alpha_2]\begin{bmatrix}\text{var}(\hat{\beta}_1) & \text{cov}(\hat{\beta}_1, \hat{\beta}_2) \\ \text{cov}(\hat{\beta}_1, \hat{\beta}_2) & \text{var}(\hat{\beta}_2)\end{bmatrix}\begin{bmatrix}\alpha_1 \\ \alpha_2\end{bmatrix}$$

$$= \alpha' V \alpha,$$

where $\alpha$ is the vector $\begin{bmatrix}\alpha_1 \\ \alpha_2\end{bmatrix}$, $\alpha'$ is the transpose of $\alpha$, and V the variance-covariance matrix of $\hat{\beta}_1$ and $\hat{\beta}_2$. That is,

$$V = \begin{bmatrix}\text{var}(\hat{\beta}_1) & \text{cov}(\hat{\beta}_1, \hat{\beta}_2) \\ \text{cov}(\hat{\beta}_1, \hat{\beta}_2) & \text{var}(\hat{\beta}_2)\end{bmatrix}.$$

Thus when V has to be estimated, we obtain an estimate $S_{\hat{h}}^2$ of $\mathrm{var}(\alpha_1\hat{\beta}_1 + \alpha_2\hat{\beta}_2)$ in the form

$$S_{\hat{h}}^2 = \alpha'\hat{V}\alpha.$$

Let us now define a (K + 1)-dimensional vector $\alpha$ by

$$\alpha' = (\alpha_0, \alpha_1, \alpha_2, \ldots, \alpha_K),$$

where the prime on $\alpha$ denotes transpose. Then the preceding example generalizes immediately to[4]

$$S_{\hat{h}}^2 = \alpha'\hat{V}\alpha \tag{4.4.3}$$

where $\hat{V}$ is given by (4.4.1). Evaluation of (4.4.3) requires a knowledge only of the rules of matrix multiplication.

EXAMPLE Let us suppose that we wish to test a hypothesis about the linear combination $\beta_0 + 2\beta_1 - 4\beta_2$. Then

$$\alpha' = (1, 2, -4)$$

Suppose, furthermore, that $\hat{V}$ is given by

$$\hat{V} = \begin{bmatrix} 1 & 3 & 2 \\ 3 & 13 & 4 \\ 2 & 4 & 6 \end{bmatrix}.$$

Then, from (4.4.3),

$$S_{\hat{h}}^2 = [1,\ 2,\ -4] \begin{bmatrix} 1 & 3 & 2 \\ 3 & 13 & 4 \\ 2 & 4 & 6 \end{bmatrix} \begin{bmatrix} 1 \\ 2 \\ -4 \end{bmatrix}$$

$$= [-1,\ 13,\ -14] \begin{bmatrix} 1 \\ 2 \\ -4 \end{bmatrix} = 81.$$

Therefore,

$$S_{\hat{h}} = 9.$$

We are now in a position to state formally the distributional result that enables us to test hypotheses about linear combinations of the $\beta_i$.

THEOREM 4.4.1 If the disturbance $u_i$ is normally distributed and $\Sigma_{k=0}^{K} \alpha_k \beta_k = h$, the ratio

$$\frac{\hat{h} - h}{S_{\hat{h}}} \sim t_{N-K-1}, \tag{4.4.4}$$

where

$$\hat{h} = \sum_{k=0}^{K} \alpha_k \hat{\beta}_k$$

and

$$S_{\hat{h}}^2 = \alpha' \hat{V} \alpha.$$

Naturally, if the model has no intercept, the ratio has a $t_{N-K}$ distribution. An example of the application of this theorem will be given shortly.

## 4.5 REPARAMETRIZATION OF A REGRESSION MODEL

The linear regression model is characterized by the equation

$$E[Y|X] = \beta_0 + \beta_1 X_{1i} + \cdots + \beta_K X_{Ki}.$$

The significant feature here is that $E[Y|X]$ is a linear function of the parameters $\beta_0, \beta_1, \ldots, \beta_K$. There are some circumstances in which it is advantageous to transform the model so that the conditional expectation becomes a linear function of different parameters, say $\gamma_0, \gamma_1, \ldots, \gamma_K$. This is known as reparametrizing the model. We will show two ways in which a model can be reparametrized together with the effects of using OLS on the transformed model.

### Replacing the $\beta$'s by Linear Combinations of the $\beta$'s

We will illustrate the procedure by a simple example before generalizing. In Example 1 of the preceding section, we saw that a test of money illusion could be formulated in terms of the hypothesis $H_0: \beta_1 + \beta_2 + \beta_3 = 0$ on the model

$$Y_i = \beta_0 + \beta_1 X_{1i} + \beta_2 X_{2i} + \beta_3 X_{3i} + u_i,$$

where

$$X_1 = \ln P,$$
$$X_2 = \ln P^*,$$
$$X_3 = \ln Y,$$
$$Y = \ln Q.$$

To test the hypothesis, the estimated variance-covariance matrix $\hat{V}$ was used. An alternative method is based on a reparametrization of the model. Let us rewrite the model in the form

$$Y_i = \beta_0 + (\beta_1 + \beta_2 + \beta_3)X_{1i} + \beta_2(X_{2i} - X_{1i}) + \beta_3(X_{3i} - X_{1i}) + u_i.$$

(What we have done is to add and subtract the terms $\beta_2 X_{1i}$ and $\beta_3 X_{1i}$ on the right-hand side.) That is,

$$Y_i = \gamma_0 + \gamma_1 X^*_{1i} + \gamma_2 X^*_{2i} + \gamma_3 X^*_{3i} + u_i,$$

where

$$\gamma_0 = \beta_0,$$
$$\gamma_1 = \beta_1 + \beta_2 + \beta_3,$$
$$\gamma_2 = \beta_2,$$
$$\gamma_3 = \beta_3$$

and

$$X^*_1 = X_1,$$
$$X^*_2 = X_2 - X_1,$$
$$X^*_3 = X_3 - X_1.$$

Each of the $\gamma$ parameters is a linear combination of all of the $\beta$ parameters. For example,

$$\gamma_2 = \alpha_0\beta_0 + \alpha_1\beta_1 + \alpha_2\beta_2 + \alpha_3\beta_3,$$

where $\alpha_2 = 1$ and $\alpha_0 = \alpha_1 = \alpha_3 = 0$. Similarly, for $\gamma_1$, $\alpha_0 = 0$ and $\alpha_1 = \alpha_2 = \alpha_3 = 1$. An essential feature of this type of reparameterization is that it always involves a transformation of the regressor variables.

Our purpose for the reparametrization in this case is that the null hypothesis now becomes $H_0 : \gamma_1 = 0$, and this can be tested using the standard t-test based on the ratio

$$t = \frac{\hat{\gamma}_1}{S_{\hat{\gamma}_1}},$$

as in (2.6.9). The need to evaluate the quadratic form (4.4.3) is avoided. When OLS is applied to the reparametrized model the quantities TSS, RSS, and ESS are all unchanged. This implies that $R^2$ also does not change. An example of this technique is given at the end of the chapter.

### Subtraction of the Same Linear Combination of Regressor Variables from Each Side of the Equation

Consider the model

$$Y_i = \beta_0 + \beta_1 X_{1i} + \beta_2 X_{2i} + u_i.$$

Let us subtract $3X_{1i} + 5X_{2i}$ from each side of the model. Then

$$Y_i - 3X_{1i} - 5X_{2i} = \beta_0 + (\beta_1 - 3)X_{1i} + (\beta_2 - 5)X_{2i} + u_i.$$

That is,

$$Y_i^* = \gamma_0 + \gamma_1 X_{1i} + \gamma_2 X_{2i} + u_i,$$

where

$$Y^* = Y - 3X_1 - 5X_2,$$

$$\gamma_0 = \beta_0,$$

$$\gamma_1 = \beta_1 - 3,$$

$$\gamma_2 = \beta_2 - 5.$$

Notice the contrast in the effect of this and the previous type of reparametrization. Here the dependent variable is transformed, but the regressor variables are not changed.

When OLS is applied to the model above we would find that

$$\hat{\gamma}_0 = \hat{\beta}_0,$$

$$\hat{\gamma}_1 = \hat{\beta}_1 - 3,$$

$$\hat{\gamma}_2 = \hat{\beta}_2 - 5,$$

and RSS is unchanged. However, because the dependent variable has been changed, TSS is not the same, and it follows from (2.7.2) that ESS also changes. This implies that $R^2$ also changes—a point that we will return to in Section 5.2. The major uses of this type of reparametrization occur in the construction of F-tests, which are discussed in Chapter 6.

## 4.6 PREDICTION

Using a multiple regression model for prediction is, in concept, identical to the simple regression case discussed in Section 2.6. We may be interested in predicting the mean of Y, conditional on a given set of values for the regressor variables, or alternatively, we may wish to predict a particular value of Y when the regressors take prescribed values.

As with simple regression, the predictor we use, $\hat{Y}_0$, is the same in both cases. If we denote by $X_1^\circ$, $X_2^\circ$, ..., $X_K^\circ$ the values of the regressor variables, then

$$\hat{Y}_0 = \hat{\beta}_0 + \hat{\beta}_1 X_1^\circ + \hat{\beta}_2 X_2^\circ + \cdots + \beta_K X_K^\circ . \tag{4.6.1}$$

Once again, it is the variance of the prediction error which is different for the two cases. We still have [compare (2.6.15)]

$$\mathrm{var}(\hat{Y}_0 - Y_0) = \sigma^2 + \mathrm{var}(\hat{Y}_0). \tag{4.6.2}$$

Now when we consider $\hat{Y}_0$, as given in (4.6.1), it is clear that it is a <u>linear combination of the $\beta$'s</u>, just as $\hat{h}$ (in the preceding section) was.

Thus, by defining

$$\alpha' = (1, X_1^\circ, X_2^\circ, \ldots, X_K^\circ), \tag{4.6.3}$$

we have immediately that

$$S_{\hat{Y}_0}^2 = \alpha'\hat{V}\alpha \tag{4.6.4}$$

and

$$S_{(\hat{Y}_0 - Y_0)}^2 = \hat{\sigma}^2 + \alpha'\hat{V}\alpha. \tag{4.6.5}$$

The results that we use for inference, analogous to (2.6.13) and (2.6.16), are

$$t = \frac{\hat{Y}_0 - E[Y_i \mid 1, X_1^\circ, X_2^\circ, \ldots, X_K^\circ]}{S_{\hat{Y}_0}} \sim t_{N-K-1} \tag{4.6.6}$$

and

$$t = \frac{\hat{Y}_0 - Y_0}{S_{(\hat{Y}_0 - Y_0)}} \sim t_{N-K-1}, \tag{4.6.7}$$

respectively. Apart from different expressions for the estimated variance of prediction error, application in the multiple regression case is identical to that of simple regression.

## 4.7 AN EXAMPLE

The following example is designed to illustrate all the techniques and concepts above. A greater understanding of the operations carried out by the MINITAB computer program will be obtained if this example is studied in detail and if reference is made to the point where each concept is discussed.

The problem we will consider is the estimation of the Cobb-Douglas production function[5]

$$Q_t = \alpha K_t^{\beta_1} L_t^{\beta_2} e^{u_t}, \tag{4.7.1}$$

where

$Q_t$ = index of output in the tth period,

$K_t$ = index of capital in the tth period,

$L_t$ = index of labor in the tth period,

$\alpha, \beta_1, \beta_2$ = coefficients to be estimated,

$e^{u_t}$ = disturbance term.

[This form of the disturbance term is needed for least squares estimates of the coefficients to be BLUE. This will become evident once logarithms of (4.7.1) are taken.]

In equation (4.7.1) $\beta_1$ is the elasticity of production with respect to capital, that is, $\beta_1 = (\partial Q/\partial K)(K/Q)$, while similarly $\beta_2 = (\partial Q/\partial L)(L/Q)$ is the elasticity of production with respect to labor. Elasticities are unit free, so $\beta_1$ and $\beta_2$ (as well as their estimates) do not depend on the units in which Q, K, and L are measured. Thus the data on each of these will be in the form of an index.

Another item of economic interest is $\beta_1 + \beta_2$, the returns to scale. If $\beta_1 + \beta_2 > 1$, increasing returns to scale exist; if $\beta_1 + \beta_2 < 1$, there are decreasing returns to scale; and $\beta_1 + \beta_2 = 1$ implies constant returns to scale. This is of interest because if constant returns to scale exist and capital and labor are paid according to their marginal products ($\partial Q/\partial K$ and $\partial Q/\partial L$, respectively), total output will be exactly exhausted. That is,

$$\frac{\partial Q}{\partial K} K + \frac{\partial Q}{\partial L} L = Q.$$

Equation (4.7.1) is not yet in the same form as (4.1.1), so must be transformed before least squares estimates can be obtained. Taking natural logarithms of both sides of (4.7.1) gives

$$\ln Q_t = \ln \alpha + \beta_1 \ln K_t + \beta_2 \ln L_t + u_t. \tag{4.7.2}$$

Let

$$Y_t = \ln Q_t,$$

$$X_{1t} = \ln K_t,$$

$$X_{2t} = \ln L_t,$$

$$\beta_0 = \ln \alpha.$$

Equation (4.7.2) can then be written as

$$Y_t = \beta_0 + \beta_1 X_{1t} + \beta_2 X_{2t} + u_t. \tag{4.7.3}$$

This is in a form that can be estimated. Instead of $\alpha$ being estimated, $\beta_0$ (= ln $\alpha$) will be estimated, but an estimate of $\alpha$ could then be obtained from

$$\hat{\alpha} = \exp(\hat{\beta}_0).$$

Fifteen observations are available and these are given in Table 4.7.1. In the construction of each index the last observation has been used as a base period. The data on Y, $X_1$, and $X_2$ (the logarithms of Q, K, and L, respectively) are also given in Table 4.7.1. It is to these data that we must apply the least squares techniques. From Table 4.7.1 we can obtain

$$N = 15 \qquad K = 2$$

$$\Sigma Y^2 = 313.5576 \qquad \bar{Y} = 4.57127$$

$$\Sigma y^2 = 313.5576 - (15)(4.57127)^2 = 0.109418$$

$$\Sigma X_1^2 = 309.8788 \qquad \bar{X}_1 = 4.54317$$

$$\Sigma x_1^2 = 309.8788 - (15)(4.54317)^2 = 0.27306$$

$$\Sigma X_2^2 = 316.9182 \qquad \bar{X}_2 = 4.59601$$

$$\Sigma x_2^2 = 316.9182 - (15)(4.59601)^2 = 0.06880$$

$$\Sigma x_1 x_2 = 313.299 - (15)(4.54317)(4.59601) = 0.092$$

$$\Sigma x_1 y = 311.665 - (15)(4.54317)(4.57127) = 0.145$$

$$\Sigma x_2 y = 315.201 - (15)(4.59601)(4.57127) = 0.057.$$

Substituting the relevant numbers from above into equations (4.3.2), (4.3.3), and (4.3.4), we obtain the estimates

$$\hat{\beta}_1 = \frac{(0.145)(0.069) - (0.057)(0.092)}{(0.273)(0.069) - (0.092)^2} = 0.460,$$

Table 4.7.1

| $t$ | $K_t$ | $L_t$ | $Q_t$ | $Y_t = \log Q_t$ | $X_{1t} = \log K_t$ | $X_{2t} = \log L_t$ |
|---|---|---|---|---|---|---|
| 1 | 105.8 | 97.0 | 100.9 | 4.6141 | 4.6616 | 4.5747 |
| 2 | 102.7 | 107.4 | 100.6 | 4.6112 | 4.6318 | 4.6766 |
| 3 | 80.3 | 94.6 | 87.3 | 4.4694 | 4.3858 | 4.5497 |
| 4 | 87.0 | 105.6 | 99.7 | 4.6022 | 4.4659 | 4.6597 |
| 5 | 137.2 | 113.7 | 119.8 | 4.7858 | 4.9214 | 4.7336 |
| 6 | 97.2 | 96.0 | 97.8 | 4.5829 | 4.5768 | 4.5643 |
| 7 | 79.6 | 95.6 | 85.7 | 4.4509 | 4.3770 | 4.5602 |
| 8 | 88.3 | 102.3 | 94.3 | 4.5465 | 4.4807 | 4.6279 |
| 9 | 98.5 | 96.1 | 98.4 | 4.5890 | 4.5901 | 4.5654 |
| 10 | 93.7 | 96.9 | 85.2 | 4.4450 | 4.5401 | 4.5737 |
| 11 | 86.6 | 99.5 | 93.3 | 4.5358 | 4.4613 | 4.6002 |
| 12 | 93.9 | 99.0 | 106.5 | 4.6681 | 4.5422 | 4.5951 |
| 13 | 94.0 | 102.8 | 96.0 | 4.5643 | 4.5433 | 4.6328 |
| 14 | 78.6 | 83.2 | 89.9 | 4.4987 | 4.3644 | 4.4212 |
| 15 | 100.0 | 100.0 | 100.0 | 4.6052 | 4.6052 | 4.6052 |

$$\hat{\beta}_2 = \frac{(0.057)(0.273) - (0.145)(0.092)}{(0.273)(0.069) - (0.092)^2} = 0.205,$$

$$\hat{\beta}_0 = 4.571 - (0.460)(4.543) - (0.205)(4.596) = 1.540.$$

Also, $\hat{\alpha} = \exp(\hat{\beta}_0) = 4.665$. Using (4.3.11) to estimate $\hat{\sigma}^2$ gives

$$\hat{\sigma}^2 = \frac{9.109 - (0.460)(0.145) - (0.205)(0.057)}{15 - 3} = 0.002616.$$

(Note: A check of some of the arithmetic might show some small discrepancies. This is because the calculations were carried out more accurately than they are reported.)

The standard errors for $\hat{\beta}_1$ and $\hat{\beta}_2$ can be found using (4.3.5) and (4.3.6), where $\sigma^2$ is replaced by $\hat{\sigma}^2$ and the square root of the whole expression is taken.

$$r_{12}^2 = \frac{(\Sigma x_1 x_2)^2}{\Sigma x_1^2 \, \Sigma x_2^2} = \frac{(0.092)^2}{(0.273)(0.069)} = 0.4509,$$

$$S_{\hat{\beta}_1} = \left[\frac{0.002616}{(0.273)(1 - 0.451)}\right]^{\frac{1}{2}} = 0.132,$$

$$S_{\hat{\beta}_2} = \left[\frac{0.002616}{(0.069)(1 - 0.451)}\right]^{\frac{1}{2}} = 0.263.$$

The covariance matrix of the $\hat{\beta}$'s [from equation (4.4.1)] is

$$\hat{V} = \begin{bmatrix} 0.848 & 0.028 & -0.212 \\ 0.028 & 0.017 & -0.023 \\ -0.212 & -0.023 & 0.069 \end{bmatrix}.$$

A 90% confidence interval for $\beta_1$, for example, is

$$\hat{\beta}_1 - (t_{0.05})(S_{\hat{\beta}_1}) \text{ to } \hat{\beta}_1 + (t_{0.05})(S_{\hat{\beta}_1}),$$

that is, 0.460 - (1.782)(0.132) to 0.476 + (1.782)(0.132) or 0.225 to 0.695. Note that the t distribution has 12 degrees of freedom.

To calculate $R^2$ we need

$$ESS = \sum_{k=1}^{2} \hat{\beta}_k \sum x_k y = (0.460)(0.145) + (0.205)(0.057)$$

$$= 0.0780,$$

$$TSS = \Sigma y^2 = 0.1094,$$

$$R^2 = \frac{0.0780}{0.1094} = 0.7131.$$

Also, RSS = TSS - ESS = 0.1094 - 0.0780 = 0.0314. The estimated production function is

$$\underset{}{\ln \hat{Q}_t} = \underset{(0.921)}{1.540} + \underset{(0.132)}{0.460} \ln K_t + \underset{(0.263)}{0.205} \ln L_t; \qquad R^2 = 0.713, \quad \hat{\sigma}^2 = 0.002616.$$

So far no mention has been made concerning what values of $R^2$ and the standard errors are regarded as good values. We would naturally like $R^2$ to be as close to 1 as possible and would also like the standard errors to be small, because this results in narrow confidence intervals. But how big should $R^2$ be, and how small should the standard error be to "keep us happy"? These are problems that will be discussed in the sections on choosing between alternative models, but a couple of guidelines will be mentioned here. First, statistical criteria such as $R^2$ and standard errors are not the only ones that should be used. Economic knowledge of which variables are important and what signs estimated coefficients should have also play an important role in choosing between models. If a model is to be used successfully for predictive purposes, an $R^2$ of greater than 0.7 or 0.8 would be desirable.

When using cross-sectional data on the behavior of some economic units, it is often extremely difficult to get an $R^2$ of greater than 0.2 or 0.3. This is because many random influences still exist and have not been averaged out as in the case of some time-series data. This low $R^2$ does not matter if one is interested primarily in estimating a particular coefficient and this estimation has been achieved accurately (the standard error is low). If one wishes to establish the importance of a particular explanatory variable, it is desirable to show that a certain coefficient is nonzero ($\beta_k \neq 0$). This can be shown fairly conclusively if the null hypothesis $\beta_k = 0$ is rejected at the 5 percent level of significance. Using the rule of thumb given in Chapter 2, this will occur if the coefficient estimate is greater than about twice the size of its standard error.

Let us now evaluate the estimated production function. The $R^2$ could be regarded as moderate. In practice, aggregate production functions estimated from time-series data usually yield a higher $R^2$. The elasticity of production with respect to capital has been estimated fairly accurately—the coefficient estimate is over three times the size of its standard error—but the labor coefficient has been estimated poorly. In fact, its standard error is greater than the estimated coefficient. The null hypothesis $\beta_2 = 0$ would be accepted at any reasonable level of significance. Despite this, our knowledge of economics would still insist that labor be included in the model, but perhaps an alternative definition should be sought.[6] Other possible reasons for a poor estimate of $\beta_2$ are the exclusion of relevant variables, multicollinearity or autocorrelation. These are future topics.

Let us now test whether constant returns to scale hold.

$$H_0: \beta_1 + \beta_2 = 1$$

$$H_1: \beta_1 + \beta_2 \neq 1.$$

Here $\hat{h} = \hat{\beta}_1 + \hat{\beta}_2 = 0.460 + 0.205 = 0.665.$

$$S^2_{\hat{h}} = \alpha'\hat{V}\alpha,$$

where

$$\alpha' = (0, 1, 1).$$

Thus

$$S^2_{\hat{h}} = [0, 1, 1]\begin{bmatrix} 0.848 & 0.028 & -0.212 \\ 0.028 & 0.017 & -0.023 \\ -0.212 & -0.023 & 0.069 \end{bmatrix}\begin{bmatrix} 0 \\ 1 \\ 1 \end{bmatrix}$$

$$= [-0.184, -0.006, 0.046]\begin{bmatrix} 0 \\ 1 \\ 1 \end{bmatrix} = 0.040,$$

giving

$$S_{\hat{h}} = (0.040)^{\frac{1}{2}} = 0.20.$$

From (4.4.4) the t-ratio is given by

$$t = \frac{0.665 - 1}{0.20} = -1.675. \qquad (4.7.4)$$

At the 5 percent level of significance, with N - K - 1 = 15 - 2 - 1 = 12 degrees of freedom, we will reject $H_0$ if $|t| > 2.179$. Clearly, $H_0$ will not be rejected and we conclude that the hypothesis of constant returns to scale is consistent with the data.

Finally, let us predict output for some given values of capital and labor. Suppose these values are $K^\circ = 93.9$ and $L^\circ = 99.0$. With these values

$$X_1^\circ = \ln K^\circ = \ln 93.9 = 4.5422$$

$$X_2^\circ = \ln L^\circ = \ln 99.0 = 4.5951$$

$$\hat{Y}_0 = \hat{\beta}_0 + \hat{\beta}_1 X_1^\circ + \hat{\beta}_2 X_2^\circ$$
$$= 1.540 + (0.460)(4.5422) + (0.205)(4.5951)$$
$$= 4.5714.$$

Thus

$$\hat{Q}_0 = \exp \hat{Y}_0 = \exp(4.5714) = 96.68.$$

This last transformation was necessary because the model is linear in the logarithms of Q, K, and L.

The estimated variance of the prediction error $(\hat{Y}_0 - Y_0)$ is given by

$$S^2_{(\hat{Y}_0 - Y_0)} = \hat{\sigma}^2 + \alpha'\hat{V}\alpha,$$

where

$$\alpha' = (1, X_1^\circ, X_2^\circ) = (1, 4.5422, 4.5951).$$

Thus[7]

$$S^2_{(\hat{Y}_0 - Y_0)} = 0.002616 + [1 \;\; 4.5422 \;\; 4.5951] \begin{bmatrix} 0.848 & 0.028 & -0.212 \\ 0.028 & 0.017 & -0.023 \\ -0.212 & -0.023 & 0.069 \end{bmatrix} \begin{bmatrix} 1 \\ 4.5422 \\ 4.5951 \end{bmatrix}$$

$$= 0.002616 + 0.000174$$

$$= 0.002790.$$

Therefore,

$$S_{(\hat{Y}_0 - Y_0)} = (0.002790)^{\frac{1}{2}} = 0.05282.$$

We are now able to construct a confidence interval for our prediction $\hat{Q}_0$. Using the result (4.5.7), a 95 percent confidence interval for $\hat{Y}_0$ is

$$4.5714 \pm (t_{0.025})(0.05282) = 4.5714 \pm (2.179)(0.05282)$$

The 95 percent confidence interval for $\hat{Y}_0$ is therefore 4.4563 to 4.6865. Taking the exponential of these two limits, we find that the 95 percent confidence interval for $\hat{Q}_0$ is 86.2 to 108.5.

## 4.8 A MINITAB PROGRAM FOR THE EXAMPLE

As a means of reinforcing the ideas discussed in Chapter 3 about the program MINITAB, we present below a program to do the example of Section 4.7, given that the first three columns of Table 4.7.1 are stored in the data file 'CD.DTA'.

```
READ 'CD.DTA' C1-C3
LET C4=LOGE(C3)
LET C5=LOGE(C1)
LET C6=LOGE(C2)
NAME C4 'LNQ' C5 'LNK' C6 'LNL'
REGR C4 2 C5 C6
STOP
```

The relevant results are shown in Display 4.8.1.

Most of the output is self-explanatory, and when compared with Section 4.7, it is seen that the results are identical. We note at this point that the three quantities ESS, RSS, and TSS are given under the heading "Analysis of Variance" and may be identified as

ESS = regression SS,

RSS = error SS,

TSS = total SS.

We will now turn our attention to testing the hypothesis of constant returns to scale, namely

$$H_0: \beta_1 + \beta_2 = 1$$

against

$$H_1: \beta_1 + \beta_2 \neq 1.$$

This test will be performed using a reparametrization of the model as discussed in Section 4.5. The null hypothesis can be written in the form

DISPLAY 4.8.1

```
MTB    READ 'CD.DTA' C1-C3
      15 ROWS READ
ROW       C1       C2       C3

  1     105.8     97.0    100.9
  2     102.7    107.4    100.6
  3      80.3     94.6     87.3
  4      87.0    105.6     99.7

MTB > LET C4=LOGE(C3)
MTB > LET C5=LOGE(C1)
MTB > LET C6=LOGE(C2)
MTB > NAME C4 'LNQ' C5 'LNK' C6 'LNL'
MTB > REGR C4 2 C5 C6

The regression equation is
LNQ = 1.54 + 0.460 LNK + 0.205 LNL

Predictor        Coef         Stdev        t-ratio
Constant       1.5396        0.9210           1.67
LNK            0.4600        0.1321           3.48
LNL            0.2049        0.2631           0.78

s = 0.05114        R-sq = 71.3%        R-sq(adj) = 66.5%

Analysis of Variance

SOURCE         DF          SS            MS
Regression      2      0.078031      0.039015
Error          12      0.031387      0.002616
Total          14      0.109418
```

$H_0$: $\beta_1 + \beta_2 - 1 = 0$, so we will transform the model so that one of the parameters (it does not matter which) is $\beta_1 + \beta_2 - 1$.

Our regression model is

$$\ln Q_i = \beta_0 + \beta_1 \ln K_i + \beta_2 \ln L_i + u_i.$$

Adding and subtracting $\beta_2 \ln K_i$ to the right-hand side yields

$$\ln Q_i = \beta_0 + (\beta_1 + \beta_2) \ln K_i + \beta_2(\ln L_i - \ln K_i) + u_i.$$

If we now substract $\ln K_i$ from each side, we have

$$\ln Q_i - \ln K_i = \beta_0 + (\beta_1 + \beta_2 - 1) \ln K_i + \beta_2(\ln L_i - \ln K_i) + u_i.$$

A test of whether the coefficient of $\ln K_i$ is zero in this model is obviously equivalent to testing $H_0$. This regression may be run by adding the following statements to our earlier program:

```
LET C7 = C6-C5 (stores lnL - lnK in C7)
LET C8 = C4-C5 (stores lnQ - lnK in C8)
REGR C8 2 C5 C7
```

The results of this regression are shown in Display 4.8.2.

The coefficient of ln $K_i$ is -0.3351, and its standard error is 0.2000. Thus, to test whether this coefficient is significantly different from zero, we form the t ratio given by [see equation (2.6.9)]

$$t = \frac{-0.3351}{0.2000} = -1.675$$

This is identical to the t-ratio (4.7.4) which we obtained using the estimated variance-covariance matrix $\hat{V}$.

A comparison of Displays 4.8.1 and 4.8.2 shows that RSS = 0.031387 in both cases but that TSS, ESS, and $R^2$ are different. This illustrates the points made in Section 4.5. A convenient method for using MINITAB to generate and analyze predictions will be given in Section 7.6.

DISPLAY 4.8.2

```
MTB > LET C7=C6-C5
MTB > LET C8=C4-C5
MTB > REGR C8 2 C5 C7

The regression equation is
C8 = 1.54 - 0.335 LNK + 0.205 C7

Predictor        Coef        Stdev      t-ratio
Constant       1.5396       0.9210         1.67
LNK           -0.3351       0.2000        -1.68
C7             0.2049       0.2631         0.78

s = 0.05114     R-sq = 66.4%     R-sq(adj) = 60.8%

Analysis of Variance

SOURCE       DF          SS          MS
Regression    2    0.062134    0.031067
Error        12    0.031387    0.002616
Total        14    0.093521
```

NOTES

1. We will use $\beta_1$ to represent any of the parameters $\beta_1, \beta_2, \ldots, \beta_K$.

2. In this case $X_1$ and $X_2$ are known as orthogonal regressors.

3. The corresponding expression for var($\hat{\beta}_0$) is not given here. The interested reader will find it in Koutsoyiannis (1977, 122).

4. An expression of this kind is known as a quadratic form in the vector $\alpha$.

5. We will follow conventional practice and use the index t instead of i when the data are time-series observations.

6. Variables such as aggregate labor (or capital) in the economy for a given period are obviously fairly difficult to define, and there are a number of possible alternatives that could be used.

7. Obviously, $\hat{V}$ must be retained to more decimal places than shown here in order to achieve the final accuracy reported. If $\hat{V}$ had been reported to eight decimal places, this accuracy could have been achieved. The necessity of carrying a large number of decimal places to avoid rounding errors and obtain an accurate final result emphasizes the need for a computer to solve multiple regression problems.

# 5

# Interpretation of the Estimated Regression Model

## 5.1 INTRODUCTION

In our analysis so far it has been taken for granted that the true model, or generating mechanism for the dependent variable, has been known. In particular, it has been assumed that

(a) The relevant regressor variables and the functional form are given;
(b) The disturbance $u_i$ satisfies the conditions (2.1.2), (2.1.3), and (2.1.4); and
(c) The regressor variables $X_1, \ldots, X_K$ are not random variables, but fixed in repeated samples.

Under these conditions OLS is BLUE and a regression analysis consists simply of applying this efficient procedure (OLS) to the data and carrying out whatever hypotheses tests the context makes necessary.

In practice, however, our prior knowledge is rarely, if ever, precise enough to tell us with certainty the true model. Needless to say, if we apply OLS to a model that is not correct, there is absolutely no guarantee that our results will be good. We must therefore examine a number of competing models to try and determine which is best. In this chapter we will be looking at some of the problems that are almost always present and which make our analysis anything but the simple exercise it has seemed so far. Our aim is to develop a framework within which to think, which will help us to determine what the effects of our inexact a priori knowledge are likely to be.

## 5.2 USES AND ABUSES OF $R^2$

A very common practice is to run a number of regressions and choose the model that yields the highest value of $R^2$. As we have seen, $R^2$ can be interpreted either as the proportion of variation in the dependent variable explained

Different Dependent Variables

Frequently, we wish to choose between two models that have different dependent variables. Consider the data shown in Table 5.2.1. Denoting consumption of pork, price of pork, and disposable personal income by $Q_t$, $P_t$, and $Y_t$, respectively, two possible models would be

$$Q_t = \beta_0 + \beta_1 P_t + \beta_2 Y_t + u_t$$

and

$$\ln Q_t = \beta_0 + \beta_1 \ln P_t + \beta_2 \ln Y_t + u_t,$$

with the second model implying constant price and income elasticities.

Estimating these models, we obtain

$$\hat{Q}_t = 56.96 - \underset{(0.316)}{1.409}P_t + \underset{(0.017)}{0.0798}Y_t; \quad R^2 = 0.583,$$

$$\ln \hat{Q}_t = 1.353 - \underset{(0.118)}{0.58} \ln P_t + \underset{(0.146)}{0.7473} \ln Y_t; \quad R^2 = 0.625.$$

Because the second model explains 62.5 percent of variation and the first 58.3 percent, we may be tempted to assume that the second model is to be preferred. However, we must note that the first model explains 58.3 percent of the variation in $Q_t$, whereas the second model explains 62.5 percent of the variation in $\ln Q_t$. The $R^2$'s are not comparable.

To obtain from the second model an $R^2$ that is comparable, we could obtain the predictions $\ln \hat{Q}_t$ of $\ln Q_t$ and exponentiate them. We would then have predictions $\hat{Q}_t^*$, say, of $Q_t$ obtained from the second model. If we then regressed $\hat{Q}_t^*$ on $Q_t$, the $R^2$ from this regression would be comparable with that obtained using the first model.

To illustrate how having different dependent variables can affect the value of $R^2$, we will consider again the Keynesian consumption function estimated in Section 2.10, namely

$$\hat{C}_t = \underset{(0.179)}{0.690} + \underset{(0.007)}{0.604}Y_t; \quad \begin{array}{l} RSS = 5.111 \\ R^2 = 0.996. \end{array}$$

If we were to use this estimated model to construct a savings function, given that $S_t = Y_t - C_t$, we would obtain

$$\hat{S}_t = Y_t - 0.690 - 0.604Y_t$$

$$= -0.690 + 0.396Y_t.$$

If we now carry out the regression of $S_t$ on $Y_t$, we obtain

Table 5.2.1

| Year | Price (cents/lb) | Consumption (lb/capita) | Income (dollars/capita) |
|---|---|---|---|
| 1922 | 26.8 | 65.7 | 541 |
| 1923 | 25.3 | 74.2 | 616 |
| 1924 | 25.3 | 74.0 | 610 |
| 1925 | 31.1 | 66.8 | 636 |
| 1926 | 33.3 | 64.1 | 651 |
| 1927 | 31.2 | 67.7 | 645 |
| 1928 | 29.5 | 70.9 | 653 |
| 1929 | 30.3 | 69.6 | 682 |
| 1930 | 29.1 | 67.0 | 604 |
| 1931 | 23.7 | 68.4 | 515 |
| 1932 | 15.6 | 70.7 | 390 |
| 1933 | 13.9 | 69.6 | 364 |
| 1934 | 18.8 | 63.1 | 411 |
| 1935 | 27.4 | 48.4 | 459 |
| 1936 | 26.9 | 55.1 | 517 |
| 1937 | 27.7 | 58.8 | 551 |
| 1938 | 24.5 | 58.2 | 506 |
| 1939 | 22.2 | 64.7 | 538 |
| 1940 | 19.3 | 73.5 | 576 |
| 1941 | 24.7 | 68.4 | 697 |

Source: Fox (1968, 284), reproduced with permission.

$$\hat{S}_t = \underset{(0.179)}{-0.690} + \underset{(0.007)}{0.396}Y_t; \qquad RSS = 5.111 \quad R^2 = 0.990.$$

We notice that the coefficients are identical to those obtained by construction from the consumption function. Furthermore, the standard errors and residual sum of squares RSS are identical. That is, these two regressions are entirely equivalent, in that one can be completely derived from the other. However, we notice that the $R^2$'s are not the same. With these data the difference is not very great but with the equivalent functions using quarterly U.S. data, Rao and Miller (1971) found that $R^2$ changed from 0.99 to 0.64! This apparent paradox of entirely equivalent regressions yielding different $R^2$'s is resolved when we recall the definition

$$R^2 = 1 - \frac{RSS}{TSS}.$$

Although RSS is the same, because there are different dependent variables, TSS will be different. This example is a case of reparametrization of the model by subtracting the variable $Y_t$ from both sides of the equation. Thus, whenever models have different dependent variables $R^2$ cannot be used as a criterion for judging which model is preferable.

## Different Number of Regressor Variables

Suppose that we fit a model of the form

$$\hat{Y}_i = \hat{\beta}_0 + \hat{\beta}_1 X_{1i}.$$

Then the least squares principle guarantees that $\hat{\beta}_0$ and $\hat{\beta}_1$ are the estimates which produce the minimum residual sum of squares. If another variable $X_2$ is now added, the residual sum of squares from OLS can never rise (otherwise, the estimated model in which $\hat{\beta}_0$ and $\hat{\beta}_1$ are unchanged and $\hat{\beta}_2 = 0$ would have a smaller RSS, which is a contradiction). In fact, when regressor variables are added to a model, it is virtually certain that RSS will fall, even if the added variables are completely irrelevant. Consequently, as

$$R^2 = 1 - \frac{RSS}{TSS}$$

and TSS is not changing, addition of variables will cause $R^2$ to rise. The problem is that as the number of regressor variables K increases, the degrees of freedom N - K - 1 decreases, leading to less precise estimates of the coefficients.

Intuitively, inclusion of a relevant variable will result in an increase in $R^2$ which is large enough to outweigh the reduction in degrees of freedom. This leads to the notion of modifying $R^2$ so as to take account not just of RSS and TSS, but of RSS and TSS relative to degrees of freedom.

Now TSS = $\Sigma y^2$ and is based on N observations on $y = Y - \bar{Y}$, subject to one restriction, namely, $\Sigma y = 0$. TSS thus has N - 1 degrees of freedom. RSS = $\Sigma e^2$ is based on N observations $e_1, e_2, \ldots, e_N$, but the residuals satisfy K + 1 restrictions, $\Sigma e = 0$, $\Sigma eX_1 = 0, \ldots, \Sigma eX_K = 0$. Thus RSS has N - (K + 1) = N - K - 1 degrees of freedom.

We now define an adjusted $R^2$ denoted by $\bar{R}^2$, as follows:

$$\bar{R}^2 = 1 - \frac{RSS/(N - K - 1)}{TSS/(N - 1)}$$

$$= 1 - \frac{RSS}{TSS}\frac{N - 1}{N - K - 1}. \tag{5.2.1}$$

That is,

$$\bar{R}^2 = 1 - (1 - R^2)\frac{N - 1}{N - K - 1}. \tag{5.2.2}$$

Equation (5.2.2) gives the simple relationship that exists between $R^2$ and $\bar{R}^2$. Whenever models having a different number of regressor variables are compared, it should be on the basis of $\bar{R}^2$ and not $R^2$. We note two properties of $\bar{R}^2$:

(i) If the number of observations N is very large compared with the number of regressor variables K, then, on dividing numerator and denominator of (5.2.2) by N, we have

$$\bar{R}^2 = 1 - (1 - R^2)\frac{1 - 1/N}{1 - (K + 1)/N}$$

$$\approx 1 - (1 - R^2) = R^2.$$

That is, in this case $\bar{R}^2$ and $R^2$ are approximately equal.

(ii) Even though $R^2$ can never be negative, if $R^2$ is small, $\bar{R}^2$ can be negative. It is easily shown that if

$$R^2 < \frac{K}{N - 1},$$

then $\bar{R}^2 < 0$.

## 5.3 SPECIFICATION ERRORS

The model that is used for regression may be misspecified in many different ways. For example, the functional form and/or the choice of regressor variables may be incorrect. We will only consider the cases of relevant variables omitted and irrelevant variables included in the analysis. An appreciation of the principles involved here can give valuable insights which are helpful in specifying the model.

## Omission of Relevant Variables

We will follow our usual practice and examine the simplest case and then generalize our conclusions. Suppose that (in deviation form)[1] the true model is

$$y_i = \beta_1 x_{1i} + \beta_2 x_{2i} + u_i, \tag{5.3.1}$$

but owing to the incorrect omission of the relevant variable $x_2$, an estimated relation of the form

$$\hat{y}_i = \hat{\beta}_1 x_{1i} \tag{5.3.2}$$

is obtained.

We know from the result (2.2.9) for simple regression that

$$\hat{\beta}_1 = \frac{\Sigma x_1 y}{\Sigma x_1^2}$$

$$= \frac{\Sigma x_1(\beta_1 x_1 + \beta_2 x_2 + u)}{\Sigma x_1^2}$$

by (5.3.1). Therefore,

$$\hat{\beta}_1 = \beta_1 + \beta_2 \frac{\Sigma x_1 x_2}{\Sigma x_1^2} + \frac{\Sigma x_1 u}{\Sigma x_1^2}.$$

On taking expectations, recalling that $E(u_i) = 0$ and $x_1$ and $x_2$ are not stochastic, we obtain

$$E(\hat{\beta}_1) = \beta_1 + \beta_2 \frac{\Sigma x_1 x_2}{\Sigma x_1^2}.$$

Also, again by (2.2.9), $\Sigma x_1 x_2 / \Sigma x_1^2$ is the slope coefficient we would obtain if we regressed $X_2$ on $X_1$. We will denote this coefficient $\hat{\beta}_{2.1}$. Thus

$$E(\hat{\beta}_1) = \beta_1 + \beta_2 \hat{\beta}_{2.1} \tag{5.3.3}$$

Equation (5.3.3) shows that provided that neither $\beta_2$ nor $\hat{\beta}_{2.1}$ is zero, the omission of the relevant variable $X_2$ will result in the parameter estimate of the included variable $X_1$ being biased. We may ignore the case $\beta_2 = 0$, as this is the same as saying that $X_2$ is irrelevant. The coefficient $\hat{\beta}_{2.1} = 0$ when $X_1$ and $X_2$ are orthogonal (uncorrelated).

Therefore, provided that the omitted variable is not orthogonal to the included variable, the OLS estimate of $\beta_1$ is no longer BLUE, as it is no longer unbiased. The magnitude of the bias depends on $\beta_2$ and the degree of correlation between $X_1$ and $X_2$, and the direction of the bias depends on the sign of $\beta_2 \hat{\beta}_{2.1}$. The sign of this term can often be deduced from prior knowledge.

Using matrix algebra, the result (5.3.4) can be generalized. Let us suppose that the true model is

$$y_i = \beta_1 x_{1i} + \cdots + \beta_r x_{ri} + \beta_{r+1} x_{r+1,i} + \cdots + \beta_K x_{Ki} + u_i$$

and that the model is misspecified, in that it is assumed to be

$$y_i = \beta_1 x_{1i} + \cdots + \beta_r x_{ri} + u_i.$$

That is, the relevant variables $X_{r+1}, \ldots, X_K$ have been omitted.

It can be shown that

$$E(\hat{\beta}_1) = \beta_1 + \beta_{r+1}\hat{\beta}_{r+1.1,2,\ldots,r} + \cdots + \beta_K\hat{\beta}_{K.1,2,\ldots,r}, \qquad (5.3.4)$$

where $\hat{\beta}_{r+1.1,2,\ldots,r}$ denotes the slope coefficient of $X_1$ that would be obtained if the omitted variable $X_{r+1}$ were regressed on all of the included variables $X_1, X_2, \ldots, X_r$. From equation (5.3.4) we can infer that unless all of the omitted variables are orthogonal to all of the included variables, OLS estimators based on the misspecified model will be biased.

ILLUSTRATIONS Suppose that we are estimating the demand $Q_t$ for some commodity, and the true model is

$$Q_t = \beta_0 + \beta_1 P_t + \beta_2 Y_t + \beta_3 P_t^* + u_t,$$

where $P_t$, $Y_t$, and $P_t^*$ are deflated price, income, and price of competitor, respectively. If the variable $P_t^*$ were omitted, the estimated model would be

$$\hat{Q}_t = \hat{\beta}_0 + \hat{\beta}_1 P_t + \hat{\beta}_2 Y_t$$

and $\hat{\beta}_1$ would be a biased estimator with

$$\text{bias} = \beta_3 \hat{\beta}_{3.1,2}.$$

As $P_t^*$ is the price of a competitor, $\beta_3 > 0$. It is probably reasonable to assume that $P_t$ and $P_t^*$ are positively correlated[2] and hence $\hat{\beta}_{3.1,2} > 0$. The estimator $\hat{\beta}_1$ will therefore have a positive bias and $\hat{\beta}_1$ will tend to overestimate $\beta_1$. It is quite conceivable that the positive bias would be sufficiently large to yield $\hat{\beta}_1 > 0$—an upward-sloping demand curve!

Consider again a constant elasticity of demand curve. Using exactly the same reasoning, we may find that our analysis indicates the demand to be inelastic when economic theory would suggest a highly elastic demand, or vice versa. These illustrations demonstrate a general fact—that when the estimated coefficients do not accord with a priori economic expectations, the reason is often, but not always, to be found in omitted relevant variables.

Inclusion of Irrelevant Variables

Let us suppose now that the true model is

$$y_i = \beta_1 x_{1i} + u_i, \tag{5.3.5}$$

and we include the irrelevant variable $X_2$ to estimate a model of the form

$$\hat{y}_i = \hat{\beta}_1 x_{1i} + \hat{\beta}_2 x_{2i}.$$

From (4.3.2),

$$\hat{\beta}_1 = \frac{\Sigma x_1 y \, \Sigma x_2^2 - \Sigma x_2 y \, \Sigma x_1 x_2}{\Sigma x_1^2 \, \Sigma x_2^2 - (\Sigma x_1 x_2)^2}, \tag{5.3.6}$$

and by substituting for y from (5.3.5).

$$\hat{\beta}_1 = \frac{\beta_1 \Sigma x_1^2 \, \Sigma x_2^2 + \Sigma x_1 u \, \Sigma x_2^2 - \beta_1 (\Sigma x_1 x_2)^2 - \Sigma x_2 u \, \Sigma x_1 x_2}{\Sigma x_1^2 \, \Sigma x_2^2 - (\Sigma x_1 x_2)^2}$$

$$= \beta_1 + \frac{\Sigma x_1 u \, \Sigma x_2^2 - \Sigma x_2 u \, \Sigma x_1 x_2}{\Sigma x_1^2 \, \Sigma x_2^2 - (\Sigma x_1 x_2)^2}.$$

Now because $E(u) = 0$ and $x_1$, $x_2$ are nonstochastic, on taking expectations we have

$$E(\hat{\beta}_1) = \beta_1.$$

That is, the inclusion of the irrelevant variable $X_2$ has not introduced any bias into $\hat{\beta}_1$.

Consider now $\hat{\beta}_2$, which from (4.3.3) is given by

$$\hat{\beta}_2 = \frac{\Sigma x_2 y \, \Sigma x_1^2 - \Sigma x_1 y \, \Sigma x_1 x_2}{\Sigma x_1^2 \, \Sigma x_2^2 - (\Sigma x_1 x_2)^2}$$

$$= \frac{\beta_1 [\Sigma x_1 x_2 \, \Sigma x_1^2 - \Sigma x_1^2 \, \Sigma x_1 x_2] + [\Sigma x_2 u \, \Sigma x_1^2 - \Sigma x_1 u \, \Sigma x_1 x_2]}{\Sigma x_1^2 \, \Sigma x_2^2 - (\Sigma x_1 x_2)^2}$$

$$= \frac{0 + \Sigma x_2 u \, \Sigma x_1^2 - \Sigma x_1 u \, \Sigma x_1 x_2}{\Sigma x_1^2 \, \Sigma x_2^2 - (\Sigma x_1 x_2)^2}.$$

Therefore,

$$E(\hat{\beta}_2) = 0.$$

Saying that $X_2$ is irrelevant is exactly the same as saying that $\beta_2 = 0$, so $\hat{\beta}_2$ is also unbiased. The general principle that emerges is that when irrelevant variables are included, no biases are introduced into the parameter estimates.

In view of this and the fact that estimates are biased when relevant variables are omitted, it would seem to be a sensible strategy to include every possible variable that may be even remotely relevant in order to minimize the risk of omitting a relevant variable. Unfortunately, there is a penalty to be paid for including irrelevant variables.

If (5.3.5) is the true model and we specify it correctly, $\beta_1$ will be estimated by

$$\hat{\beta}_1 = \frac{\Sigma x_1 y}{\Sigma x_1^2},$$

and by (2.4.2),

$$\text{var}(\hat{\beta}_1) \equiv V_1 = \frac{\sigma^2}{\Sigma x_1^2}. \tag{5.3.7}$$

If, on the other hand, $X_2$ is incorrectly included in the model, we will estimate $\beta_1$ by (5.3.6) and by (4.3.5),

$$\text{var}(\hat{\beta}_1) \equiv V_2 = \frac{\sigma^2}{\Sigma x_1^2(1 - r_{12}^2)}, \tag{5.3.8}$$

where $r_{12}$ is the sample correlation between $X_1$ and $X_2$. This expression for the variance of $\hat{\beta}_1$ holds regardless of whether $X_2$ is relevant, because it does not involve the parameter $\beta_2$.

It follows from (5.3.7) and (5.3.8) that unless $X_1$ and $X_2$ are orthogonal (i.e., $r_{12} = 0$)

$$V_2 > V_1$$

and the inclusion of an irrelevant variable has resulted in $\beta_1$ being estimated with less precision. We say that there has been a loss of efficiency. OLS is no longer BLUE, because it is no longer best. The percentage increase in the variance of $\hat{\beta}_1$ incurred by including the irrelevant variable $X_2$ is given by

$$\frac{100(V_2 - V_1)}{V_1} = \frac{100 r_{12}^2}{1 - r_{12}^2}$$

Table 5.3.1 Percentage Increase in Variance of $\hat{\beta}_1$

| $r_{12}^2$ | Increase (%) | $r_{12}^2$ | Increase (%) |
|---|---|---|---|
| 0.1 | 11.1 | 0.7 | 233.3 |
| 0.2 | 25.0 | 0.8 | 400.0 |
| 0.3 | 42.9 | 0.9 | 900.0 |
| 0.4 | 66.7 | 0.95 | 1900.0 |
| 0.5 | 100.0 | 0.99 | 9900.0 |
| 0.6 | 150.0 | | |

Table 5.3.1 gives some idea of when this misspecification is likely to be serious. In terms of standard deviations, $r_{12}^2 = 0.1$ would increase the standard deviation of $\hat{\beta}_1$ by about 5 percent, $r_{12}^2 = 0.5$ by about 40 percent, and $r_{12}^2 = 0.95$ by about 350 percent. As we shall see when we discuss multicollinearity, values of $r_{12}^2$ of the order of 0.95 are the rule rather than the exception when dealing with economic data.

This effect of "blowing up" standard deviations results in t-values becoming small, which in turn may convince us that highly relevant variables are, in fact, irrelevant. Dropping these variables would, of course, result in misspecification bias being introduced.

## 5.4 JUDGING THE RELEVANCE OF ADDED VARIABLES

In practical situations there would usually be certain variables which are suggested on theoretical grounds as being obviously important in a regression relationship. In addition, there may be other variables whose relevance requires investigation. In the preceding sections we have considered three criteria that bear on the relevance of these added variables. They were:

(a) The change in the value of $\bar{R}^2$; if $\bar{R}^2$ rises, this is an indication that the additions to the regressors are worthwhile.

(b) If there is only one additional variable, its statistical significance can be assessed by evaluating the t-ratio given by $t = \hat{\beta}_i / S_{\hat{\beta}_i}$.[3] We remind the reader of the rule of thumb which states that $\hat{\beta}_i$ is significant at the 5 percent level if the magnitude of the t-ratio is larger than 2.

(c) The changes in the estimated coefficients when variables are added can often provide valuable clues on the relevance of added variables.

It is important that all of these three criteria be considered as attempts are made to choose the best model. Finally, we note that (a) and (b) above are quite closely connected. Haitovsky (1969) has shown that when a variable is added to the model, $\bar{R}^2$ must rise if the magnitude of the t-ratio associated with the variable is greater than 1. Conversely, if $|t| < 1$, $\bar{R}^2$ will fall. From this result we can deduce that the criterion (b) is stronger than (a)—if the coefficient of the added variable is significant at the 5 percent level, $\bar{R}^2$ must have risen. On the other hand, $\bar{R}^2$ may rise even when the coefficient is not statistically significant.

EXAMPLE We have already estimated a Keynesian consumption function, using the data from Table 2.10.1, by

$$C_t = \underset{(0.18)}{0.69} + \underset{(0.01)}{0.60}Y_t; \qquad \bar{R}^2 = 0.996.$$

From an economic point of view, an essential feature of this model is that it is static. That is, any changes in income Y will result in an instantaneous and complete response in consumption. When we say that the marginal propensity to consume is 0.6, we mean that a 10-dollar increase in income will induce an instantaneous increase of 6 dollars in consumption. This is an exercise in comparative statics. It may, however, be far more realistic to assume that a change in income will induce changes in consumption which may continue for many time periods. One reason for this is that habitual consumption patterns do not change instantaneously. One possible model that may enable us to capture these dynamic effects is

$$C_t = \beta_0 + \beta_1 Y_t + \beta_2 C_{t-1} + u_t. \tag{5.4.1}$$

In this model there is not just one multiplier, but an infinite number. Economists usually distinguish the short-run (or impact) multiplier $\beta_1$ and the long-run multiplier $\beta_1/(1 - \beta_2)$. This second multiplier gives the total response to a change in income after all adjustments have taken place. The parameter $\beta_2$ is related to the speed of adjustment. Clearly, if $\beta_2$ is close to zero, the dynamic model is quite similar to the static model. That is, adjustment takes place very rapidly. On the other hand, if $\beta_2$ is large, adjustments take place slowly and a static model would not be appropriate. To make economic sense, the multipliers must be positive and less than 1. That is, $0 < \beta_1 < 1$ and $0 < \beta_1/(1 - \beta_2) < 1$. This second condition implies that $\beta_2 > 0$ and $\beta_1 + \beta_2 < 1$.

From a regression point of view, one way of finding out whether a dynamic model is appropriate is to examine the relevance of the variable $C_{t-1}$. Estimating the model (5.4.1), we obtain

$$\hat{C}_t = \underset{(0.08)}{0.26} + \underset{(0.03)}{0.22}Y_t + \underset{(0.05)}{0.66}C_{t-1}; \qquad \bar{R}^2 = 0.999.$$

We notice first that the coefficient of $C_{t-1}$ is highly significant with a t-ratio given by

$$t = \frac{0.66}{0.05} = 13.0.$$

This implies that $\bar{R}^2$ must have risen, and indeed we see that $\bar{R}^2$ has risen from 0.996 to 0.999.

Also, if $C_{t-1}$ is relevant, the static model is misspecified and $\hat{\beta}_1$ in that model would be a biased estimator, the bias being given by $\beta_2\hat{\beta}_{2.1}$. Now we know that $\beta_2 > 0$. Also, as $Y_t$ and $C_{t-1}$ are positively correlated ($r_{12}$ = 0.995), $\hat{\beta}_{2.1} > 0$.

Thus $\hat{\beta}_1$ in the static model would be positively biased if $C_{t-1}$ is relevant, and we would expect the value of $\hat{\beta}_1$ to decrease when $C_{t-1}$ is added to the regression. In fact, $\hat{\beta}_1$ decreases from 0.60 to 0.22. It is clear, then, that all criteria point to the fact that $C_{t-1}$ is a worthwhile addition and consequently, that a dynamic model allowing for gradual adjustment is more appropriate than a static model.

Finally, we notice that the "speed of adjustment" parameter $\hat{\beta}_2$ has a value 0.66 which is quite large (the maximum possible value is $1 - \hat{\beta}_1 = 0.78$), implying a slow rate of adjustment. The impact multiplier is 0.22 and the long-run multiplier $\hat{\beta}_1/(1 - \hat{\beta}_2) = 0.65$, which implies that only about one-third of the total adjustment to an income change takes place instantaneously.

Whenever this process of assessing the relevance of additional variables is being carried out, it should be remembered that conclusions are necessarily tentative until we have almost zeroed-in on the true model. Until all important variables have been included in the regression, misspecification bias will mask the true situation. An inherent difficulty is that we will never know for certain when the model specification is close to being correct. Judgment on this point should improve with experience. It is part of the art of being a good applied econometrician.

## 5.5 DATA DEFICIENCY AND MULTICOLLINEARITY

A further problem that frequently arises in econometric work to cloud the interpretation of a regression analysis is that of deficient data. Our data may not have enough information to enable us to distinguish the various partial effects influencing the dependent variable. In such cases we may be, as Johnston (1972) remarks, "in the statistical position of not being able to make bricks without straw."

Consider the general model

$$Y_i = \beta_0 + \beta_1 X_{1i} + \beta_2 X_{2i} + \cdots + \beta_K X_{Ki} + u_i.$$

It can be shown that the variance of $\hat{\beta}_1$ is given by the generalization of (4.3.5), in the form

$$\operatorname{var}(\hat{\beta}_1) = \frac{\sigma^2}{\Sigma x_1^2(1 - R_1^2)}, \qquad (5.5.1)$$

where $R_1^2$ is the $R^2$ that would be obtained if $X_1$ were regressed on the other regressors $X_2, \ldots, X_K$. The term $\Sigma x_1^2(1 - R_1^2)$ can be interpreted in exactly the same way as in Section 4.3 as the autonomous information content in the observations on $X_1$.

A successful analysis will be characterized by estimates with small variances. Conversely, large variances mean that the parameters have been estimated with very little precision and are therefore unreliable. There are two types of data characteristics which are likely to give rise to a large variance of $\hat{\beta}_1$ due to the denominator $\Sigma x_1^2(1 - R_1^2)$ being small relative to $\sigma^2$.

(a) $\Sigma x_1^2$ may be very small. That is, there may be very little variation in the observations on $X_1$. As $\beta_1$ gives the marginal effect of $X_1$ on Y, holding the other variables fixed, a small amount of variation in $X_1$ will make it very difficult to isolate this marginal effect.

(b) $R_1^2$ may be close to 1. In this case $1/(1 - R_1^2)$ would be large, leading to a large variance in $\hat{\beta}_1$. This large $R_1^2$ means that the variation in $X_1$ is almost completely explained by the other variables $X_2, \ldots, X_K$. There is very little autonomous information contained in $X_1$. Thus, if $X_2, \ldots, X_K$ were held fixed, $X_1$ also would be substantially fixed and it would be difficult to estimate its marginal effect with any precision. It is this sort of characteristic that is known as multicollinearity.

If we are only interested in prediction of the dependent variable Y, this problem may not matter very much. It is when the analysis requires precise estimates of the individual parameters, for example to test hypotheses, that it becomes serious.

## Overcoming Data Deficiency

(a) If we are in the position of gathering our own data, it may be possible to ensure that the explanatory variables exhibit a lot of variation.

(b) It may be possible to increase the sample size. This will rarely be possible when dealing with time-series data, but may be possible when the data are cross-sectional. When the sample size is increased, it can be shown that $\Sigma x_1^2$ cannot decrease and will, with almost complete certainty, increase. This in turn decreases the variance of $\hat{\beta}_1$.

(c) A very common practice is to drop a variable (or variables) that has a large standard error. We have already seen that when a variable is added to a regression, $R^2$ will rise. Conversely, if a variable is dropped, $R^2$ will fall. Thus, if one or several of the variables $X_2, \ldots, X_K$ are omitted, $R_1^2$ will fall, which by (5.5.1) will cause var $(\hat{\beta}_1)$ to decrease. This effect will be most marked if the variable that is dropped is highly correlated with $X_1$. This procedure must be used with caution. If the variable that is dropped is relevant, $\hat{\beta}_1$ will suffer from misspecification bias. It is important, then, to try to assess (using the principles of Section 5.4) whether a variable is relevant and not simply omit it because of a large standard error.

(d) There may be some a priori information with which to augment the data. Suppose that

$$Y_t = \beta_0 + \beta_1 X_{1t} + \beta_2 X_{2t} + \beta_3 X_{3t} + u_t$$

and $X_{2t}$ is highly correlated with $X_{1t}$. This will probably mean that the standard errors of $\hat{\beta}_1$ and $\hat{\beta}_2$ are unacceptably large. It may be known (perhaps from other similar studies) that $\beta_2$ is probably close to $\beta_2^*$ in value, where $\beta_2^*$ is some known number. Then, approximately,

$$Y_t = \beta_0 + \beta_1 X_{1t} + \beta_2^* X_{2t} + \beta_3 X_{3t} + u_t,$$

that is,

$$Y_t - \beta_2^* X_{2t} = \beta_0 + \beta_1 X_{1t} + \beta_3 X_{3t} + u_t.$$

Regressing $Y_t - \beta_2^* X_{2t}$ on $X_{1t}$ and $X_{3t}$ will very likely result in much more precise estimates of $\beta_1$ and $\beta_3$. This technique would be most effective if $X_1$ and $X_3$ were not highly correlated.

(e) It may be possible to augment time-series data with cross-sectional data. Suppose the demand for a product is a function of its price and per capita income. In time-series data, price and per capita income will usually move together and hence multicollinearity will be a problem. If it is possible to obtain cross-sectional data, price will be effectively constant, and the partial regression coefficient of demand with respect to income can be estimated. This can then be used to modify the model as in (d) above, and the time-series data used to estimate the price effect.

(f) There are techniques available which are purely statistical; that is, they make no attempt to incorporate extra information into the analysis in order to overcome multicollinearity. Two such methods are ridge regression, first suggested by Hoerl and Kennard (1970), and the method of principal components. We will briefly sketch the main features of the latter method, using a three-variable model,

$$Y_i = \beta_0 + \beta_1 X_{1i} + \beta_2 X_{2i} + \beta_3 X_{3i} + u_i,$$

and assuming that there is severe multicollinearity among the regressors.

It is always possible to construct another set of variables $Z_1$, $Z_2$, and $Z_3$ which are linear combinations of the original variables. That is,

$$\begin{aligned} Z_{1i} &= \alpha_{11} X_{1i} + \alpha_{12} X_{2i} + \alpha_{13} X_{3i}, \\ Z_{2i} &= \alpha_{21} X_{1i} + \alpha_{22} X_{2i} + \alpha_{23} X_{3i}, \\ Z_{3i} &= \alpha_{31} X_{1i} + \alpha_{32} X_{2i} + \alpha_{33} X_{3i}, \end{aligned} \tag{5.5.2}$$

where the $\alpha$'s are known numbers. Clearly, there are an infinite number of ways in which the $\alpha$'s could be specified, and consequently, an infinite number of ways in which the $Z_1$, $Z_2$, and $Z_3$ could be constructed.

The Z variables are called principal components when the $\alpha$'s are chosen in such a way that

(i) $Z_1$, $Z_2$, and $Z_3$ are mutually orthogonal;
(ii) $Z_1$ accounts for most of the variation in the X's followed next by $Z_2$ and then by $Z_3$.

Two remarks could be made at this stage. First, (ii) is rather vague, but it is difficult to make it more precise without introducing some matrix algebra. Second, we have not given any idea of how the $\alpha$'s can be chosen to produce Z variables with these properties. For our purposes it is sufficient to accept that it can be done.

Let us suppose now that Y is regressed on $Z_1$, $Z_2$, and $Z_3$, giving

$$\hat{Y}_i = \hat{\gamma}_0 + \hat{\gamma}_1 Z_{1i} + \hat{\gamma}_2 Z_{2i} + \hat{\gamma}_3 Z_{3i}.$$

Substituting from (5.5.2) yields

$$\hat{Y}_t = \hat{\gamma}_0 + \hat{\gamma}_1(\alpha_{11}X_{1t} + \alpha_{12}X_{2t} + \alpha_{13}X_{3t}) + \cdots$$

$$= \hat{\gamma}_0 + (\alpha_{11}\hat{\gamma}_1 + \alpha_{21}\hat{\gamma}_2 + \alpha_{31}\hat{\gamma}_3)X_{1t} + \cdots.$$

Thus

$$\hat{Y}_t = \hat{\beta}_0^* + \hat{\beta}_1^* X_{1t} + \hat{\beta}_2^* X_{2t} + \hat{\beta}_3^* X_{3t},$$

where

$$\hat{\beta}_1^* = \alpha_{11}\hat{\gamma}_1 + \alpha_{21}\hat{\gamma} + \alpha_{31}\hat{\gamma}_3, \text{ etc.} \qquad (5.5.3)$$

It can be shown that $\hat{\beta}_1^* = \hat{\beta}_1$, where $\hat{\beta}_1$ is the OLS estimate of $\beta$, obtained by regressing Y on $X_1$, $X_2$, and $X_3$. It appears that we have gone to a lot of trouble to achieve exactly the same result as a straight regression on the original variables!

The usefulness of this approach comes at the next stage. By construction, the variable $Z_1$ accounts for most of the variation (or information) contained in the original set $X_1$, $X_2$, and $X_3$. Suppose then that we drop the variables $Z_2$ and $Z_3$. Because the Z variables are orthogonal, $\hat{\gamma}_1$ would not change, and we would obtain another estimator of $\hat{\beta}_1$, given by

$$\hat{\beta}_1^{**} = \alpha_{11}\hat{\gamma}_1. \qquad (5.5.4)$$

It can be shown that $\hat{\beta}_1^{**}$ is a biased estimator of $\beta_1$, but the variance of $\hat{\beta}_1^{**}$ will usually be far smaller than that of the OLS estimator $\hat{\beta}_1$. Whether the method of principal components will be useful in any given context will depend on whether the reduction in variance more than offsets the bias that is introduced.

We conclude our discussion of multicollinearity with a simple analogy. Let us suppose that we wish to make orange juice and we have been able to obtain the best juice extractor that modern technology can produce. If the oranges we have available are dry, even with this extractor we will not be able to produce much juice. We may, however, be able to obtain a satisfactory result using many more oranges.

The method of least squares is a good information extractor (under the usual assumptions it is BLUE). However, if the available data do not contain much information (we have multicollinearity), we will not, even with this good extractor, be able to get good results. By adding information, either a priori or more sample data, we may be able to get a satisfactory result.

## 5.6 PROXY VARIABLES

Let us suppose that in deviation form, the true model is

$$y_i = \beta_1 x_{1i} + \beta_2 x_{2i} + u_i, \tag{5.6.1}$$

and that we are particularly interested in estimating the parameter $\beta_1$. For example, this parameter may be the own-price elasticity in a demand study or the response to a particular type of labor in a production function. We will also assume that no data are available on the variable $X_2$. Given that our interest lies in the parameter $\beta_1$ and not $\beta_2$, we may be tempted to drop $X_2$ from the analysis and simply regress Y on $X_1$. We have already seen in (5.3.3) that this result in a biased estimator of $\beta_1$, and in fact the bias may be so great as to render $\hat{\beta}_1$ useless.

Even though it may not be possible to obtain observations on $X_2$, there may be another variable $X_2^*$ which is correlated with $X_2$, and for which observations are available. If $X_2^*$ is used in the regression instead of $X_2$, it is known as a proxy variable.

When Y is regressed on $X_1$ and $X_2^*$, using (4.3.2) the estimate of $\beta_1$ is given:

$$\hat{\beta}_1 = \frac{\Sigma x_1 y \, \Sigma x_2^{*2} - \Sigma x_2^* y \, \Sigma x_1 x_2^*}{\Sigma x_1^2 \, \Sigma x_2^{*2} - (\Sigma x_1 x_2^*)^2}. \tag{5.6.2}$$

In order to find the bias in $\hat{\beta}_1$, we substitute (5.6.1) into (5.6.2) to eliminate y and take expectations, to obtain

$$E(\hat{\beta}_1) = \beta_1 + \beta_2 \frac{\Sigma x_1 x_2 \, \Sigma x_2^{*2} - (\Sigma x_2^* x_2)(\Sigma x_1 x_2^*)}{\Sigma x_1^2 \, \Sigma x_2^{*2} - (\Sigma x_1 x_2^*)^2}. \tag{5.6.3}$$

By using (4.3.2) again, we see that the factor multiplying $\beta_2$ in the expression above is the estimated coefficient of $x_1$ we would obtain if we were able to regress the unobserved variable $X_2$ on $X_1$ and the proxy $X_2^*$. Thus

$$E(\hat{\beta}_1) = \beta_1 + \beta_2 \hat{\gamma}_1, \tag{5.6.4}$$

where $\hat{\gamma}_1$ is defined by the estimated regression

$$\hat{x}_2 = \hat{\gamma}_1 x_1 + \hat{\gamma}_2^* x_2^*.$$

We see then that whether we drop $X_2$ altogether or include a proxy variable $X_2^*$ to represent it, $\beta_1$ is a biased estimator of the parameter of interest, $\beta_1$.

The question then naturally arises as to whether it is worthwhile including the proxy. Let us suppose, to start with, that the proxy variable $X_2^*$ is perfectly correlated with $X_2$. In this case we have $x_2^* = kx_2$, where k is some constant. Substituting into (5.6.3), we immediately obtain $\hat{\gamma}_1 = 0$, implying that $\hat{\beta}_1$ is an unbiased estimator. In practice, of course, although $X_2^*$ may be highly correlated with $X_2$, it would never be perfectly correlated.

By assuming that $x_2^* = kx_2 + \epsilon$, it can be shown that provided that $\epsilon$ is uncorrelated with both $X_1$ and $X_2$, then, at least for reasonably large samples, the bias from including the proxy will be less than that incurred by dropping $X_2$ altogether. This is the conclusion reached by McCallum (1972) and Wickens (1972). However, Aigner (1974) has pointed out that the inclusion of a proxy will increase the variance of $\hat{\beta}_1$. Once again, as with principal components, we are faced with a trade-off between bias and variance.

## Inclusion of the Variable t in a Regression

Perhaps the most common form of proxy variable in regressions using time-series data is a "trend variable" t. To illustrate why this variable may be introduced, we will again consider a demand study. Let us suppose that the true model is of the form

$$\ln Q_t = \beta_0 + \beta_1 \ln P_t + \beta_2 \ln Y_t + \sum_{i=1}^{K} \beta_i^* \ln P_{it}^* + u_t.$$

The term $\Sigma_{i=1}^{K} \beta_i^* \ln P_{it}^*$ represents the effects of all the substitutes and complements whose price variations affect the demand for the commodity in question. We will assume, furthermore, that our interest is in estimating $\beta_1$, the own-price elasticity. If data on the price movements of the $P_i^*$ are not available, it may be perfectly reasonable to assume that they are all growing linearly with time. That is,

$$\beta_i^* \ln P_{it}^* = \alpha_{0i} + \alpha_{1i} t.$$

In this case,

$$\Sigma \beta_i^* \ln P_{it}^* = \Sigma \alpha_{0i} + t \Sigma \alpha_{1i}$$

and the model becomes

$$\begin{aligned} \ln Q_t &= (\beta_0 + \Sigma \alpha_{0i}) + \beta_1 \ln P_t + \beta_2 \ln Y_t + (\Sigma \alpha_{1i})t + u_t \\ &= \beta_0^* + \beta_1 \ln P_t + \beta_2 \ln Y_t + \gamma_3 t + u_t. \end{aligned}$$

The variable t is a proxy for all the omitted prices which if the assumption of linear growth is appropriate, should reduce the bias in the estimates of own-price and income elasticity.

We should note that the intercept $\beta_0^*$ and the coefficient of t, $\gamma_3$, are capturing many different effects and will therefore not be capable of interpretation. A very useful discussion on the use of t as a variable is given in Rao and Miller (1971, 99-104).

NOTES

1. Consider the model $Y_i = \beta_0 + \beta_1 X_{1i} + \cdots + \beta_K X_{Ki} + u_i$. If we average over all sample values of the Y's and X's, we have $\bar{Y} = \beta_0 + \beta_1 \bar{X}_1 + \cdots + \beta_K \bar{X}_K + \bar{u}$. Subtracting gives us

$$Y_i - \bar{Y} = (\beta_0 - \beta_0) + \beta_1(X_{1i} - \bar{X}) + \cdots + \beta_K(X_{Ki} - \bar{X}) + v_i$$

or

$$y_i = \beta_1 x_{1i} + \cdots + \beta_K x_{Ki} + v_i.$$

This is known as the deviation form of the model. Its essential features are that (a) all variables are expressed as deviations from their sample means; and (b) there is no intercept term.

2. If data are available on $P_t^*$, this can always be checked.

3. If there is more than one additional variable, the equivalent test involves the F-ratio. This will be considered in detail subsequently.

# 6

# Hypothesis Testing with the Regression Model

## 6.1 THE t-TEST

In Sections 2.6, 4.3, and 4.4 we saw how certain hypotheses can be tested with a suitable t-test. This type of testing procedure has some appeal in that it represents a minor extension of the sort of reasoning we carry out in everyday life. Let us suppose, for example, that we believe (hypothesize) that $\beta_1 = 3$. If we run the regression and observe that $\hat{\beta}_1 = 2.99$ or $\hat{\beta}_1 = 3.01$, we are likely to conclude that our beliefs seem to be reasonable. We have no reason to reject our hypothesis. On the other hand, if $\hat{\beta}_1 = 1500$, we are likely to seriously doubt our null hypothesis $\beta_1 = 3$. Viewed slightly differently, we seek evidence about the truth or falsity of the null hypothesis by considering the difference between the estimated value $\hat{\beta}_1$ and the hypothesized value $\beta_1^*$. If this difference, given by $\hat{\beta}_1 - \beta_1^*$, is large in magnitude, we would probably reject the null hypothesis, whereas if it is small, we would accept the null.

Although this criterion seems to be reasonable, what do we mean by "large" and "small"? Obviously, we need to have some standard of measurement with which to compare the difference $\hat{\beta}_1 - \beta_1^*$.

When considering heights of adult males, we would describe someone of height 1.95 meters as "tall" or "very tall," and someone of 1.45 meters as "short." We are able to do this because we have an intuitive idea of a standard height (probably about 1.75 meters) and 1.95 is tall (relative to the average) and 1.45 is short (relative to the average). In order, then, to determine whether $\hat{\beta}_1 - \beta_1^*$ is large or small we need some measure of the average size of this difference if $\beta_1 = \beta_1^*$ (i.e., the null hypothesis is true). The standard error $S_{\hat{\beta}_1}$ is precisely this required average. This then leads us to consider the t-ratio

$$t = \frac{\hat{\beta}_1 - \beta_1^*}{S_{\beta_1}}, \qquad (6.1.1)$$

which is simply the difference between $\hat{\beta}_1$ (estimated) and $\beta_1^*$ (hypothesized), relative to the average difference we could expect if the null hypothesis were true. At the risk of laboring the point, we can interpret a value $t = 5$ (say) to mean that the difference between estimated value and hypothesized value is five times the average difference that would occur if the null were true.

We have already used the result which says that if $H_0$ is true, this t-ratio has a t distribution, with N - K - 1 degrees of freedom (or N - K if the model has no intercept). This enables us to make probability statements about the computed t-ratio, and becomes the basis of the hypothesis-testing procedures we have already considered. For example, if $H_0$ is true, there is (approximately)[1] a probability of 0.05 that $|t| > 2$. Thus if we use the rule of thumb "reject $H_0$ if $|t| > 2$," we will only incorrectly reject $H_0$ about 5 percent of times.

The t-test has the advantage that its basis is conceptually easily grasped. It has, however, a major drawback. It cannot be used to carry out several tests simultaneously. Before we introduce a procedure that can be used more widely, we will give another interpretation of the t-test.

## An Alternative Interpretation of the t-Test

Let us consider the simplest form of a t-test—testing $H_0: \beta_1 = 0$ against $H_1: \beta_1 \neq 0$ in the context of the simple regression model

$$Y_i = \beta_0 + \beta_1 X_i + u_i.$$

From (2.5.8) and (2.5.10), the square of the t-ratio is given by

$$t^2 = \frac{\hat{\beta}_1^2}{S^2_{\hat{\beta}_1}} = \hat{\beta}_1^2 \frac{\Sigma x^2}{\hat{\sigma}^2}$$

The numerator, by (2.7.1), is

$$\hat{\beta}_1^2 \Sigma x^2 = \Sigma y^2 - \Sigma e^2 = \text{TSS} - \text{RSS}.$$

It will be shown in Section 6.4 that TSS is identical to the RSS that would arise from OLS on the model

$$Y_i = \beta_0 + u_i.$$

This is the model that we obtain when $H_0$ is assumed to be true. Accordingly, we will denote this residual sum of squares by RSS($H_0$). Also, by (2.5.5),

$$\hat{\sigma}^2 = \frac{\text{RSS}}{N - 2}.$$

As RSS here refers to the model in which $\beta_1$ is assumed to be nonzero (i.e.,

$H_1$ assumed true), we will denote it by $RSS(H_1)$. We therefore have

$$t^2 = \frac{TSS - RSS}{RSS/(N - 2)}$$

or

$$t^2 = \frac{RSS(H_0) - RSS(H_1)}{RSS(H_1)/(N - 2)}. \qquad (6.1.2)$$

Thus the square of the t-ratio is seen to depend on the proportional change that occurs to RSS when the null hypothesis $H_0$ is applied. This interpretation of a t-ratio does not just apply to this specific example, but is true in general, and leads us to a new principle of hypothesis testing.

## 6.2 THE F-TEST

The F-test procedure can be used for a wide variety of hypothesis tests, but whatever the test, they all have in common an examination of changes in the residual sum of squares RSS. To illustrate the general principles involved, we will return to the simple Keynesian consumption function, estimated in Section 2.10 using the data of Table 2.10.1. The estimated consumption function was

$$\hat{C}_i = \underset{(0.179)}{0.690} + \underset{(0.007)}{0.604}Y_i. \qquad (6.2.1)$$

The residual sum of squares is easily calculated using (2.7.1) as

$$\begin{aligned} RSS &= \Sigma c^2 - \hat{\beta}_1 \Sigma cy \\ &= 1184.1193 - 0.604413 \times 1950.6672 \\ &= 5.1107. \end{aligned}$$

As (6.2.1) gives the line of best fit, the RSS of 5.1107 must be smaller than the RSS associated with any other line involving the same variables.

Let us suppose now that we constrain the line to pass through the origin. That is, we set $\hat{\beta}_0 = 0$. Using OLS on the model

$$C_i = \beta_1 Y_i + u_i,$$

we obtain

$$\hat{C}_i = 0.630Y_i,$$

with

$$\begin{aligned} RSS &= \Sigma C^2 - \hat{\beta}_1 \Sigma CY \\ &= 8711.732776 - 0.629482 \times 13{,}827.641483 \\ &= 7.4814. \end{aligned}$$

Notice the RSS has risen from 5.1107 to 7.4814.

We can utilize this change in RSS to test the null hypothesis $\beta_0 = 0$. If this hypothesis is true, we would expect that $\hat{\beta}_0$ in the unconstrained model would be quite close to zero. Thus, constraining $\hat{\beta}_0 = 0$ would not change the estimated line by very much and hence the RSS, although it must rise, would not rise by very much. On the other hand, if the null hypothesis is not true, then $\hat{\beta}_0$ in the unconstrained model would probably be nowhere near zero. Applying the constraint $\beta_0 = 0$ would, in this case, produce a line quite different from the unconstrained line, which would be characterized by a greatly increased RSS.

We now have the basis of a testing procedure which we can set out as follows:

(a) Run the unconstrained model, that is, the full model which does not constrain the parameters in any way. We obtain the residual sum of squares, denoted from now on by $RSS(H_1)$.
(b) Run the model that is constrained by assuming the null hypothesis $H_0$ to be true. The residual sum of squares from this regression will be denoted by $RSS(H_0)$.
(c) Consider the increase in residual sum of squares $[(RSS(H_0) - RSS(H_1)]$. If this increase is small, this constitutes evidence that $H_0$ is true, whereas if it is large, $H_0$ would be rejected.[2]

In constrast to the t-test, this procedure can be applied to a wide variety of situations. All we need is to be able to constrain the model by assuming $H_0$ to be true, and run the constrained regression.

EXAMPLE 1

$$Y_i = \beta_0 + \beta_1 X_{1i} + \beta_2 X_{2i} + \beta_3 X_{3i} + \beta_4 X_{4i} + u_i.$$

Test the hypothesis

$$H_0: \beta_1 = \beta_2; \quad \beta_3 + 2\beta_4 = 1.$$

For the unconstrained regression, we simply regress Y on $X_1$, $X_2$, $X_3$, $X_4$ and obtain $RSS(H_1)$. For the constrained regression we assume $H_0$ to be true. We can now eliminate two $\beta$'s from the regression. Under $H_0$, $\beta_1 = \beta_2$; $\beta_3 = 1 - 2\beta_4$, so

$$Y_i = \beta_0 + \beta_2(X_{1i} + X_{2i}) + (1 - 2\beta_4)X_{3i} + \beta_4 X_{4i} + u_i,$$

giving

DISPLAY 6.2.1

```
MTB > LET C4=LOGE(C3)
MTB > LET C5=LOGE(C1)
MTB > LET C6=LOGE(C2)
MTB > LET C7=C4-C6
MTB > LET C8=C5-C6
MTB > REGR C4 2 C5 C6

The regression equation is
C4 = 1.54 + 0.460 C5 + 0.205 C6

Predictor         Coef          Stdev      t-ratio
Constant        1.5396         0.9210         1.67
C5              0.4600         0.1321         3.48
C6              0.2049         0.2631         0.78

s = 0.05114      R-sq = 71.3%      R-sq(adj) = 66.5%

Analysis of Variance

SOURCE         DF          SS          MS
Regression      2    0.078031    0.039015
Error          12    0.031387    0.002616
Total          14    0.109418



SOURCE         DF      SEQ SS
C5              1    0.076445
C6              1    0.001586

Continue? NO
MTB > REGR C7 1 C8

The regression equation is
C7 = - 0.0030 + 0.411 C8

Predictor         Coef          Stdev      t-ratio
Constant      -0.00303        0.01585        -0.19
C8              0.4107         0.1374         2.99

s = 0.05458      R-sq = 40.7%      R-sq(adj) = 36.2%

Analysis of Variance

SOURCE         DF          SS          MS
Regression      1    0.026611    0.026611
Error          13    0.038727    0.002979
Total          14    0.065339
```

$$Y_i - X_{3i} = \beta_0 + \beta_2(X_{1i} + X_{2i}) + \beta_4(X_{4i} - 2X_{3i}) + u_i.$$

This is the constrained model. We thus regress $Y - X_3$ (dependent variable) on $X_1 + X_2$, $X_4 - 2X_3$ and obtain $RSS(H_0)$.

Notice that we obtained the constrained equation by eliminating $\beta_1$ and $\beta_3$. Suppose that we had used $H_0$ to eliminate $\beta_1$ and $\beta_4$. In this case we would have obtained the constrained model

$$Y_i - \frac{1}{2}X_{4i} = \beta_0 + \beta_2(X_{1i} + X_{2i}) + \beta_3(X_{3i} - \frac{1}{2}X_{4i}) + u_i.$$

This is different from the earlier constrained model. However, the RSS's will be identical. We will illustrate this point in the next example.

EXAMPLE 2 Consider the Cobb-Douglas production function

$$\ln Q_t = \beta_0 + \beta_1 \ln K_t + \beta_2 \ln L_t + u_t,$$

which was estimated in Section 4.7 using the data of Table 4.7.1. If we want to test the hypothesis of constant returns to scale, that is,

$$H_0 : \beta_1 + \beta_2 = 1,$$

we need to constrain the model by assuming that $H_0$ is true. Let us use $H_0$ to eliminate $\beta_2$.

Then, because $\beta_2 = 1 - \beta_1$,

$$\ln Q_t = \beta_0 + \beta_1 \ln K_t + (1 - \beta_1)(\ln K_t - \ln L_t) + u_t.$$

That is,

$$\ln Q_t - \ln L_t = \beta_0 + \beta_1(\ln K_t - \ln L_t) + u_t.$$

Thus, to obtain $RSS(H_0)$ we would regress (ln Q - ln L) on (ln K - ln L). A MINITAB® program to obtain both $RSS(H_1)$ and $RSS(H_0)$ is as follows (assuming that K, L, and Q are stored in Columns C1, C2, and C3, respectively):

```
LET C4 = LOGE(C3)
LET C5 = LOGE(C1)
LET C6 = LOGE(C2)
```

(ln Q, ln K, and ln L are now stored in C4, C5, and C6, respectively.)

```
LET C7 = C4-C6
LET C8 = C5-C6
```

((ln Q - ln L) and (ln K - ln L) are stored in C7 and C8, respectively.)

```
REGR C4 2 C5 C6
```

(This is the unconstrained regression.)

```
REGR C7 1 C8
```

(This is the constrained regression.) The results of these two regressions are shown in Display 6.2.1.

Suppose now that $H_0$ had been used to eliminate $\beta_1$ instead of $\beta_2$. In this case it is easily seen that the constrained model is

$$\ln Q_t - \ln K_t = \beta_0 + \beta_2(\ln L_t - \ln K_t) + u_t.$$

A MINITAB® program to accomplish this regression is

```
LET C9  = C4-C5
LET C10   C6-C5
REGR C9 1 C10
```

The results of this regression are shown in Display 6.2.2.

As noted in Section 4.8, in the MINITAB® output the residual sum of squares is shown under the heading "Analysis of Variance" beside the word "Error." Thus, referring to Display 6.2.1, for the unconstrained model we have $RSS(H_1) = 0.031387$, and for the constrained model $RSS(H_0) = 0.038727$. Once again, we notice that constraining the model results in a rise in residual sum of squares. When we turn to Display 6.2.2, which records the result of a constrained regression in which $\beta_2$ is eliminated, we see that $RSS(H_0) = 0.038727$, which is identical to the previous value of $RSS(H_0)$. This confirms that even if a constraint can be applied in more than one way, the resulting RSS is unique.

We have seen that the fundamental concept in this new approach to

DISPLAY 6.2.2

```
-- SUBT C5 C4 C9
-- SUBT C5 C6 C10
-- REGR C9  1 C10
THE REGRESSION EQUATION IS
Y = 0 .0030 + 0.589 X1
                                           ST. DEV.      T-RATIO =
       COLUMN      COEFFICIENT             OF COEF.      COEF/S.D.
        --            -0.00303             0.01585         -0.19
X1     C10             0.5893              0.1374           4.29

THE ST. DEV. OF Y ABOUT REGRESSION LINE IS
S = 0.05458
WITH (  15- 2) =  13 DEGREES OF FREEDOM
R-SQUARED = 58.6 PERCENT
R-SQUARED = 55.4 PERCENT, ADJUSTED FOR D.F.

ANALYSIS OF VARIANCE
 DUE TO       DF          SS          MS=SS/DF
REGRESSION     1      0.054794       0.054794
RESIDUAL      13      0.038727       0.002979
TOTAL         14      0.093522
```

hypothesis testing is the examination of the increase in RSS when a model is constrained according to a null hypothesis. We will now refine this concept a little in order to make it operational. The increase in RSS not only reflects the truth or falsity of the null hypothesis, but the following three factors:

(a) The units of measurement of the dependent variable Y. If Y is some quantity measured in kilograms, and the unit of measurement was changed to grams, the increase in RSS would change by a factor of $10^6$. We can achieve freedom from units of measurement by considering the ratio $[RSS(H_0) - RSS(H_1)]/RSS(H_1)$ instead of simply the increment in RSS. That is, we consider the proportional increase.

(b) The number of constraints implied by the null hypothesis. It stands to reason that the more constraints that are imposed, the larger will be the increase in RSS. We should therefore consider this increase relative to the number of constraints, and thus use $[RSS(H_0) - RSS(H_1)]/M$, where M is the number of constraints.

(c) The number of degrees of freedom in the unconstrained model. We have already seen that when a variable is added to a model, RSS will always fall even if the variable is irrelevant. In such cases, the fall in RSS is due to the reduction in degrees of freedom. In the same way, when a constraint is imposed on the model, RSS will rise, whether or not the constraint reflects a hypothesis which is true, because the degrees of freedom have been increased. We can adjust our measure of increase in RSS in (a) to take account of degrees of freedom by using $RSS(H_1)/(N - K - 1)$ in the denominator instead of $RSS(H_1)$.

We are thus led to measure the magnitude of increase in RSS when a model is constrained by a hypothesis $H_0$ by the "F-ratio," defined by

$$F = \frac{[RSS(H_0) - RSS(H_1)]/M}{RSS(H_1)/(N - K - 1)}. \qquad (6.2.2)$$

If F is small, we will accept $H_0$, and if F is large, $H_0$ will be rejected. As with the t-ratio, we can only decide on whether F is large or small when we know how it is distributed when $H_0$ is true. The following theorem gives us this information.

THEOREM 6.2.1 Provided that the disturbance is normally distributed, the F-ratio defined in (6.2.2) has a F-distribution with parameters M and N - K - 1 when $H_0$ is true. That is, when $H_0$ is true,

$$F \sim F_{M, N-K-1}. \qquad (6.2.3)$$

If the calculated F value is too large to be consistent (at some preassigned level of significance) with the appropriate F-distribution, the null hypothesis is rejected.

EXAMPLE 1 We use the results in Display 6.2.1 and/or 6.2.2 to test for constant returns to scale in the Cobb-Douglas function. Here $H_0: \beta_1 + \beta_2 = 1$.

Table 6.2.1 Time Series Dealing with the Consumption of Textiles in the Netherlands, 1923-1939

| Year | (1) Volume of textile consumption per capita[a] | (2) Real income per capita[a] | (3) Relative price of textiles[a] |
|---|---|---|---|
| 1923 | 99.2 | 96.2 | 101.0 |
| 1924 | 99.0 | 98.1 | 100.1 |
| 1925 | 100.0 | 100.0 | 100.0 |
| 1926 | 111.6 | 104.9 | 90.6 |
| 1927 | 122.2 | 104.9 | 86.5 |
| 1928 | 117.6 | 109.5 | 89.7 |
| 1929 | 121.1 | 110.8 | 90.6 |
| 1930 | 136.0 | 112.3 | 82.8 |
| 1931 | 154.2 | 109.3 | 70.1 |
| 1932 | 153.6 | 105.3 | 65.4 |
| 1933 | 158.5 | 101.7 | 61.3 |
| 1934 | 140.6 | 95.4 | 62.5 |
| 1935 | 136.2 | 96.4 | 63.6 |
| 1936 | 168.0 | 97.6 | 52.6 |
| 1937 | 154.3 | 102.4 | 59.7 |
| 1938 | 149.0 | 101.6 | 59.5 |
| 1939 | 165.5 | 103.8 | 61.3 |

[a]Index base 1925 = 100.

Source: Theil (1971, 102), reproduced with permission.

$$\text{RSS}(H_0) = 0.038727,$$
$$\text{RSS}(H_1) = 0.031387,$$
$$M = 1,$$
$$N - K - 1 = 12.$$

Thus

$$F = \frac{(0.038727 - 0.031387)/1}{0.031387/12}$$
$$= 2.806.$$

At the 5 percent level of significance, the critical point for $F_{1,12}$ is 4.75. Therefore, because

$$F < F_{1,12}(0.05),$$

$H_0$ is not rejected, and the data are shown to be consistent with the hypothesis of constant returns to scale.

EXAMPLE 2 Consider the data relating to the consumption of textiles in the Netherlands,shown in Table 6.2.1. A constant-elasticity model for consumption is

$$\ln Y_t = \beta_0 + \beta_1 \ln X_{1t} + \beta_2 \ln X_{2t} + u_t,$$

where

$Y_t$ = volume of textile consumption per capita,

$X_{1t}$ = real income per capita,

$X_{2t}$ = relative price of textiles.

There are 17 observations. Let us suppose that we wish to test the hypothesis that the income and price elasticities are equal, having the value unity. That is,

$$H_0: \beta_1 = -\beta_2 = 1$$

A MINITAB® program to carry out the unconstrained regression, assuming that Y, $X_1$, and $X_2$ are stored in C1, C2, and C3, respectively, is

```
LET C4 = LOGE(C1)
LET C5 = LOGE(C2)
LET C6 = LOGE(C3)
REGR C4 2 C5 C6
```

The results of this regression are shown in Display 6.2.3.

DISPLAY 6.2.3

```
MTB > LET C4=LOGE(C1)
MTB > LET C5=LOGE(C2)
MTB > LET C6=LOGE(C3)
MTB > REGR C4 2 C5 C6

The regression equation is
C4 = 3.16 + 1.14 C5 - 0.829 C6

Predictor        Coef        Stdev     t-ratio
Constant       3.1636       0.7048        4.49
C5             1.1432       0.1560        7.33
C6           -0.82884      0.03611      -22.95

s = 0.03118      R-sq = 97.4%      R-sq(adj) = 97.1%

Analysis of Variance

SOURCE        DF          SS          MS
Regression     2     0.51733     0.25867
Error         14     0.01361     0.00097
Total         16     0.53094
```

Reprinted with permission of Minitab, Inc.

To obtain the constrained equation, we assume $H_0$ to be true and obtain

$$\ln Y_t = \beta_0 + \ln X_{1t} - \ln X_{2t} + u_t.$$

That is,

$$Y_t^* = \beta_0 + u_t, \tag{6.2.4}$$

where

$$Y_t^* = \ln Y_t - \ln X_{1t} + \ln X_{2t}. \tag{6.2.5}$$

This regression model is different from any that we have met so far, in that it has no regressor variables. Although, in theory, this presents no problems, running such a regression on MINITAB® is not completely straightforward. It should be obvious that the best estimator we can obtain for $\beta_0$, given the model (6.2.4), is $\hat{\beta}_0 = \bar{Y}^*$, and the predictor of $Y_t^*$ will also be $\bar{Y}^*$. Thus, by the definition of residual sum of squares,

$$RSS(H_0) = \Sigma(Y_t^* - \bar{Y}^*)^2. \tag{6.2.6}$$

The standard deviation of $Y_t^*$, denoted by S and obtained by the MINITAB® command STDEV, is defined as

$$S = \left[\frac{\Sigma(Y_t^* - \bar{Y}^*)^2}{N - 1}\right]^{\frac{1}{2}},$$

so, by (6.2.6),

$$RSS(H_0) = (N - 1)S^2. \tag{6.2.7}$$

A MINITAB® program to obtain the constrained residual sum of squares is

LET C7 = C4-C5+C6

(The variable $Y_t^*$ is now stored in C7.)

LET K1 = 16*STDEV(C7)*STDEV(C7)

[$(N - 1)S^2$, where $N = 17$, is now in K1.]

PRINT K1

The results of these computations are shown in Display 6.2.4. From the output we have

$$RSS(H_0) = 0.03943,$$

$$RSS(H_1) = 0.01361,$$

$$M = 2,$$

$$N - K - 1 = 17 - 2 - 1 = 14.$$

Therefore,

$$F = \frac{(0.03943 - 0.01361)/2}{0.01361/14} = 13.28.$$

At the 5 percent level of significance, the critical value for $F_{2,14}$ is 3.74. Because $F > F_{2,14}(0.05)$, the null hypothesis is rejected.

DISPLAY 6.2.4

```
SOURCE          DF        SEQ SS
C5               1       0.00508
C6               1       0.51225


MTB > LET C7=C4-C5+C6
MTB > LET K1=16*STDEV(C7)*STDEV(C7)
MTB > PRINT K1
K1          0.0394298
```

Reprinted with permission of Minitab, Inc.

Before leaving this general discussion of the F-test to consider some special formulations, we note the fundamental connection between the t- and F-tests. In Example 1, we tested the hypothesis of constant returns to scale using the F-test. In Section 4.7 we carried out the same test using a t-test. In each case the null was accepted. Will this always be the case? Can a t-test always be used instead of an F-test?

We notice that in the constant-returns-to-scale case, there is only one constraint. In general, whenever the null hypothesis imposes only one constraint (i.e., $M = 1$), either a t-test or an F-test can be used. Furthermore, the tests are equivalent in the sense that the calculated F-ratio is equal to the square of the t-ratio. We have already seen an example of this in the preceding section. In equation (4.7.4) we have

$$t = -1.675,$$

and from Example 1 of this section,

$$F = 2.806 = (-1.675)^2.$$

We should note, however, that it is not possible to carry out a one-tail test using the F-test procedure. Whenever a null hypothesis implies more than one constraint ($M > 1$) (see Example 2), a t-test cannot be used. Finally, although the F-test can be used to deal with more general situations, it involves carrying out two regressions,[3] whereas the t-test involves only one.

## 6.3 F-TESTS IN TERMS OF CHANGE IN $R^2$

We may consider hypotheses as being either homogeneous or nonhomogeneous in the $\beta$ parameters. A homogeneous hypothesis is one in which the right-hand side of the constraint equation is zero. For example,

$$\beta_1 = \beta_2 = 0, \qquad \beta_1 + \beta_2 = 0$$

are homogeneous relationships. On the other hand, $\beta_1 = -1$ or $\beta_1 + \beta_2 = 1$ are nonhomogeneous. Again, $\beta_1/\beta_2 = 3$ is homogeneous, because it can be rewritten in the form $\beta_1 - 3\beta_2 = 0$. Whenever a regression model is constrained by homogeneous relationships, the dependent variable of the constrained model is the same as that of the unconstrained model. On the other hand, a nonhomogeneous constraint will always result in a constrained model with a different dependent variable.

We have already seen in Section 5.2 that models can validly be compared using $R^2$ or $\bar{R}^2$ when (but only when) they have the same dependent variable. This may lead us to suspect that when the hypotheses are homogeneous, the F-test may be expressible in terms of changes in $R^2$ or $\bar{R}^2$.

Let us assume then that $H_0$ consists only of homogeneous relationships in the $\beta$'s. Consequently, the constrained model has the same dependent variable, so the total sum of squares TSS is the same for both constrained and unconstrained models. We have

$$F = \frac{[RSS(H_0) - RSS(H_1)]/M}{RSS(H_1)/(N - K - 1)}.$$

Dividing numerator and denominator by TSS, and using

$$R^2 = 1 - \frac{RSS}{TSS},$$

we obtain

$$F = \frac{[(1 - R_0^2) - (1 - R_1^2)]/M}{(1 - R_1^2)/(N - K - 1)}, \tag{6.3.1}$$

where $R_0^2$ and $R_1^2$ are the $R^2$ values for the constrained and unconstrained regressions, respectively. Therefore,

$$F = \frac{\Delta R^2/M}{(1 - R_1^2)/(N - K - 1)}, \tag{6.3.2}$$

where

$$\Delta R^2 = R_1^2 - R_0^2$$

and is the decrease in $R^2$ that occurs when the model is constrained.

Equation (6.3.2) can be given an intuitively appealing interpretation. $(1 - R_1^2)$ is the proportion of variation in the dependent variable which is not explained by the unconstrained regression. $\Delta R^2$ measures the increase in the unexplained variation that occurs when the model is constrained. If this increase is large (relative to the unexplained variation of the unconstrained model), the constraint is probably false and $H_0$ would be rejected.

Finally, we note that the result in Section 5.4, concerning the relationship between the t-value of a variable and the change in $\bar{R}^2$ which occurs when that variable is added, can be generalized. Edwards (1969) has shown that if a group of variables is added, $\bar{R}^2$ will increase if $F > 1$, where F is the F-ratio that would be obtained by regarding the original model as the constrained model and the expanded model as the unconstrained model.

## 6.4 A ROUTINE F-TEST

Whenever a regression analysis is carried out, a fundamental question is whether the group of regressors that have been chosen have any relevance in explaining variation in the dependent variable. Clearly, if they are irrelevant, there is no sense in using the model for prediction or inference. We should, therefore, always test the null hypothesis

$$H_0: \beta_1 = \beta_2 = \cdots = \beta_K = 0, \tag{6.4.1}$$

which simply states that the regressors are irrelevant. If the regressors have been sensibly chosen, we would, of course, expect that this hypothesis would be rejected.

To consider this special test, we recall that

$$TSS = ESS + RSS, \tag{6.4.2}$$

where

$$TSS = \text{total sum of squares},$$

$$ESS = \text{explained sum of squares},$$

$$RSS = \text{residual sum of squares}.$$

When we constrain the model by (6.4.1), we obtain

$$Y_i = \beta_0 + u_i,$$

and it has already been shown [see equation (6.2.6)] that for this model

$$RSS(H_0) = \Sigma(Y_i - \bar{Y})^2 = \Sigma y_i^2 = TSS.$$

Thus, because M = K in this case,

$$F = \frac{[RSS(H_0) - RSS(H_1)]/K}{RSS(H_1)/(N - K - 1)}$$

$$= \frac{(TSS - RSS)/K}{RSS/(N - K - 1)}.$$

That is, by (6.4.2),

$$F = \frac{ESS/K}{RSS/(N - K - 1)}. \tag{6.4.3}$$

The numerator ESS/K is often called the <u>explained mean square</u> or <u>regression mean square</u>. Similarly, the denominator RSS/(N - K - 1) is usually called the <u>residual mean square</u> or <u>error mean square</u>. This enables us to write this particular F-ratio, to test the hypothesis (6.4.1), as

$$F = \frac{\text{regression MS}}{\text{error MS}}. \tag{6.4.4}$$

In a MINITAB® output, this information is contained under the heading "Analysis of Variance." This table shows how the TOTAL sum of squares partitions into the two components REGRESSION sum of squares (which we call the explained sum of squares) and ERROR sum of squares (which we call residual sum of squares). Divided by the appropriate degrees of freedom K and N - K - 1, the mean squares are computed.

We illustrate by reference to Display 6.2.3. Here

regression MS = 0.25867

error MS = 0.00097,

yielding

$$F = \frac{0.25867}{0.00097} = 266.7.$$

The critical value for $F_{2,14}$ at the 1 percent level is 6.51. Clearly, the null hypothesis (6.4.1) is rejected and the regressors $\ln X_1$ and $\ln X_2$ are accepted as relevant.

Returning to (6.4.3), we can reformulate the expression for the F-ratio. Now

$$ESS = R^2 \; TSS,$$

$$RSS = (1 - R^2) \; TSS,$$

so

$$F = \frac{R^2/K}{(1 - R^2)/(N - K - 1)}. \qquad (6.4.5)$$

Again using the data in Display 6.2.3, where $R^2 = 0.974$, $K = 2$, and $N - K - 1 = 14$,

$$F = \frac{0.974/2}{(1 - 0.974)/14} = 262.2.$$

This value differs from the previous value only because of the rounding of $R^2$ to three significant figures. We emphasize that the expressions (6.4.3) and (6.4.5) apply only to testing the null hypothesis that all the regressors are irrelevant, that is, the hypothesis (6.4.1).

## 6.5 TESTING BETWEEN LINEAR AND LOG-LINEAR MODELS

Economic theory often provides guidance about the variables that should be included in a regression relationship, but rarely about the functional form of the relationship. For example, we may postulate that in a demand equation, price and income should be included. However, we may have little a priori knowledge to tell us which of

$$Q_i = \beta_0 + \beta_1 P_i + \beta_2 Y_i + u_i$$

or

$$\ln Q_i = \beta_0 + \beta_1 \ln P_i + \beta_2 \ln Y_i + u_i$$

is more appropriate. The second relationship implies that both price and income elasticities are constant. Because the log-linear model has important applications in economics, we will now present a method for discriminating between the linear and log-linear models. However, it should be noted that the test has wider applicability in that it is essentially a test about whether the dependent variable should be expressed in logarithmic form.

## Theoretical Considerations

We have already seen in Section 5.2 that when two relationships have different dependent variables, $R^2$ cannot be used as a basis for assessing the relative merits of the models. When two models do have the same dependent variable, because $R^2 = 1 - \text{RSS/TSS}$ and TSS is the same for both models, using $R^2$ to discriminate between models is essentially a convenient way of comparing residual sums of squares. To show why it is that we cannot validly compare residual sums of squares in this situation, we will consider the following two simple regression models:

$$Y_i = \beta_0 + \beta_1 X_i + u_i, \tag{6.5.1}$$

$$Z_i \equiv \ln Y_i = \gamma_0 + \gamma_1 X_i + u_i, \tag{6.5.2}$$

denoting the residual sums of squares by $\text{RSS}_1$ and $\text{RSS}_2$, respectively. Let us suppose now that the units of measurement of $Y_i$ are changed. In new units, we will denote $Y_i$ by $Y_i^*$, and there will be a relationship of the form

$$Y_i^* = \frac{Y_i}{K}, \tag{6.5.3}$$

where K is some constant. For example, if $Y_i$ is measured in grams and $Y_i^*$ is the same quantity measured in kilograms, K = 1000. Then, from equation (2.2.9),

$$\hat{\beta}_1^* = \frac{\Sigma xy^*}{\Sigma x^2} = K^{-1}\frac{\Sigma xy}{\Sigma x^2} = \frac{\hat{\beta}_1}{K}.$$

That is, when the units of measurement of Y change, the OLS estimates change in the same way.

Also, from (2.5.6),

$$\begin{aligned} RSS_1^* &= \Sigma y^{*2} - \hat{\beta}_1^* \Sigma xy^* \\ &= K^{-2} \Sigma y^2 - K^{-1} \hat{\beta}_1 \Sigma x(K^{-1}y) \\ &= K^{-2}(\Sigma y^2 - \hat{\beta}_1 \Sigma xy) = \frac{RSS_1}{K^2}. \end{aligned}$$

That is,

$$RSS_1^* = \frac{RSS_1}{K^2}. \tag{6.5.4}$$

Let us turn now to the model (6.5.2) in which the dependent variable is $Z_i$ (= ln $Y_i$). Now

$$Z_i^* = \ln Y_i^* = \ln \frac{Y_i}{K} = \ln Y_i - \ln K.$$

Thus

$$\bar{Z}^* = \overline{\ln Y} - \ln K.$$

Subtracting these two relationships yields

$$z_i^* = Z_i^* - \bar{Z}^* = \ln Y_i - \overline{\ln Y_i} = Z_i - \bar{Z}.$$

Thus

$$z_i^* = z_i.$$

Then, using (2.2.9) again, we have

$$\hat{\gamma}_1^* = \frac{\Sigma xz^*}{\Sigma x^2} = \frac{\Sigma xz}{\Sigma x^2} = \hat{\gamma}_1.$$

That is, the OLS estimates do <u>not</u> change when the units of measurement are changed.[4] Also,

$$\begin{aligned} RSS_2^* &= \Sigma z^{*2} - \hat{\gamma}_1^* \Sigma z^* y \\ &= \Sigma z^2 - \hat{\gamma}_1 \Sigma zy = RSS_2. \end{aligned}$$

That is,

$$RSS_2^* = RSS_2. \tag{6.5.5}$$

Comparing (6.5.4) and (6.5.5), we have contrasting results on the effect of a unit change on the residual sum of squares. When Y is the dependent variable, a change of units changes RSS in accordance with (6.5.4), whereas when ln Y is the dependent variable, RSS is invariant with respect to change of units of measurement. It should be obvious from these simple cases that this conclusion will hold in the multiple regression case regardless of what regressor variables are included in the models.

The significance of this conclusion is that we cannot meaningfully compare RSS to discriminate between the models (6.5.1) and (6.5.2). For example, let us assume that Y is measured in grams and we find $RSS_1 = 5000$ and $RSS_2 = 5$. We may be tempted to assume that the log-linear model explains the data far better. However, if the dependent variable were measured in kilograms (and units of measurement are, of course, completely arbitrary), then by (6.5.4) and (6.5.5), we would find $RSS_1 = 0.005$ and $RSS_2 = 5$ (unchanged). In summary, we can make the relative sizes of $RSS_1$ and $RSS_2$ whatever we like by a suitable choice of units of measurement.

To overcome this problem, we will define a new variable $Y_i'$ by

$$Y_i' = \frac{Y_i}{C}, \tag{6.5.6}$$

where

$$C = \exp\left(\frac{\Sigma \ln Y_i}{N}\right). \tag{6.5.7}$$

The main feature of the new variable $Y_i'$ is that it is invariant with respect to changes of unit in $Y_i$. Consequently, regressions with $Y_i'$ as a dependent variable will result in residual sums of squares that are not dependent on units of measurement. We establish this property as follows: Let $(Y_i^*)' = Y_i^*/C^*$, where $Y_i^*$ is given by (6.5.3) for some arbitrary constant K, and $C^*$ is defined by (6.5.7) with $Y_i$ replaced by $Y_i^*$. Then $Y_i'$ will be independent of units of measurement if

$$Y_i' = (Y_i^*)'.$$

Now

$$(Y_i^*)' = \frac{Y_i^*}{C^*} = \frac{Y_i}{KC^*}, \tag{6.5.8}$$

and

$$\begin{aligned} C^* &= \exp\left(\frac{\Sigma \ln Y_i^*}{N}\right) \\ &= \exp\left[\frac{\Sigma(\ln(Y_i/K))}{N}\right] \\ &= \exp\left[\frac{\Sigma(\ln Y_i)}{N} - \frac{\Sigma(\ln K)}{N}\right] \\ &= \exp\left[\frac{\Sigma(\ln Y_i)}{N} - \ln K\right] \end{aligned}$$

$$= \frac{\exp[\Sigma(\ln Y_i)/N]}{K}.$$

Thus $C^* = C/K$. Substituting into (6.5.8), we have

$$(Y_i^*)' = \frac{Y_i}{K(C/K)} = \frac{Y_i}{C} = Y_i'.$$

Thus a change in measurement units never affects the value of the variable $Y_i'$ defined by equation (6.5.6). If we now denote by $RSS_1'$ and $RSS_2'$ the residual sums of squares obtained when $Y_i'$ and $\ln Y_i'$ are taken as dependent variables, a comparison of $RSS_1'$ and $RSS_2'$ is a valid way of discriminating between the two models.

Box and Cox (1964) have derived a statistical test based on this comparison. The null hypothesis is that the two models are empirically equivalent. We define the $\ell$-statistic by

$$\ell = \frac{N}{2}\left|\ln \frac{RSS_1'}{RSS_2'}\right|, \tag{6.5.9}$$

and if the null hypothesis is true, $\ell \sim \chi^2(1)$. Thus, at the 5 percent level of significance, we reject $H_0$ if $\ell > 3.84$, and at the 1 percent level, if $\ell > 6.63$. If $H_0$ is rejected, then naturally the model for which RSS' is smaller is recognized as being a more appropriate description of the data.

## Practical Application

From (6.5.9) it appears that to discriminate between the two alternative models it is necessary to transform each observation on the dependent variable using (6.5.6) and then rerun the regressions to compute the value of $\ell$. In fact, this is not necessary. Using (6.5.9), (6.5.4), and (6.5.5) we have

$$\ell = \frac{N}{2}\left|\ln \frac{RSS_1/C^2}{RSS_2}\right| \tag{6.5.10}$$

and it is clear that once C is computed, the $\ell$-statistic can be calculated from the original output.

EXAMPLE In the analysis of production Q, economic theory suggests that both capital K and labor L should be included among the regressors. Two possible functional forms would be

$$Q_i = \beta_0 + \beta_1 K_i + \beta_2 L_i + u_i$$

and

$$\ln Q_i = \beta_0 + \beta_1 \ln K_i + \beta_2 \ln L_i + u_i.$$

We will use the data of Table 4.7.1 to see which of these models (if either) is the more appropriate.

Carrying out the regression of Q on K and L, we obtain

$$\hat{Q}_i = \underset{(19.48)}{32.02} + \underset{(0.122)}{0.450}K_i + \underset{(0.256)}{0.244}L_i; \quad RSS_1 = 277.07. \tag{6.5.11}$$

Referring to Display 4.8.1, we have

$$\ln \hat{Q}_i = \underset{(0.921)}{1.540} + \underset{(0.132)}{0.460} \ln K_i + \underset{(0.263)}{0.205} \ln L_i; \quad RSS_2 = 0.078031. \tag{6.5.12}$$

Before we can compare $RSS_1$ and $RSS_2$, we must divide $RSS_1$ by $C^2$. Now

$$\begin{aligned} C &= \exp\left(\frac{\Sigma \ln Q_i}{N}\right) \\ &= \exp(\text{average value of } \ln Q_i) \\ &= \exp(4.5713) \\ &= 96.6671. \end{aligned}$$

Thus

$$C^2 = 9344.5375.$$

A MINITAB® program to find $C^2$ would be (assuming that $\ln Q_i$ is stored in C4)

```
LET K1 = EXP(MEAN(C4))
LET K2 = K1**2
PRINT K2
```

Thus

$$\begin{aligned} \ell &= \frac{15}{2}\left|\ln \frac{277.07/9344.5375}{0.078031}\right| \\ &= \frac{15}{2}\left|\ln \frac{0.029651}{0.078031}\right| \\ &= 7.257. \end{aligned}$$

As $\ell > 3.84$, we reject $H_0$ at the 5 percent level. As $RSS_1' = 0.0296$ and $RSS_2' = 0.0780$, we judge that the model (6.5.11) is preferable to the Cobb-Douglas model (6.5.12).

## 6.6 LACK-OF-FIT TEST

In most situations of econometric analysis, there is one observation on the dependent variable Y for each observation on the set of regressors $X_1$, $X_2$, ..., $X_K$. In such cases our guide to the choice of an adequate model is mainly the value of $R^2$, augmented, of course, by a priori knowledge on the signs of coefficients.

Sometimes, however, a researcher may analyze experimental data, which are usually characterized by several observations on Y corresponding to each observation on $X_1$, $X_2$, ..., $X_K$. These observations are known as replications. Let us consider, for example, an experiment to determine how crops respond to nitrogen and phosphorus fertilization, denoting the quantities of nitrogen and phosphorus fertilizers which are applied by $X_1$ and $X_2$. It is almost certain that each combination of $X_1$ and $X_2$ will be applied to several plots. Thus there will be several values of Y (crop yield), corresponding to each observation $(X_1, X_2)$. The important thing about having replications is that they give valuable information which can be utilized in validating a model. Unfortunately, this extra information is usually ignored (see Battese, 1977). We will now sketch the basis for the lack-of-fit test. We have already seen [see equation (4.3.8)] that the disturbance variance $\sigma^2$ can be estimated by $\hat{\sigma}^2$, given by

$$\hat{\sigma}^2 = \frac{\text{RSS}}{N - K - 1}$$

and this estimator can be used whether or not we have replicated data. It is an unbiased estimator if, and only if, the model is correctly specified.

The most significant feature of replicated data is that it enables us to obtain two independent estimators of $\sigma^2$ which are both different from $\hat{\sigma}^2$. One of these, which we will call $\hat{\sigma}_1^2$, is always an unbiased estimator of $\sigma^2$. The other, denoted by $\hat{\sigma}_2^2$, is unbiased only if the model is correctly specified. If the model is misspecified, $\hat{\sigma}_2^2$ tends to overestimate $\sigma^2$. This simple fact forms the basis for a model specification test.

If $\hat{\sigma}_1^2$ and $\hat{\sigma}_2^2$ are approximately equal, it is probably because $\hat{\sigma}_2^2$ is unbiased (remember, $\hat{\sigma}_1^2$ is always unbiased), and hence the model is probably correctly specified. On the other hand, if $\hat{\sigma}_2^2$ is considerably larger than $\hat{\sigma}_1^2$, it is probably because $\hat{\sigma}_2^2$ is positively biased—that is, the model is misspecified.

We will now formally derive expressions for $\hat{\sigma}_1^2$ and $\hat{\sigma}_2^2$. Let us suppose that there are t different combinations of the regressor (or control) variables. We will denote the ith combination by $(X_{1i}, X_{2i}, \ldots, X_{Ki})$, and refer to the replicated observations on Y corresponding to this combination as the ith group. In the cases of just one and two control variables, the ith groups are as depicted in Figure 6.6.1.

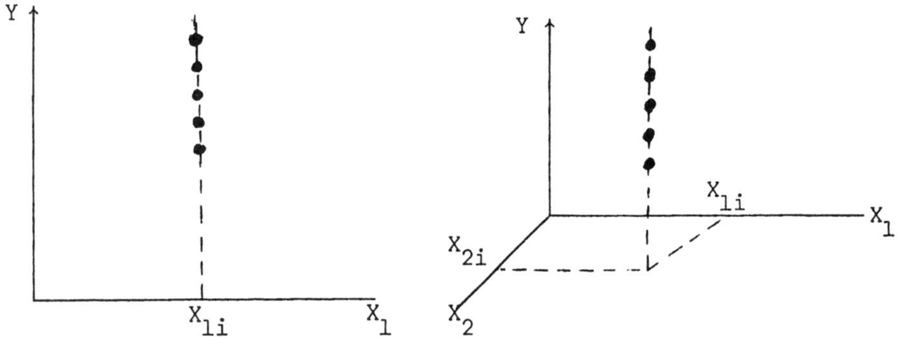


Figure 6.6.1. Group of replicated observations in the cases of one and two control variables.

Let $Y_{ij}$ be the jth observation on Y in the ith group, and let $n_i$ be the total number of observations in the ith group. Obviously,

$$n_1 + n_2 + \cdots + n_t \equiv \sum_{i=1}^{t} n_i = N. \tag{6.6.1}$$

The value of the replications is that the variation within each group can be used to obtain an estimate of $\sigma^2$ which is independent of the model specification. Take, for example, the first group. The variance of this group can be estimated in the usual way by

$$\hat{\sigma}_{11}^2 = \frac{1}{n_1 - 1} \sum_{j=1}^{n_1} (Y_{1j} - \bar{Y}_1)^2,$$

where $\bar{Y}_1$ is the sample mean of the first group. Similarly, for the second group, we can estimate $\sigma^2$ by $\hat{\sigma}_{12}^2$, given by

$$\hat{\sigma}_{12}^2 = \frac{1}{n_2 - 1} \sum_{j=1}^{n_2} (Y_{2j} - \bar{Y}_2)^2.$$

Proceeding in this way with the t groups, we will obtain t different estimators of $\sigma^2$, namely, $\hat{\sigma}_{11}^2, \hat{\sigma}_{12}^2, \ldots, \hat{\sigma}_{1t}^2$. Each of these estimators is unbiased and does not depend on any specification of a relationship between Y and $X_1, X_2, \ldots, X_K$.

It should be obvious that rather than using any of the individual $\hat{\sigma}_{1i}^2$ as an estimator of $\hat{\sigma}^2$, it would be much better to use a weighted average of all of them. Thus we define $\hat{\sigma}_1^2$ as

$$\hat{\sigma}_1^2 = \omega_1\hat{\sigma}_{11}^2 + \omega_2\hat{\sigma}_{12}^2 + \cdots + \omega_t\hat{\sigma}_{1t}^2,$$

where $\Sigma\omega_i = 1$. A sensible choice of weights would depend on the size of the groups.[5] This leads us to define $\hat{\sigma}_1^2$ by

$$\hat{\sigma}_1^2 = \frac{(n_1 - 1)\hat{\sigma}_{11}^2 + (n_2 - 1)\hat{\sigma}_{12}^2 + \cdots + (n_t - 1)\hat{\sigma}_{1t}^2}{(n_1 - 1) + (n_2 - 1) + \cdots + (n_t - 1)}$$

$$= \frac{(n_1 - 1)\hat{\sigma}_{11}^2 + (n_2 - 1)\hat{\sigma}_{12}^2 + \cdots + (n_t - 1)\hat{\sigma}_{1t}^2}{(n_1 + n_2 + \cdots + n_t) - t}.$$

That is,

$$\hat{\sigma}_1^2 = \frac{\Sigma_{i=1}^{t} (n_i - 1)\hat{\sigma}_{1i}^2}{N - t}. \qquad (6.6.2)$$

This estimator of $\sigma^2$ shares the same property as the $\hat{\sigma}_{1i}^2$—namely, it is unbiased[6] <u>regardless of any model specification</u>.

From the expressions above defining $\hat{\sigma}_{11}^2$ and $\hat{\sigma}_{12}^2$, it is clear that

$$(n_i - 1)\hat{\sigma}_{1i}^2 = \sum_{j=1}^{n_i} (Y_{ij} - \bar{Y}_i)^2,$$

and hence, substituting into (6.6.2),

$$\hat{\sigma}_1^2 = \frac{\Sigma_{i=1}^{t} \Sigma_{j=1}^{n_i} (Y_{ij} - \bar{Y}_i)^2}{N - t}. \qquad (6.6.3)$$

The numerator of this expression is often called the <u>pure error sum of squares</u> and denoted by PSS. That is,

$$PSS = \sum_{i=1}^{t} \sum_{j=1}^{n_i} (Y_{ij} - \bar{Y}_i)^2 \qquad (6.6.4)$$

and

$$\hat{\sigma}_1^2 = \frac{PSS}{N - t}. \qquad (6.6.5)$$

In Chapter 2 we showed how the total sum of squares can be partitioned into two further sums of squares—the explained (or regression) sum of squares and the residual sum of squares.[7]

It can be shown that <u>when there are replications</u>, the residual sum of squares can itself be partitioned into two further sums of squares: the pure error sum of squares, defined above, and the lack-of-fit sum of squares (LSS). That is,

$$\text{RSS} = \text{PSS} + \text{LSS}. \tag{6.6.6}$$

Before exploiting this partitioning, we note the following:

(a) If there are no replications, that is, $n_i = 1$ for all i, then $Y_{ij} = \bar{Y}_i$ (the average of one observation is itself) and thus by (6.6.4), PSS $\equiv$ 0. Furthermore, $N = \Sigma n_i = t$ in this case, so $N - t = 0$. Thus, both numerator and denominator in (6.6.5) are zero, so $\hat{\sigma}_1^2$ is undefined. It is only only when there are replications that the estimator $\hat{\sigma}_1^2$ is available.

(b) We know that RSS has (N - K - 1) degrees of freedom. Also, PSS being a weighted average of $\hat{\sigma}_{11}^2, \hat{\sigma}_{12}^2, \ldots, \hat{\sigma}_{1t}^2$ has $(n_1 - 1) + (n_2 - 1) + \cdots + (n_t - 1) = N - t$ degrees of freedom. Thus, by equation (6.6.6), LSS = RSS - PSS must have $(N - K - 1) - (N - t) = t - K - 1$ degrees of freedom (or $t - K$ if there is no intercept).

(c) RSS is, of course, related to the specification of a model. As PSS is independent of the choice of any model, then by (6.6.6) LSS must be related to a model. In fact, it can be shown that LSS is simply the residual sum of squares that would be obtained if each group of observations on Y were replaced by its sample mean, and these means were regressed on the X's.[8] Thus LSS is a measure of how well the model predicts the group means.

We now define a <u>second</u> estimator of $\sigma^2$, denoted by $\hat{\sigma}_2^2$ and given by

$$\hat{\sigma}_2^2 = \frac{\text{LSS}}{t - K - 1} = \frac{\text{RSS} - \text{PSS}}{t - K - 1}. \tag{6.6.7}$$

This estimator is unbiased only if the model is correctly specified. If there is misspecification, then $\hat{\sigma}_2^2$ is <u>positively biased</u>; that is, it will tend to overestimate $\sigma^2$.

At this point we pause to clarify the relationship between $\hat{\sigma}^2$ (the estimator which, so far, has always been used) and the two estimators $\hat{\sigma}_1^2$ and $\hat{\sigma}_2^2$ just introduced. Using (6.6.6) gives us

$$\hat{\sigma}^2 \equiv \frac{\text{RSS}}{N - K - 1} = \frac{\text{PSS}}{N - K - 1} + \frac{\text{LSS}}{N - K - 1}$$

$$= \frac{N - t}{N - K - 1}\,\frac{\text{PSS}}{N - t} + \frac{t - K - 1}{N - K - 1}\,\frac{\text{LSS}}{t - K - 1}.$$

If we set $\alpha = (N - t)/(N - K - 1)$, it is easily shown that $(t - K - 1)/(N - K - 1) = 1 - \alpha$. Furthermore, if the number of parameters to be estimated (K + 1) is less than the number of groups (t), then $0 < \alpha < 1$. Then

$$\hat{\sigma}^2 = \alpha\hat{\sigma}_1^2 + (1 - \alpha)\hat{\sigma}_2^2, \tag{6.6.8}$$

showing that $\hat{\sigma}^2$ is a <u>weighted average</u> of $\hat{\sigma}_1^2$ and $\hat{\sigma}_2^2$. If there are no replications, $N = t$ and so $\alpha = 0$, which implies that $\hat{\sigma}^2 = \hat{\sigma}_2^2$.

We may now compare $\hat{\sigma}_1^2$ and $\hat{\sigma}_2^2$ or, equivalently, consider the ratio $\hat{\sigma}_2^2/\hat{\sigma}_1^2$. If this ratio is significantly larger than 1, there is probably model misspecification. As we have seen on a number of occasions already, we can make a judgment about whether a ratio is significantly large only if we know the distribution of the ratio. The following theorem provides the necessary result.

THEOREM 6.6.1 If the $Y_{ij}$ are normally distributed, and if the model is correctly specified, then

$$F = \frac{\hat{\sigma}_2^2}{\hat{\sigma}_1^2}$$

has an F-distribution, with parameters t - K - 1 and N - t. Thus, if this F-ratio is larger than the critical value (at some preassigned level of significance), the hypothesis of correct specification is rejected.

For future reference, we observe that

$$F = \frac{\hat{\sigma}_2^2}{\hat{\sigma}_1^2} = \frac{LSS/(t - K - 1)}{PSS/(N - t)} .$$

That is, by (6.6.6),

$$F = \frac{(RSS - PSS)/(t - K - 1)}{PSS/(N - t)} . \tag{6.6.9}$$

EXAMPLE [Data from Whan (1972)] Wheat plants were subjected to differing amounts of nitrogen fertilizer (N) and the grain yield (Y) was examined. There

Table 6.6.1

| N (lb/acre) | Grain yield, Y: Replication 1 | Replication 2 | $\Sigma(Y_{ij} - \bar{Y}_i)^2$ |
|---|---|---|---|
| 0 | 770 | 821 | 1,300.5 |
| 46 | 1,251 | 1,432 | 16,380.5 |
| 97 | 2,653 | 2,700 | 1,104.5 |
| 144 | 4,086 | 3,706 | 72,200 |
| 178 | 3,816 | 3,212 | 182,408 |
| 238 | 3,895 | 3,903 | 32 |
| 265 | 3,858 | 3,617 | 29,040.5 |
| 312 | 3,805 | 3,646 | 12,640.5 |

were two replications at each of eight levels of application. The results are shown in Table 6.6.1, together with $\Sigma(Y_{ij} - \bar{Y}_i)^2$ for each group. For the first group, for example,

$$\Sigma(Y_{1j} - \bar{Y}_1)^2 = \Sigma Y_{1j}^2 - \frac{(\Sigma Y_{1j})^2}{2} = 1{,}266{,}941 - \frac{(1591)^2}{2}$$

$$= 1300.5.$$

Once again we emphasize that these terms reflect the within-group variation and do not depend in any way on the specification about how Y and N are related.

The pure error sum of squares is obtained by adding the contributions $\Sigma(Y_{ij} - \bar{Y}_i)^2$ from the eight groups. That is,

$$\text{PSS} = 1300.5 + 16{,}380.5 + \cdots + 12{,}640.5$$

$$= 315{,}106.5$$

DISPLAY 6.6.1

```
MTB > REGR C2 2 C1 C11;
SUBC> PURE.

The regression equation is
C2 = 593 + 27.6 C1 - 0.0570 C11

Predictor        Coef          Stdev      t-ratio
Constant        593.3          196.2         3.02
C1             27.551          2.912         9.46
C11         -0.057002       0.008982        -6.35

s = 325.9         R-sq = 93.6%      R-sq(adj) = 92.6%

Analysis of Variance

SOURCE         DF          SS          MS
Regression      2    20212121    10106060
Error          13     1381162      106243
Total          15    21593282

SOURCE         DF      SEQ SS
C1              1    15933593
C11             1     4278527

Unusual Observations
Obs.       C1        C2        Fit  Stdev.Fit  Residual   St.Resid
  7       144    4086.0     3378.6      123.9     707.4      2.35R

R denotes an obs. with a large st. resid.

Pure error test - F = 5.41  P =  DF(pure error) = 8
```

Reprinted with permission of Minitab, Inc.

We will now postulate the model

$$Y_{ij} = \beta_0 + \beta_1 N_i + \beta_2 N_i^2 + u_{ij}.$$

Carrying out the regression, we obtain

$$\hat{Y}_{ij} = \underset{(196)}{593} + \underset{(2.9)}{27.6}N_i - \underset{(0.009)}{0.057}N_i^2; \quad R^2 = 0.936 \quad RSS = 1,381,162.$$

From (6.6.6),

$$LSS = 1,381,162 - 315,106.5 = 1,066,055.5.$$

Therefore,

$$F = \frac{1,066,055.5/(8 - 2 - 1)}{315,106.5/(16 - 8)} = 5.41.$$

The critical value at the 5 percent level for $F_{5,8}$ is 3.69. Thus the model shows evidence of lack of fit. Even though $R^2$ is quite high, coefficients have the signs we would expect and all coefficients are significant, the valuable information conveyed by the replications has enabled us to detect model misspecification. The lack-of-fit test does not, however, help us decide how the model specification should be changed.

We have shown how the pure error SS can be calculated directly. In MINITAB® the subcommand PURE causes the lack-of-fit F-statistic to be calculated. Display 6.6.1 illustrates this and gives F=5.41 as above. In the next chapter we show another method for computing the test, both as one illustration of the use of dummy variables, and for users whose regression package does not have this facility.

## NOTES

1. The actual probability depends, of course, on the degrees of freedom.
2. At this stage we have not quantified the notions of "small" and "large."
3. There is one exception to this statement, which we will consider shortly.
4. This statement does not apply to the estimate of the intercept. In this case the reader can easily show that $\hat{\gamma}_0^* = \hat{\gamma}_0 - \ln K$.
5. We choose the weight $\omega_i$ proportional to $(n_i - 1)$ because this is the number of degrees of freedom of $\hat{\sigma}_{1i}^2$.
6. A weighted average of unbiased estimators must itself be unbiased.
7. The reader may be interested to verify this relationship numerically by considering some of the computer output displays.
8. Actually, all observations on both Y and the X's have to be weighted by $n_i^{\frac{1}{2}}$, but this is an unimportant technical detail.

# 7

# Extension of the Regression Model: Dummy Variables

## 7.1 INTRODUCTION

There are many situations in which observations on a dependent variable Y and regressor variables $X_1$, $X_2$, ..., $X_K$ can be regarded as coming from different sources. These differences may arise in many ways, and we list several below to give the general idea.

(a) Observations on time-series data may be obtained from different eras, for example, wartime and peacetime.
(b) In a cross-sectional study, observations may be taken on people having different sociological characteristics, for example, ethnic groupings and levels of education.
(c) Observations may be obtained from different countries, or different geographical regions within a country.
(d) Observations may be drawn from different seasons of the year—the March quarter, September quarter, and so on.

To highlight the problem that different sources may pose, let us look at two extreme situations:

(i) The model

$$Y_i = \beta_0 + \beta_1 X_{1i} + \cdots + \beta_K X_{Ki} + u_i$$

is the same for all sources of data. That is, the parameters $\beta_0$, $\beta_1$, ..., $\beta_K$ and $\sigma^2$ are identical for each source. Clearly, in this situation the distinction between sources is irrelevant, and the obvious procedure is to pool the data from all sources and run a single regression.

(ii) The model is completely different across sources. That is, the parameters $\beta_0$, $\beta_1$, ..., $\beta_K$ and $\sigma^2$ are all different across sources. In such a situation the model for each source would be estimated using data from that source alone.

The two approaches of ignoring the different sources and thus pooling the data, or of running a separate regression for each source are the only options we have at this stage. The usual situation belongs to neither (i) nor (ii), but somewhere in between. The models are likely to have some parameters in common, and others different across the sources. In this case, if we proceed as in (i) and simply pool the data, ignoring the differences across sources, we will obtain a sort of hybrid model that is not appropriate for any source. The estimators for the parameters of any particular source would be biased.

On the other hand, if we proceed as in (ii) and run a separate regression for each source, although we will not introduce any bias, we will be ignoring the information that some parameters have a common value. Consequently, our estimators will be inefficient, that is, not as precise as they could be if we incorporated this information into the analysis. What is needed, then, is a method that will enable us to build into our analysis the a priori information about which parameters are common and which differ across data sources. This may be achieved by introducing into the model artificial variables of a special kind, known as dummy variables.

## 7.2 DUMMY VARIABLES

This concept will be introduced by way of a simple example.

### Example

To analyze labor force participation, a researcher postulates the model

$$Y_i = \beta_0 + \beta_1 X_{1i} + \beta_2 X_{2i} + u_i, \tag{7.2.1}$$

where Y is the labor force participation, $X_1$ the unemployment rate, and $X_2$ the wage rate. Data are obtained from three cities: Sydney, Melbourne, and Adelaide.

Case I: Different intercepts. Suppose it can be reasonably assumed that $\beta_1$, $\beta_2$, and $\sigma^2$ are the same for the three cities. In this context, we are assuming that for any given values of $X_1$ and $X_2$ the levels of labor force participation differ across cities, but the marginal effects of changes in $X_1$ and $X_2$ are the same in the three cities. Furthermore, it is assumed that for given values of $X_1$ and $X_2$, the variation in labor force participation is the same in all cities.

We adopt the following procedure: Select one city (say Sydney) as base and express the intercepts for Melbourne and Adelaide relative to the intercept for Sydney. The model (7.2.1) may now be written as follows:

Sydney: $$Y_i = \beta_0 + \beta_1 X_{1i} + \beta_2 X_{2i} + u_i; \tag{7.2.2}$$

Melbourne: $$Y_i = (\beta_0 + \gamma_M) + \beta_1 X_{1i} + \beta_2 X_{2i} + u_i; \tag{7.2.3}$$

Adelaide: $$Y_i = (\beta_0 + \gamma_A) + \beta_1 X_{1i} + \beta_2 X_{2i} + u_i. \tag{7.2.4}$$

Notice that these three equations incorporate all our a priori knowledge. The same parameters, namely $\beta_1$ and $\beta_2$, are associated with the variables $X_1$ and $X_2$. The parameter $\gamma_M$ represents the amount by which the intercept for Melbourne differs from that for the base Sydney. If $\gamma_M > 0$, Melbourne has a greater intercept; if $\gamma_M < 0$, it is less; and if $\gamma_M = 0$, Sydney and Melbourne have the same intercept. The parameter $\gamma_A$ is interpreted in exactly the same way.

We now define two <u>dummy variables</u> $D_{1i}$ and $D_{2i}$ as follows:

$D_{1i} = 1$ whenever Melbourne observations are used;
$\phantom{D_{1i}} = 0$ otherwise.

$D_{2i} = 1$ whenever Adelaide observations are used;
$\phantom{D_{2i}} = 0$ otherwise.

With these definitions, the <u>three</u> equations (7.2.2), (7.2.3), and (7.2.4) can be written as a <u>single</u> equation:

$$Y_i = \beta_0 + \gamma_M D_{1i} + \gamma_A D_{2i} + \beta_1 X_{1i} + \beta_2 X_{2i} + u_i, \tag{7.2.5}$$

and we can estimate the parameters $\beta_0$, $\beta_1$, $\beta_2$, $\gamma_M$, $\gamma_A$, and $\sigma^2$ in the usual way by regressing $Y_i$ on $D_{1i}$, $D_{2i}$, $X_{1i}$, and $X_{2i}$, pooling the data from the three sources: Sydney, Melbourne, and Adelaide. The following points should be noted:

(a) To run separate regressions for each of the three cities would be inefficient. To see this, assume that $N_1$, $N_2$, and $N_3$ observations are available from Sydney, Melbourne, and Adelaide, respectively. Using a separate regression, $\hat{\beta}_1$ for Sydney would be estimated with $N_1 - 3$ degrees of freedom. In the combined model (7.2.5) there are $N_1 + N_2 + N_3 - 5$ degrees of freedom available. Thus, provided that $N_2 + N_3 > 2$, there are more degrees of freedom from the combined model, and hence, greater efficiency.

(b) If the differences in intercept were ignored and the data simply pooled to estimate (7.2.1), the estimators of <u>all</u> parameters would be <u>biased</u> because the relevant variables $D_1$ and $D_2$ had been omitted. Of course, the degree of bias would depend on the strength of correlation between the omitted variables $D_1$ and $D_2$, and the included variables $X_1$ and $X_2$.

Probably in this case this correlation would be quite small, so the bias in $\beta_1$ and $\beta_2$ would also be small. However, this need not always be the case.

(c) There are in this example three different sources of data, leading to two dummy variables. In general, S different sources will give rise to S - 1 dummy variables.

(d) Suppose now that instead of assuming that there is a difference in intercepts we wish to test whether this is the case. The unconstrained model is (7.2.5), as this allows the intercepts to differ. This regression is run and $RSS(H_1)$ obtained. If the model is constrained so that the intercepts are the same (i.e., $\gamma_M = \gamma_A = 0$), then

$$Y_i = \beta_0 + \beta_1 X_{1i} + \beta_2 X_{2i} + u_i.$$

This regression will yield $RSS(H_0)$. The F-ratio is then computed as in (6.2.2).

Case II: Different slope coefficients. Consider now the situation in which $\beta_0$, $\beta_2$, and $\sigma^2$ are the same for all cities, but the slope parameter $\beta_1$ changes. This case is more difficult to justify in this context, but can be used to show how we can deal with such cases.

As before, we nominate one source (say Sydney) as base and express the models relative to Sydney.

Sydney: $$Y_i = \beta_0 + \beta_1 X_{1i} + \beta_2 X_{2i} + u_i; \quad (7.2.6)$$

Melbourne: $$Y_i = \beta_0 + (\beta_1 + \gamma_M) X_{1i} + \beta_2 X_{2i} + u_i; \quad (7.2.7)$$

Adelaide: $$Y_i = \beta_0 + (\beta_1 + \gamma_A) X_{1i} + \beta_2 X_{2i} + u_i. \quad (7.2.8)$$

Defining $D_{1i}$ and $D_{2i}$ in exactly the same way as before, these three equations may be expressed as the single equation

$$Y_i = \beta_0 + \beta_1 X_{1i} + \gamma_M (D_{1i} X_{1i}) + \gamma_A (D_{2i} X_{1i}) + \beta_2 X_{2i} + u_i. \quad (7.2.9)$$

If we now compare (7.2.5) with (7.2.9) we should be able to generalize and see how a dummy variable equation may be set up to incorporate our prior beliefs about which parameters change across data sources, and which parameters remain fixed. The method is set out below in steps.

Step 1: Identify the different data sources. If there are S of them, S - 1 dummy variables are defined. Notice that the number of dummy variables is not affected by the number of parameters which change, but determined only by the number of different sources.

Step 2: Identify the parameters that change across sources.

Step 3: For each parameter that changes, S - 1 new variables are added to the model. For example, if $\beta_1$ changes, the model will include the additional variables

$$D_1X_1,\ D_2X_1,\ \ldots,\ D_{S-1}X_1.$$

If the intercept changes, we add the variables $D_1$, $D_2$, ..., $D_{S-1}$.

Case III: Models in which all parameters are assumed to change. We will continue with our simple example in which

$$Y_i = \beta_0 + \beta_1 X_{1i} + \beta_2 X_{2i} + u_i,$$

and data are obtained from three cities. If every parameter is assumed to change, then following the three steps above, the model takes the form

$$Y_i = \beta_0 + \gamma_0 D_{1i} + \delta_0 D_{2i} + \beta_1 X_{1i} + \gamma_1(D_{1i}X_{1i}) + \delta_1(D_{2i}X_{1i})$$
$$+ \beta_2 X_{2i} + \gamma_2(D_{1i}X_{2i}) + \delta_2(D_{2i}X_{2i}) + u_i. \tag{7.2.10}$$

Notice, three data sources means two dummy variables. Three changing parameters means $3 \times 2 = 6$ extra variables. Hence $3 + 6 = 9$ terms altogether.

The appropriate parameter estimates are set out in Table 7.2.1. The question now arises: If three separate regressions were run, one for each city, would the parameter estimates be the same as the above? For example, would the coefficient of $X_1$ for the Melbourne regression be the same as $\hat{\beta}_1 + \hat{\gamma}_1$? The answer is "yes." However, the standard errors of the estimates would not be the same. The reason for this is that the two methods are based on different assumptions about the variance of the disturbance, $\sigma^2$. A dummy variable approach would be adopted if it were assumed that $\sigma^2$ was the same for each city, and separate regressions otherwise. [See Rao and Miller (1971, 90-91) for a fuller discussion of this point.]

Finally, for carrying out hypothesis tests it is important to realize that when a dummy variable regression of the type (7.2.10) is carried out, the residual sum of squares is identical to the sum of the RSSs obtained from

Table 7.2.1

| | Sydney | Melbourne | Adelaide |
|---|---|---|---|
| Intercept | $\hat{\beta}_0$ | $\hat{\beta}_0 + \hat{\gamma}_0$ | $\hat{\beta}_0 + \hat{\delta}_0$ |
| $X_1$ | $\hat{\beta}_1$ | $\hat{\beta}_1 + \hat{\gamma}_1$ | $\hat{\beta}_1 + \hat{\delta}_1$ |
| $X_2$ | $\hat{\beta}_2$ | $\hat{\beta}_2 + \hat{\gamma}_2$ | $\hat{\beta}_2 + \hat{\delta}_2$ |

doing separate regressions. That is, if we denote by $RSS_i$ the residual sum of squares from the regression on the ith source, then

$$RSS = \sum_i RSS_i, \tag{7.2.11}$$

where RSS refers to the dummy variable regression.

## Seasonal Dummy Variables

A particular case of the use of dummy variables is to be found in the way seasonality is often treated. Suppose that the consumption function is being estimated with quarterly data, in which case data come from four different sources: March quarter, June quarter, September quarter, and December quarter. We will assume that the marginal propensity to consume stays the same for all quarters but the level of consumption changes. For example, the level of consumption could be expected to be highest in the December quarter because of Christmas. We would postulate a model

$$C_t = \beta_0 + \beta_1 Y_t + u_t,$$

where $\beta_1$ is the same for all quarters but $\beta_0$ varies.

We define three seasonal dummy variables:

$$\begin{aligned} S_{1t} &= 1 \quad \text{for March quarter;} \\ &= 0 \quad \text{otherwise.} \\ S_{2t} &= 1 \quad \text{for June quarter;} \\ &= 0 \quad \text{otherwise.} \\ S_{3t} &= 1 \quad \text{for September quarter;} \\ &= 0 \quad \text{otherwise.} \end{aligned}$$

By implication we have taken the December quarter as base. The model that incorporates our assumptions would then be of the form

$$C_t = \beta_0 + \gamma_1 S_{1t} + \gamma_2 S_{2t} + \gamma_3 S_{3t} + \beta_1 Y_t + u_t. \tag{7.2.12}$$

The estimates of the intercepts in this case would be:

March: $\hat{\beta}_0 + \hat{\gamma}_1$,

June: $\hat{\beta}_0 + \hat{\gamma}_2$,

September: $\hat{\beta}_0 + \hat{\gamma}_3$,

December: $\hat{\beta}_0$.

Probably, $\hat{\gamma}_1$, $\hat{\gamma}_2$, and $\hat{\gamma}_3$ would all be negative. One often reads of a model using quarterly data being set up in which the variables are specified as $Y, X_1, X_2, \ldots, X_K$, "together with three seasonal dummy variables." The underlying assumption here is that only the intercept changes with the seasons. Again, if we were to assume that both the marginal propensity to consume and the intercept change across seasons, the model would be

$$C_t = \beta_0 + \gamma_1 S_{1t} + \gamma_2 S_{2t} + \gamma_3 S_{3t} + \beta_1 Y_t + \delta_1(S_{1t}Y_t) + \delta_2(S_{2t}Y_t) + \delta_3(S_{3t}Y_t) + u_t. \tag{7.2.13}$$

## 7.3 THE CHOW TEST

There are situations in which it is desirable to divide the data set into two groups and test whether the same regression model applies to each group. The most obvious application is when there is reason to suspect that a different model may apply for the two groups. For example, it would be reasonable to assume that the consumption function may be different in wartime from peacetime. Again, in an investigation on the determinants of government expenditure on education, it may be supposed that a change of government would give rise to a different philosophy on educational expenditure.

Another very important application is to test the stability of the estimated model. If a model is to be used for forecasting, it is desirable to be sure that at least over the sample period, the model appears to be stable. A method of testing this is to divide the sample data into two (or more) groups and see whether models estimated for the separate groups differ significantly.

Let us assume, for the moment, that the two groups contain $N_1$ and $N_2$ observations and that both $N_1$ and $N_2$ are greater than $K + 1$, where K is the number of regressor variables. Chow (1960) suggested the following procedure:

(a) Carry out the separate regressions and denote the residual sums of squares by $RSS_1$ and $RSS_2$.
(b) Carry out the regression over the entire data set, denoting the residual sum of squares by RSS.
(c) Construct an F-ratio, defined by

$$F = \frac{[RSS - (RSS_1 + RSS_2)]/(K + 1)}{(RSS_1 + RSS_2)/[N_1 + N_2 - 2(K + 1)]} \tag{7.3.1}$$

This F-ratio is compared with $F_{K+1, N_1+N_2-2(K+1)}$ and if F is significantly large, a different model applies across the two groups.

This test is in fact identical to the following test using dummy variables. Define

$D_i = 1$ for observations on group 1

$= 0$ otherwise.

(a) Regress Y on $X_1, \ldots, X_K, D, DX_1, \ldots, DX_K$. As usual, denote the residual sum of squares by $RSS(H_1)$. Then, by (7.2.11),

$$RSS(H_1) = RSS_1 + RSS_2.$$

(b) Regress Y on $X_1, \ldots, X_K$. This is a constrained regression (as coefficients of $D, DX_1, \ldots, DX_K$ are assumed to be equal to zero). Denote the residual sum of squares by $RSS(H_0)$. This is of course identical to the residual sum of squares we obtain in (b). Thus

$$RSS(H_0) = RSS.$$

(c) The number of constraints is clearly K + 1. Thus the F-ratio is

$$F = \frac{[RSS(H_0) - RSS(H_1)]/(K + 1)}{RSS(H_1)/[N_1 + N_2 - 2(K + 1)]},$$

which is identical to (7.3.1).

For completeness, we note that if either $N_1$ or $N_2$ is less than or equal to K + 1, the Chow test as described above cannot be computed. Let us now suppose that $N_1 > K + 1$, but $N_2 \le K + 1$. In this case Chow showed that the ratio defined by

$$F = \frac{(RSS - RSS_1)/N_2}{RSS_1/(N_1 - K - 1)} \tag{7.3.2}$$

is appropriate and should be compared with $F_{N_2,(N_1-K-1)}$.

## 7.4 CONSTRUCTING DUMMY VARIABLES IN MINITAB®

When we apply OLS to models containing dummy variables, it is necessary to be able to construct them. In MINITAB® this is usually done with the SET and INDIcator commands, either separately or together. In this section we discuss, by way of several simple examples, how to use these commands.

### The SET Command

As we saw in Chapter 3, the SET command can be used to introduce data into the computer. A very powerful feature of SET is that it enables convenient entry of repeated and patterned data. A number of examples of the use of SET follow to illustrate its capabilities.

EXAMPLE 1 Data: 0, 0, ..., 0, 5, 5, ..., 5, 8, 8, ..., 8 where there are 6 zeros, 10 fives, and 7 eights.

```
SET C1
6(0) 10(5) 7(8)
END
```

EXAMPLE 2 Data: 1, 2, 3, ..., 20.

```
SET C1
(1:20)
END
```

EXAMPLE 3 Data: 1, 5, 9, 13, ..., 29. Here the numbers are increasing sequentially in steps of 4.

```
SET C1
(1:29/4)
END
```

EXAMPLE 4 Data: The sequence of numbers 1, 2, 3, 4 is repeated six times. That is, we have 1, 2, 3, 4, 1, 2, 3, 4, ..., 1, 2, 3, 4.

```
SET C1
6(1:4)
END
```

The commands

```
SET C1
6(1:29/4)
END
```

would repeat the data of Example 3 six times.

EXAMPLE 5 Data: The numbers 1, 2, 3, 4 are used again, but now 1 is repeated six times followed by 2 repeated six times, and so on. That is, we have 1, 1, ..., 1, 2, 2, ..., 2, ..., 4, 4, ..., 4.

```
SET C1
(1:4)6
END
```

Here the repeat number (6) is placed after the parentheses.

## The INDIcator Command

This command uses a column of integers which has already been defined (the indicator variable) to create a number of other columns which are dummy variables, that is, take only the values 0 and 1. A simple example will illustrate its use. Suppose that the column C1 (the indicator variable) contains the numbers 1, 2, 3, 4. The command

```
INDIcator C1 C2-C5
```

creates four new columns as follows:

| C1 | C2 | C3 | C4 | C5 |
|---|---|---|---|---|
| 1 | 1 | 0 | 0 | 0 |
| 2 | 0 | 1 | 0 | 0 |
| 3 | 0 | 0 | 1 | 0 |
| 4 | 0 | 0 | 0 | 1 |

Thus C2 contains 1 every time 1 occurs in C1 and zero otherwise. C3 contains 1 every time 2 occurs in C1 and zero otherwise, and so on. Many different types of dummy variables can be created by using SET to form an appropriate indicator variable, followed by INDIcator.

## A Further Example of Dummy Variable Construction

We will continue with our introductory example of data from the cities Sydney, Melbourne, and Adelaide. Let us assume that $N_1 = 10$, $N_2 = 15$, and $N_3 = 12$ and that the data have been sorted so that all Sydney observations come first, followed by Melbourne observations, and then Adelaide observations. We want to construct three dummy variables corresponding to the three cities, which we will store in C1, C2, and C3.

<u>Method 1</u>

```
SET C1
10(1) 27(0)
END

SET C2
10(0) 15(1) 12(0)
END

SET C3
25(0) 12(1)
END
```

The meaning of these commands is fairly obvious in the context of our example. This particular method would become very tedious if we wanted to set up a large number of dummy variables.

Method 2. This method is much more elegant and is carried out in two steps. The first step is to construct an indicator variable which takes the value 1 for Sydney observations and 2 and 3 for Melbourne and Adelaide observations, respectively. This indicator variable, stored in C1, is constructed by

```
SET C1
10(1) 15(2) 12(3)
END
```

(i.e., 10 ones, followed by 15 twos, followed by 12 threes).

We now use the INDIcator command as follows:

```
INDIcator C1 C2-C4
```

This command defines three variables from C1 and stores them in C2, C3, and C4, as follows:

C2 = 1 when C1 = 1
= 0 otherwise

(Sydney dummy variable)

C3 = 1 when C1 = 2
= 0 otherwise

(Melbourne dummy variable)

C4 = 1 when C1 = 3
= 0 otherwise

(Adelaide dummy variable)

We started this example by assuming that the observations had been sorted into city groups. A far more usual situation is where there is a coding variable to indicate the data source. In such cases this variable is taken as the indicator variable and we then proceed as before.

## Seasonal Dummy Variables

Let us suppose that we have a number of years of quarterly data, and we wish to construct the four seasonal dummy variables $S_1$, $S_2$, $S_3$, and $S_4$, which correspond to the March, June, September, and December quarters, respectively. We will assume that we have 20 years of data. Then

```
SET C5
20(1:4)
END
INDIcator C5 C1-C4
```

has created $S_1, \ldots, S_4$ in columns C1 through C4. The SET command has stored the indicator variable 1, 2, 3, 4, ..., 1, 2, 3, 4 (representing the 20 years of quarterly data) in C5. The INDIcator command has used C5 to create the required dummy variables.

## 7.5 EXAMPLES ON THE USE OF DUMMY VARIABLES

### Example 1

A special case of dummy variable application is when the a priori information indicates that different variables influence the dependent variable across sources. Let us suppose that for one country a variable Y is assumed to depend on the variables $X_1$ and $X_2$. For a second country, Y is believed to be influenced by $X_1$ and $X_3$. Defining $D = 0$ for country 1 and $D = 1$ for country 2, we obtain the dummy variable model

$$Y = \beta_0 + \gamma_0 D + \beta_1 X_1 + \gamma_1 DX_1 + \beta_2 X_2 + \gamma_2 DX_2 + \beta_3 X_3 + \gamma_3 DX_3 + u,$$

where $\beta_3 = 0$, $\beta_2 + \gamma_2 = 0$. That is,

$$Y = \beta_0 + \gamma_0 D + \beta_1 X_1 + \gamma_1 DX_1 + \beta_2 (1 - D) X_2 + \gamma_3 DX_3 + u.$$

The only point in setting up a dummy variable model is if either the intercept or the coefficient of $X_1$ is common to both countries.[1] Let us assume now that the coefficient of $X_1$ is the same for both countries. That is, $\gamma_1 = 0$. The final model is then

$$Y = \beta_0 + \gamma_0 D + \beta_1 X_1 + \beta_2 (1 - D) X_2 + \gamma_3 DX_3 + u.$$

### Example 2

In a study of unemployment in Great Britain, Gujarati (1972, 1978) estimated the relationship between the unemployment rate and the job-vacancy rate, using quarterly data from 1958-IV to 1971-II. The author believed that a change in the National Insurance Act in 1966-IV meant that the data could be regarded as coming from two different sources. Observations 1958-IV to 1966-III (32 observations) comprise the first group and 1966-IV to 1971-II

Table 7.5.1

| Time | Unemployment, $U_t$ | Vacancy, $V_t$ | Time | Unemployment, $U_t$ | Vacancy, $V_t$ |
|---|---|---|---|---|---|
| 1958-IV | 1.915 | 0.510 | 1965-I | 1.201 | 0.997 |
| 1959-I | 1.876 | 0.541 | -II | 1.192 | 1.035 |
| -II | 1.842 | 0.590 | -III | 1.259 | 1.040 |
| -III | 1.750 | 0.690 | -IV | 1.192 | 1.086 |
| -IV | 1.648 | 0.771 | 1966-I | 1.089 | 1.101 |
| 1960-I | 1.450 | 0.836 | -II | 1.101 | 1.058 |
| -II | 1.393 | 0.908 | -III | 1.243 | 0.987 |
| -III | 1.322 | 0.968 | -IV | 1.623 | 0.819 |
| -IV | 1.260 | 0.998 | 1967-I | 1.821 | 0.740 |
| 1961-I | 1.171 | 0.968 | -II | 1.990 | 0.661 |
| -II | 1.182 | 0.964 | -III | 2.114 | 0.660 |
| -III | 1.221 | 0.952 | -IV | 2.115 | 0.698 |
| -IV | 1.340 | 0.849 | 1968-I | 2.150 | 0.695 |
| 1962-I | 1.411 | 0.748 | -II | 2.141 | 0.732 |
| -II | 1.600 | 0.658 | -III | 2.167 | 0.749 |
| -III | 1.780 | 0.562 | -IV | 2.107 | 0.800 |
| -IV | 1.941 | 0.510 | 1969-I | 2.104 | 0.783 |
| 1963-I | 2.178 | 0.510 | -II | 2.056 | 0.800 |
| -II | 2.067 | 0.544 | -III | 2.170 | 0.794 |
| -III | 1.942 | 0.568 | -IV | 2.161 | 0.790 |
| -IV | 1.764 | 0.677 | 1970-I | 2.225 | 0.757 |
| 1964-I | 1.532 | 0.794 | -II | 2.241 | 0.746 |
| -II | 1.455 | 0.838 | -III | 2.366 | 0.739 |
| -III | 1.409 | 0.885 | -IV | 2.324 | 0.707 |
| -IV | 1.296 | 0.978 | 1971-I | 2.516[a] | 0.583[a] |
| | | | -II | 2.909[a] | 0.524[a] |

<u>Source</u>: Gujarati (1972), reproduced with permission.
[a]Provisional estimates.

(19 observations) the second. These data are shown in Table 7.5.1. We will use these data to illustrate a number of features of dummy variable analysis.

Following Gujarati, we will denote the unemployment rate by $UN_t$ and the job-vacancy rate by $V_t$. As there are two sources for the data, we define one dummy variable $D_t$, defined by

$$D_t = 0 \quad \text{for periods 1958-IV to 1966-III}$$
$$= 1 \quad \text{for periods 1966-IV to 1971-II.}$$

We will assume that a model of the form

$$UN_t = \beta_0 + \beta_1 V_t + u_t$$

is appropriate, and will carry out the following four regressions:

(a) $UN_t$ on $V_t$ (observations 1-51);
(b) $UN_t$ on $D_t$, $V_t$, $D_tV_t$ (observations 1-51);
(c) $UN_t$ on $V_t$ (observations 1-32);
(d) $UN_t$ on $V_t$ (observations 33-51).

Our first task in setting up a MINITAB® program to do these regressions is to construct the necessary data columns. We assume that UN and V are stored in columns C1 and C2, respectively.

To set up the dummy variable, we may proceed as follows:

```
SET C3
32(0) 19(1)
END
```

The dummy variable $D_t$ is now stored in column C3.

```
LET C4 = C2*C3
```

This command places $D_tV_t$ in column C4. Finally, we must set up four additional columns containing the observations on UN and V from the separate data sources. This is achieved by the commands

```
COPY C1 C2 C5 C6;
USE rows 1 to 32.
COPY C1 C2 C7 C8;
USE rows 33 to 51.
```

The results of regressions (a) and (b) are shown in Display 7.5.1 and those from (c) and (d) in Display 7.5.2. We will analyze the estimated models entirely in terms of regressions (a) and (b) and then verify our results using regressions (c) and (d).

DISPLAY 7.5.1

```
MTB > REGR C1 1 C2

The regression equation is
C1 = 3.32 - 2.00 C2

Predictor        Coef          Stdev       t-ratio
Constant       3.3186         0.1921        17.28
C2            -2.0033         0.2402        -8.34

s = 0.2856        R-sq = 58.7%        R-sq(adj) = 57.8%

Analysis of Variance

SOURCE        DF          SS          MS
Regression     1      5.6739      5.6739
Error         49      3.9960      0.0816
Total         50      9.6699

MTB > REGR C1 3 C2-C4

The regression equation is
C1 = 2.75 - 1.53 C2 + 1.15 C3 - 0.851 C4

Predictor        Coef          Stdev       t-ratio
Constant       2.7491         0.1022        26.90
C2            -1.5294         0.1218       -12.56
C3             1.1507         0.3171         3.63
C4            -0.8511         0.4294        -1.98

s = 0.1339        R-sq = 91.3%        R-sq(adj) = 90.7%

Analysis of Variance

SOURCE        DF          SS          MS
Regression     3      8.8268      2.9423
Error         47      0.8431      0.0179
Total         50      9.6699
```



From Display 7.5.1 we have (using all 51 observations)

(a) $$\underset{}{UN_t} = \underset{(0.1921)}{3.3186} - \underset{(0.2402)}{2.0033V_t}; \quad \bar{R}^2 = 0.578, \quad RSS = 3.9960.$$

(b) $$UN_t = \underset{(0.1022)}{2.7491} - \underset{(0.1218)}{1.5294V_t} + \underset{(0.3171)}{1.1507D_t} - \underset{(0.4294)}{0.8511D_tV_t}; \quad \bar{R}^2 = 0.907, \quad RSS = 0.8431.$$

A glance at the values of $\bar{R}^2$ certainly suggests that adding the regressors $D_t$ and $D_tV_t$ is worthwhile. We may test this as follows: Regression (b) is the unconstrained model, so

$$RSS(H_1) = 0.8431.$$

Regression (a) constrains the coefficients of $D_t$ and $D_tV_t$ to be zero—it is the constrained model corresponding to the null hypothesis

$$H_0 : \beta_2 = \beta_3 = 0.$$

Thus, $RSS(H_0) = 3.9960$, $M = 2$, and $K = 3$. Then, by equation (6.2.2),

$$F = \frac{(3.9960 - 0.8431)/2}{0.8431/(51 - 3 - 1)} = 87.9.$$

At the 5 percent level of significance, the critical value for $F_{2,47}$ is approx-

DISPLAY 7.5.2

```
MTB > REGR C5 1 C6

The regression equation is
C5 = 2.75 - 1.53 C6

Predictor        Coef         Stdev      t-ratio
Constant      2.74912       0.06603        41.63
C6           -1.52941       0.07869       -19.43

s = 0.08652       R-sq = 92.6%       R-sq(adj) = 92.4%

Analysis of Variance

SOURCE        DF          SS           MS
Regression     1      2.8278       2.8278
Error         30      0.2246       0.0075
Total         31      3.0524

MTB > REGR C7 1 C8
1

The regression equation is
C7 = 3.90 - 2.38 C8

Predictor        Coef         Stdev      t-ratio
Constant       3.8998        0.4275         9.12
C8            -2.3805        0.5865        -4.06

s = 0.1907        R-sq = 49.2%       R-sq(adj) = 46.2%

Analysis of Variance

SOURCE        DF          SS           MS
Regression     1     0.59948      0.59948
Error         17     0.61852      0.03638
Total         18     1.21800
```

imately 3.20. Thus as F is greater than 3.20, $H_0$ is rejected, and we take model (b) as being more appropriate than model (a).

Following Section 7.2, we may interpret model (b) as estimating a different model for the two time periods 1-32 and 33-51. In fact, model (b) implies that

$$UN_t = 2.7491 - 1.5294V_t \quad \text{(observations 1-32);}$$

$$\begin{aligned} UN_t &= (2.7491 + 1.1507) - (1.5294 + 0.8511)V_t \\ &= 3.8998 - 2.3805V_t \quad \text{(observations 33-51).} \end{aligned}$$

It can be seen from Display 7.5.2 that these are precisely the estimated models that have been obtained by running a separate regression for the two periods. However, although it cannot be demonstrated from this output, the standard errors would not be the same. We leave it as an exercise for the reader to verify that applying a Chow test using the results of regressions (a), (c), and (d) yields exactly the same F-ratio as the one we have obtained using a dummy variable regression.

## 7.6 USING DUMMY VARIABLES TO COMPUTE THE LACK-OF-FIT TEST

In equation (6.6.9) the lack-of-fit statistic was given in the form

$$F = \frac{(RSS - PSS)/(t - K - 1)}{PSS/(N - t)}.$$

This statistic is so similar to the conventional F-statistic, given by

$$F = \frac{[RSS(H_0) - RSS(H_1)]/M}{RSS(H_1)/(N - K - 1)},$$

that it suggests that the lack-of-fit test is really a test of $M = t - K - 1$ constraints.

As before, we will denote the jth observation in the ith group by $Y_{ij}$. It will be assumed that

$$E(Y_{ij}) = \mu_i \tag{7.6.1}$$

and

$$\text{var}(Y_{ij}) = \sigma^2. \tag{7.6.2}$$

These two equations imply that the mean of each group is different but the variances of all groups are the same. A completely equivalent way of expressing these two assumptions about $Y_{ij}$ is

$$Y_{ij} = \mu_i + v_{ij} \quad (i = 1, 2, \ldots, t), \tag{7.6.3}$$

where

$$E(v_{ij}) = 0, \quad \text{var}(v_{ij}) = \sigma^2. \tag{7.6.4}$$

We now define t dummy variables, one for each group, such that $D_{ij} = 1$ whenever observations on Y come from the ith group, and $D_{ij} = 0$ otherwise. The t equations (7.6.3) can now be written in the form

$$Y_{ij} = \mu_1 D_{1j} + \mu_2 D_{2j} + \cdots + \mu_t D_{tj} + v_{ij}. \tag{7.6.5}$$

The unknown means $\mu_1, \mu_2, \ldots, \mu_t$ can be estimated by regressing Y on the t dummy variables, $D_1, D_2, \ldots, D_t$ (with no intercept). The estimates that are obtained will be

$$\hat{\mu}_i = \bar{Y}_i \quad (i = 1, 2, \ldots, t),$$

and the residual sum of squares will therefore be given by

$$\text{RSS} = \sum_{i=1}^{t} \sum_{j=1}^{n_i} (Y_{ij} - \bar{Y}_i)^2, \tag{7.6.6}$$

which is simply the pure error sum of squares. This is an <u>unconstrained</u> regression, because we are not constraining the means $\mu_i$ in any way. As always, the number of degrees of freedom is given by the number of observations less the number of parameters which are estimated. Here we are estimating the t parameters $\mu_1, \mu_2, \ldots, \mu_t$, so the number of degrees of freedom is N - t.

If we now <u>constrain</u> the $\mu_i$ to be linearly related to the control variables $X_1, X_2, \ldots, X_K$, we have

$$\mu_i = \beta_0 + \beta_1 X_{1i} + \cdots + \beta_K X_{Ki}, \tag{7.6.7}$$

so, by (7.6.3),

$$Y_{ij} = \beta_0 + \beta_1 X_{1i} + \cdots + \beta_K X_{Ki} + v_{ij}. \tag{7.6.8}$$

The constraint (7.6.7) implies that the t parameters $\mu_1, \mu_2, \ldots, \mu_t$ can be expressed in terms of only K + 1 parameters $\beta_0, \beta_1, \ldots, \beta_K$. That is, there are t - (K + 1) = t - K - 1 constraints implied. The residual sum of squares RSS, obtained from regressing Y on $X_1, X_2, \ldots, X_K$, is thus the constrained residual sum of squares $\text{RSS}(H_0)$.

Using the usual F-test, we can test whether the constraints are true [i.e., whether the model (7.6.8) is true] by evaluating

$$F = \frac{[\text{RSS}(H_0) - \text{RSS}(H_1)]/M}{\text{RSS}(H_1)/(N - K - 1)},$$

Table 7.6.1[a]

| N | Y | $D_1$ | $D_2$ | $D_3$ | $D_4$ | $D_5$ | $D_6$ | $D_7$ | $D_8$ |
|---|---|---|---|---|---|---|---|---|---|
| 0 | 770 | 1 | | | | | | | |
| 0 | 821 | 1 | | | | | | | |
| 46 | 1251 | | 1 | | | | | | |
| 46 | 1432 | | 1 | | | | | | |
| 97 | 2653 | | | 1 | | | | | |
| 97 | 2700 | | | 1 | | | | | |
| 144 | 4086 | | | | 1 | | | | |
| 144 | 3706 | | | | 1 | | | | |
| 178 | 3816 | | | | | 1 | | | |
| 178 | 3212 | | | | | 1 | | | |
| 238 | 3895 | | | | | | 1 | | |
| 238 | 3903 | | | | | | 1 | | |
| 265 | 3858 | | | | | | | 1 | |
| 265 | 3617 | | | | | | | 1 | |
| 312 | 3805 | | | | | | | | 1 |
| 312 | 3646 | | | | | | | | 1 |

[a]The zero values of the $D_i$ ($i = 1, 2, \ldots, 8$) have been omitted in order to display the structure of the dummy variables more clearly.

DISPLAY 7.6.1

```
MTB > REGR C2 8 C3-C10;
SUBC> NOCO.

The regression equation is
C2 = 796 C3 + 1342 C4 + 2677 C5 + 3896 C6 + 3514 C7 + 3899 C8 + 3737 C9
           + 3726 C10

Predictor          Coef          Stdev       t-ratio
Noconstant
C3                795.5          140.3          5.67
C4               1341.5          140.3          9.56
C5               2676.5          140.3         19.07
C6               3896.0          140.3         27.76
C7               3514.0          140.3         25.04
C8               3899.0          140.3         27.78
C9               3737.5          140.3         26.63
C10              3725.5          140.3         26.55

s = 198.5
Analysis of Variance

SOURCE         DF           SS            MS
Regression      8    160347128      20043391
Error           8       315107         39388
Total          16    160662234
```

where

$$\begin{aligned} RSS(H_0) &= RSS \text{ from } (7.6.8), \\ RSS(H_1) &= RSS \text{ from } (7.6.5) \\ &= PSS, \\ M &= t - K - 1, \\ N - K - 1 &= N - t. \end{aligned}$$

Thus

$$F = \frac{[RSS - PSS]/(t - K - 1)}{PSS/(N - t)} = \frac{\hat{\sigma}_2^2}{\hat{\sigma}_1^2} \qquad (7.6.9)$$

The lack-of-fit test can therefore be computed by performing the following three steps:

(a) Regress Y on $D_1, D_2, \ldots, D_t$. This is the unconstrained regression and the residual sum of squares is PSS.
(b) Regress Y on $X_1, \ldots, X_K$ in the usual way. The residual sum of squares from this regression is the constrained RSS.
(c) F is found from (7.6.9).

EXAMPLE We return to the example used in Section 6.6, in which we discovered significant lack of fit for the model

$$Y_{ij} = \beta_0 + \beta_1 N_i + \beta_2 N_i^2 + u_{ij}.$$

To compute the pure error sum of squares as the residual sum of squares of an unconstrained model, we set up the data matrix as shown in Table 7.6.1. The dummy variables $D_1$ to $D_8$ can be constructed as follows:

```
SET C20
(1:8)2
END
INDIcator C20 C3-C10
```

The output obtained from regressing Y on $D_1$, $D_2$, ..., $D_8$ (with no intercept) is shown in Display 7.6.1. It is apparent that for this regression RSS = 315,107, which is identical to the pure error sum of squares calculated in Section 6.6. The F-test is now easily obtained from (7.6.9).

To conclude, we note two interesting features from the output of this regression.

(a) The estimated coefficients 795.5, ..., 3725.5 are simply the sample means of the yields of the eight groups. For example, for the first group the yields were 770 and 821, having an average of 795.5.

(b) The standard deviations of the coefficients are all the same, being equal to 140.3. The reason for this can be found in terms of equation (4.3.15), namely,

$$S^2_{\hat{\beta}_1} = \frac{\hat{\sigma}^2}{\Sigma x_1^2(1 - R_1^2)} .$$

For this particular regression, the regressors $D_1$, $D_2$, ..., $D_8$ are identical except for the ordering of the observations. As OLS results are independent of ordering, it follows that $\Sigma x_1^2(1 - R_1^2) = \Sigma x_2^2(1 - R_2^2) = \cdots = \Sigma x_8^2(1 - R_8^2)$. Hence all standard errors are equal.

## 7.7 USING DUMMY VARIABLES TO OBTAIN PREDICTIONS AND PREDICTION VARIANCES

We have seen in Section 4.6 that when Y is to be predicted at the values $X_1^\circ$, $X_2^\circ$, ..., $X_K^\circ$,[2] the prediction is given by

$$\hat{Y}_0 = \hat{\beta}_0 + \hat{\beta}_1 X_1^\circ + \hat{\beta}_2 X_2^\circ + \cdots + \hat{\beta}_K X_K^\circ$$

and the estimated variance of prediction error by [see equation (4.6.5)]

$$S^2_{(\hat{Y}_0 - Y_0)} = \hat{\sigma}^2 + \alpha'\hat{V}\alpha .$$

While computing the prediction (given $X_1^\circ$, $X_2^\circ$, ..., $X_K^\circ$) does not present any problems, obtaining the estimated error variance can be something of a burden. Computing the estimated error variance in the example of Section 4.7 required 12 multiplications and additions. In general, if there are K regressors, the number of multiplications and additions would be $K(K + 1)$. Clearly, as K increases, this number would become very large, and apart from the labor involved, the probability of making an error would be quite considerable. Salkever (1976) has suggested a method, using dummy variables, which enables the computations to be done on a standard regression package such as MINITAB.®

Let us suppose that we wish to obtain predictions and standard errors of prediction at r different points $(1, X_{1i}^\circ, \ldots, X_{Ki}^\circ)$ $(i = 1, 2, \ldots, r)$. Salkever's method consists of running the regression on an artificially constructed set of data which is shown schematically in Figure 7.7.1. As can be seen from the figure, the data are modified in two ways:

(a) New observations are added to the observations on the existing variables. For the dependent variable, $\bar{Y}$ is added r times. In the case of the regressor variables, observations are added that correspond to the points at which prediction is required. Thus, when r predictions are required, r observations are added.

| | | | | | | | |
|---|---|---|---|---|---|---|---|
| $Y_1$ | $X_{11}$ | $X_{21}$ | $X_{K1}$ | 0 | 0 | | 0 |
| $Y_2$ | $X_{12}$ | $X_{22}$ | $X_{K2}$ | 0 | 0 | | 0 |
| ⋮ | ⋮ | ⋮ | ⋮ | ⋮ | ⋮ | | ⋮ |
| $Y_N$ | $X_{1N}$ | $X_{2N}$ | $X_{KN}$ | 0 | 0 | | 0 |
| $\bar{Y}$ | $X_{11}^0$ | $X_{21}^0$ | $X_{K1}^0$ | 1 | 0 | | 0 |
| $\bar{Y}$ | $X_{12}^0$ | $X_{22}^0$ | $X_{K2}^0$ | 0 | 1 | | 0 |
| ⋮ | ⋮ | ⋮ | ⋮ | ⋮ | ⋮ | | ⋮ |
| $\bar{Y}$ | $X_{1r}^0$ | $X_{2r}^0$ | $X_{Kr}^0$ | 0 | 0 | | 1 |

Figure 7.7.1. Schematic representation of constructed data.

(b) New variables are added, in the form of r dummy variables. These variables have zeros everywhere except for one position. The first dummy has 1 in the row corresponding to the first prediction observation, and so on.

Having constructed new observations and variables in this way, the regression is run, and the result interpreted as follows:

(a) The first K + 1 coefficients $\hat{\beta}_0, \hat{\beta}_1, \ldots, \hat{\beta}_K$ are simply the estimates we would obtain if we regressed Y on $X_1, \ldots, X_K$ using the original observations. The standard errors are also the same.

(b) The next r parameter estimates bear a very simple relationship to the required predictions. In fact,

$$\bar{Y} - \hat{\beta}_{K+i} = \hat{Y}_{0i}, \tag{7.7.1}$$

where $\hat{Y}_{0i}$ is the predictor of Y at the ith point.

(c) The standard errors of $\hat{\beta}_{K+i}$ $(i = 1, 2, \ldots, r)$ are the required prediction standard errors.

(d) RSS and $R^2$ are identical to those we would obtain if we regressed Y on $X_1, X_2, \ldots, X_K$.

Thus with a single regression it is possible to obtain all the usual information on $\hat{\beta}$'s, standard errors, RSS, and $R^2$, together with predictions and standard errors of predictions.

Table 7.7.1

| t | $K_t$ | $L_t$ | $Q_t$ | $Y_t = \log Q_t$ | $X_{1t} = \log K_t$ | $X_{2t} = \log L_t$ |
|---|---|---|---|---|---|---|
| 1 | 105.8 | 97.0 | 100.9 | 4.6141 | 4.6616 | 4.5747 |
| 2 | 102.7 | 107.4 | 100.6 | 4.6112 | 4.6318 | 4.6766 |
| 3 | 80.3 | 94.6 | 87.3 | 4.4694 | 4.3858 | 4.5497 |
| 4 | 87.0 | 105.6 | 99.7 | 4.6022 | 4.4659 | 4.6597 |
| 5 | 137.2 | 113.7 | 119.8 | 4.7858 | 4.9214 | 4.7336 |
| 6 | 97.2 | 96.0 | 97.8 | 4.5829 | 4.5768 | 4.5643 |
| 7 | 79.6 | 95.6 | 85.7 | 4.4509 | 4.3770 | 4.5602 |
| 8 | 88.3 | 102.3 | 94.3 | 4.5465 | 4.4807 | 4.6279 |
| 9 | 98.5 | 96.1 | 98.4 | 4.5890 | 4.5901 | 4.5654 |
| 10 | 93.7 | 96.9 | 85.2 | 4.4450 | 4.5401 | 4.5737 |
| 11 | 86.6 | 99.5 | 93.3 | 4.5358 | 4.4613 | 4.6002 |
| 12 | 93.9 | 99.0 | 106.5 | 4.6681 | 4.5422 | 4.5951 |
| 13 | 94.0 | 102.8 | 96.0 | 4.5643 | 4.5433 | 4.6328 |
| 14 | 78.6 | 83.2 | 89.9 | 4.4987 | 4.3644 | 4.4212 |
| 15 | 100.0 | 100.0 | 100.0 | 4.6052 | 4.6052 | 4.6052 |

Table 7.7.2

| Y | $X_1$ | $X_2$ | $D_1$ |
|---|---|---|---|
| 4.6141 | 4.6616 | 4.5747 | 0 |
| 4.6112 | 4.6318 | 4.6776 | 0 |
| 4.4694 | 4.3858 | 4.5497 | 0 |
| ⋮ | ⋮ | ⋮ | ⋮ |
| 4.6052 | 4.6052 | 4.6052 | 0 |
| 4.5713 | 4.5422 | 4.5951 | 1 |

EXAMPLE In Section 4.7 we investigated a Cobb-Douglas production function

$$\ln Q_t = \beta_0 + \beta_1 \ln K_t + \beta_2 \ln L_t + u_t$$

using 15 observations on $Q_t$, $K_t$, and $L_t$. These data are reproduced in Table 7.7.1 for convenience.

It was required to estimate the model and predict output when $K° = 93.9$ and $L° = 99.0$. That is,

$$X_1^\circ = \ln K^\circ = \ln 93.9 = 4.5422,$$

$$X_2^\circ = \ln K^\circ = \ln 99.0 = 4.5951.$$

In the terminology we have been using, we want to predict Y (= ln Q) at the point (1,4.5422,4.5951).

To estimate the model and carry out the predictions, the data would be set up as in Table 7.7.2, where we have used $\bar{Y} = 4.5713$, denoted the extra dummy variable by $D_1$, and shown the new observation below the dashed line. Notice that because we are only predicting Y at one point, there is only one dummy variable and one extra observation. Assuming now that the 15 observations on K, L, and Q are stored in columns C1, C2, and C3, respectively, a MINITAB® program to achieve the analysis is as follows:

```
LET C4 = LOGE(C3)
LET C5 = LOGE(C1)
LET C6 = LOGE(C2)
```

DISPLAY 7.7.1

```
MTB > REGR C4 3 C5-C7

The regression equation is
C4 = 1.54 + 0.460 C5 + 0.205 C6 + 0.0007 C7

Predictor         Coef         Stdev     t-ratio
Constant        1.5396        0.9210        1.67
C5              0.4600        0.1321        3.48
C6              0.2049        0.2631        0.78
C7             0.00066       0.05282        0.01

s = 0.05114      R-sq = 71.3%      R-sq(adj) = 64.1%

Analysis of Variance

SOURCE        DF          SS          MS
Regression     3    0.078031    0.026010
Error         12    0.031387    0.002616
Total         15    0.109418
```

(We now have ln Q = Y in C4, ln K = $X_1$ in C5, and ln L = $X_2$ in C6.)

```
INSERT C4-C6
4.5713 4.5422 4.5951
END
SET C7
15(0) 1
END
REGR C4 3 C5-C7
```

The output from this regression is shown in Display 7.7.1.

The prediction of Y at the point (1,4.5422,4.5951) is given by

$$\bar{Y} - \hat{\beta}_4 = 4.5713 - 0.00066 = 4.5706$$

and the standard error of prediction by

$$S_{\hat{\beta}_4} = 0.05282.$$

Comparison with Section 4.7 shows that the results are identical in every respect.

NOTES

1. We have implicitly assumed that the variance of u is the same for both countries. If this is not so, the dummy variable model is heteroscedastic. This problem will be dealt with subsequently.

2. Using terminology taken from geometry, the vector $(1, X_1^\circ, \ldots, X_K^\circ)$ is often called a point.

# 8

# Analysis of Residuals, Heteroscedasticity, and Autocorrelation

## 8.1 INTRODUCTION

The basic tool that is used in regression analysis is the method of ordinary least squares (OLS). We have seen that provided that

(a) the model is correctly specified;
(b) the regressor variables are fixed in repeated samples; and
(c) the disturbance $u_i$ has the properties

$$\text{(i)}\quad E(u_i) = 0,$$

$$\text{(ii)}\quad E(u_i^2) = \sigma^2,$$

$$\text{(iii)}\quad E(u_i u_j) = 0 \qquad (i \neq j),$$

then OLS is unbiased and has a smaller variance than that of any other linear unbiased estimator. In the shorthand notation we have often used, OLS is BLUE. Furthermore, if in addition to the properties in (c), $u_i$ is also normally distributed, it is possible to construct t- and F-tests to carry out a range of inferences on the regression model.

In Chapter 5, we considered some techniques for checking whether the model has been correctly specified. Another very valuable source for diagnosing possible misspecifications is the residuals that are generated when a model is fitted. In this chapter we first examine some of the types of model misspecification that may be revealed by an analysis of the residuals. Following this we consider how OLS should be modified when the assumptions in (c) no longer give an adequate representation of reality.

## 8.2 ANALYSIS OF RESIDUALS

When analyzing the residuals, it is important to realize that misspecification errors will often not be reflected in the residuals. Let us consider the case in which a relevant variable has been omitted. Suppose that the true model is

$$Y_i = \beta_0 + \beta_1 X_{1i} + \beta_2 X_{2i} + u_i,$$

and we misspecify the model by omitting the variable $X_{2i}$. That is, we assume that

$$Y_i = \beta_0 + \beta_1 X_{1i} + v_i.$$

Clearly,

$$v_i = u_i + \beta_2 X_{2i},$$

and it might be expected that the residuals would reflect the presence of $X_{2i}$ in $v_i$ and give a clue to the misspecification. In practice, however, the residuals will not often show the effect of omitted variables. The reason for this is to be found in the theory of Section 5.3. If $X_{2i}$ is linearly related to $X_{1i}$, then $\hat{\beta}_1$ will be biased. In fact,

$$E(\hat{\beta}_1) = \beta_1 + \beta_2 \hat{\beta}_{2.1}$$

and the effect of the omitted variable $X_{2i}$ will be soaked up by the term $\beta_2 \hat{\beta}_{2.1}$, and it is very likely that the residuals will be quite well behaved. We now consider some ways in which examination of the residuals may help us to improve the model specification.

### Outliers

Sometimes one or two residuals are very much larger in absolute value than all the others. Such residuals are known as outliers, and although it is possible that they relate to observations with exceptionally large disturbance values, it is also conceivable that some errors have been made in transcribing particular observations onto the data file. This possibility should be checked out before simply accepting the residuals as they have been generated. If observations have been recorded incorrectly, obviously a new regression estimate will be required, which will give rise to a corrected residual plot.

Outliers may arise for reasons other than transcription errors. For example, in the seasons 1942-43 wheat production in New Zealand was substantially higher than one would expect in terms of a regression relationship relating supply to the relative prices of agricultural products. The large positive outliers that turn up in those particular years may be rationalized after considering certain special factors which operated at that time. Those were the years of the Pacific War and a large publicity campaign was directed at farmers by the New Zealand government to try and ensure that the country was reasonably self-sufficient in wheat. Knowing these facts, a dummy variable taking nonzero values only in those years, and interpreted as "patriotic response," could be added to the regression relationship to explain these outliers.

It is important that dummy variables be added to the model only if sound reasons can be advanced to support the belief that a different model applied at the observations corresponding to the outliers. MINITAB® identifies both outliers and observations whose regressor values exert large influence on the estimation. Such observations should be examined.

## Seasonality

An adequate model to account for the variation in the dependent variable $Y_i$ should include all the systematic effects, and the residuals should therefore not exhibit any recognizable pattern. If, for example, quarterly observations were used and a plot of residuals against time showed a distinct seasonal pattern, this would be a clear indication that seasonal effects should have been explicitly accounted for in the model. This could probably be achieved by adding three seasonal dummy variables. The regression would then be rerun and examined to see whether there had been any improvement. This would be assessed by observing whether there had been an increase in $\bar{R}^2$ and whether the seasonality in the residuals had been removed. When using MINITAB,® the residuals can be stored by using the subcommand RESI.

## Clustered Residuals

If a histogram is constructed from the residuals, there should be a peak around zero and a distinct tailing-off on either side, as depicted in Figure 8.2.1. Any marked departure from this type of histogram may alert us that something has not been accounted for. Suppose, for example, that the histogram appeared as shown in Figure 8.2.2. This would indicate that the residuals are generally either well above zero or well below zero. This situation could arise if the data were really drawn from two different sources, in which the true relationships differed in their intercepts as shown in Figure 8.2.3. The estimated line would not adequately represent either group, and this would lead to a clustering of residuals in the tails of the histogram. By examining the observations corresponding to these two groups of residuals, it may be possible to identify the characteristic that separates the two

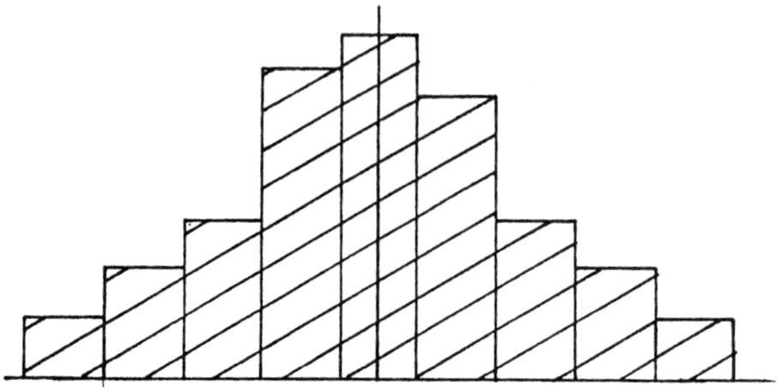

Figure 8.2.1. Typical histogram of residuals.

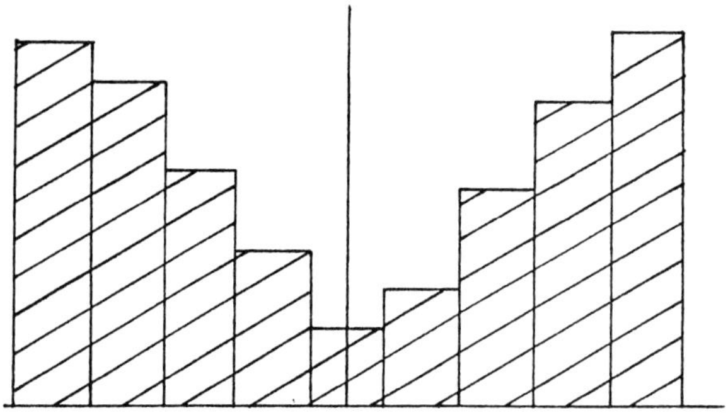

Figure 8.2.2. Residual histogram indicating model misspecification.

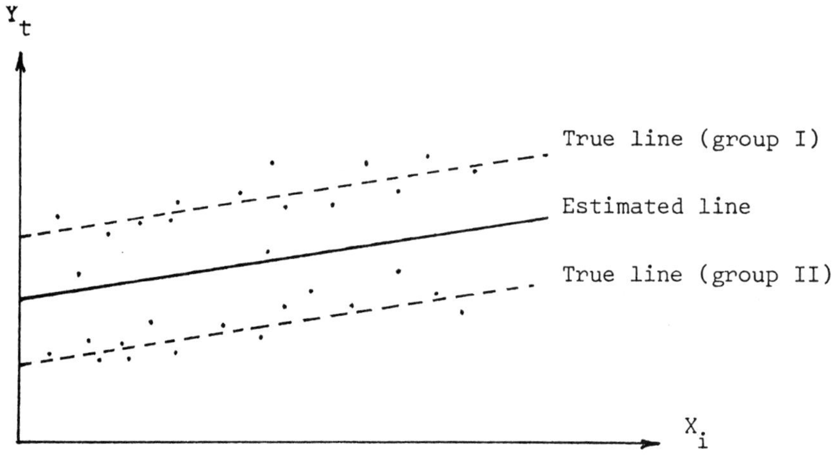


Figure 8.2.3. Observations drawn from two groups.

groups. A dummy variable could then be defined to account for this characteristic, and included in the model.

We illustrate this effect by again referring to the data taken from Gujarati (1972) and used in Section 7.5 (see Table 7.5.1). Let us suppose that we carry out the simple regression of $UN_t$ (stored in C1) on $V_t$ (C2) by the MINITAB® command

```
REGR C1  1  C2;
RESI C10.
```

The residuals are then stored in column C10. A histogram of the residuals is obtained by

```
HIST C10
```

and is shown in Display 8.2.1. This histogram shows definite bimodel characteristics. This suggests that an appropriate intercept dummy variable may have been omitted. Notice, however, that it does not tell us how the dummy variable should be defined. When we now define the dummy $D_t$ by

$$D_t = 0 \quad \text{for periods 1958-IV to 1966-III}$$
$$= 1 \quad \text{for periods 1966-IV to 1971-II}$$

(for the reasons discussed in Section 7.4) and include it as a regressor in column C3, we have

```
REGR C1  2  C2  C3;
RESI C11.
HIST C11
```

DISPLAY 8.2.1

```
-- HIST C10
  C10
   MIDDLE OF    NUMBER OF
   INTERVAL     OBSERVATIONS
       -1.5       6      ******
       -1.0       3      ***
       -0.5      16      ****************
        0.0      10      **********
        0.5       2      **
        1.0       4      ****
        1.5       8      ********
        2.0       1      *
        2.5       1      *
```

Reprinted with permission of Minitab, Inc.

DISPLAY 8.2.2

```
-- HIST C11
  C11
   MIDDLE OF    NUMBER OF
   INTERVAL     OBSERVATIONS
      -3.0        1    *
      -2.5        1    *
      -2.0        1    *
      -1.5        1    *
      -1.0        5    *****
      -0.5        7    *******
       0.0       12    ************
       0.5       12    ************
       1.0        8    ********
       1.5        2    **
       2.0        0
       2.5        0
       3.0        1    *
```

Reprinted with permission of Minitab, Inc.

The results are shown in Display 8.2.2.

This histogram is now unimodal, as required, although there is evidence of skew. It is of interest to note that when the variable $D_tV_t$ is also included among the regressors, the residuals, as shown in Display 8.2.3, now give rise to a symmetric histogram but indicate four outliers, corresponding to observations 33, 34, 35, and 51. The nonrandom position of these observations (three together, and the last observation of the first group) suggests that there may be some systematic effect here which should be investigated.

DISPLAY 8.2.3

```
-- HIST C12
  C12
   MIDDLE OF    NUMBER OF
   INTERVAL     OBSERVATIONS
      -2.5        3    ***
      -2.0        0
      -1.5        2    **
      -1.0        4    ****
      -0.5        6    ******
       0.0       14    **************
       0.5       10    **********
       1.0        8    ********
       1.5        3    ***
       2.0        0
       2.5        1    *
```

Reprinted with permission of Minitab, Inc.

## Polynomial Relationships

There are many relationships which cannot be adequately represented by models that are linear in variables. For example, if we were investigating the relationship between crop yield and fertilizer application, we would expect to find the marginal effect on crop yield from added fertilizer diminishing (at least after a certain level of application). Similarly, it is usually accepted that a total cost curve, as a function of output, would not be linear but as shown in Figure 8.2.4.

In both of these examples, the marginal change is not constant, but a function of one of the regressor variables. One way of modeling such a situation is by using a polynomial function of the form

$$Y_i = \beta_0 + \beta_1 X_i + \beta_2 X_i^2 + \cdots + \beta_p X_i^p + \gamma Z_i + u_i, \tag{8.2.1}$$

where we have used $\gamma Z_i$ as a shorthand representation of other relevant regressor variables and parameters.

If now we assumed a model of the form

$$Y_i = \beta_0 + \beta_1 X_i + \gamma Z_i + u_i,$$

we would have omitted variables $X_i^2$, $X_i^3$, ..., $X_i^p$ which bear a nonlinear relationship with the included variable $X_i$. Anscombe (1973) has suggested that a plot of residuals against predicted Y (i.e., $\hat{Y}_i$) should always be part of a regression analysis. Such a plot will often reveal the omission of polynomial variables.

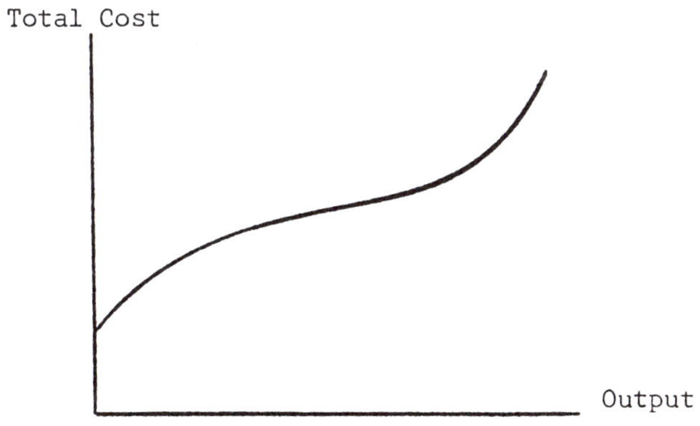


Figure 8.2.4. Typical total cost curve.

Table 8.2.1

| Observation | Total cost, Y (thousands of dollars) | Total output, X (hundreds of physical units) |
|---|---|---|
| 1 | 10 | 1 |
| 2 | 28.6 | 3 |
| 3 | 19.5 | 2 |
| 4 | 32.9 | 4 |
| 5 | 52.4 | 6 |
| 6 | 42.4 | 5 |
| 7 | 62.9 | 7 |
| 8 | 86.3 | 9 |
| 9 | 74.1 | 8 |
| 10 | 100.0 | 10 |
| 11 | 133.9 | 12 |
| 12 | 115.7 | 11 |
| 13 | 154.8 | 13 |
| 14 | 178.7 | 14 |
| 15 | 203.1 | 15 |

Source: Koutsoyiannis (1977, 136), reproduced with permission.

EXAMPLE Consider the data on total cost and output in Table 8.2.1. In general, if a model has been specified correctly, a plot of $e_i$ against $\hat{Y}_i$ should reveal a random scatter of points about the horizontal line $e_i = 0$.

Suppose now that using the data in Table 8.2.1 we regress $Y_i$ (in C1) on $X_i$ (in C2) with the following commands:

```
REGR C1  1  C2  C9  C11;
RESI  C10.
```

This results in the estimated model

$$\hat{Y}_i = \underset{(6.8)}{-18.8} + \underset{(0.8)}{13.1}X_i \ ; \quad \bar{R}^2 = 0.956 \ .$$

The residuals $e_i$ are stored in C10 and the predictions $\hat{Y}_i$ in C11. A plot

DISPLAY 8.2.4

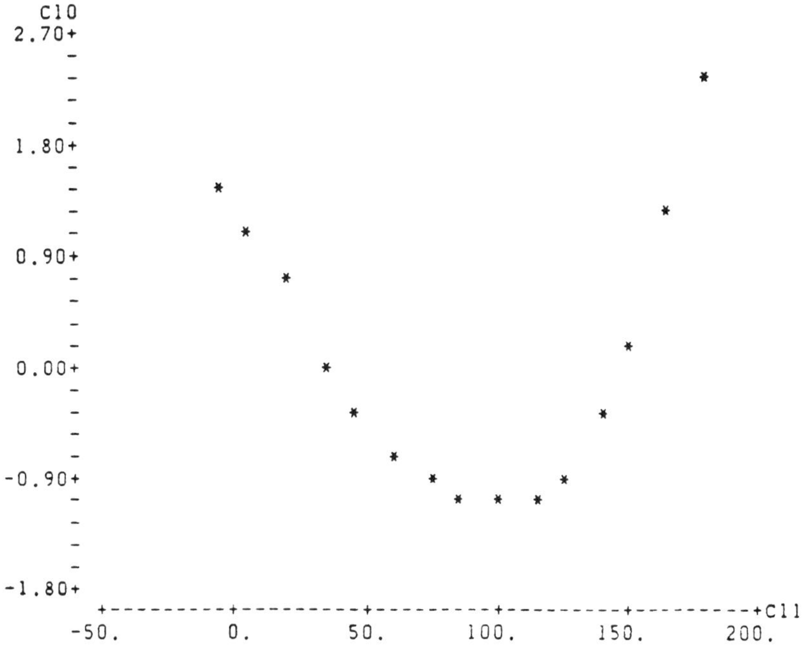


of $e_i$ against $\hat{Y}_i$ is accomplished by

```
PLOT C10 C11
```

and this plot is shown in Display 8.2.4.

Obviously, the residuals are not randomly scattered about the line $e_i = 0$, but show a very strong quadratic relationship. This suggests adding the regressor $X_i^2$ to the model. Thus

```
LET C3 = C2**2
REGR C1 2 C2 C3 C50 C13;
RESI C12.
```

The results of this regression are

$$\hat{Y}_i = \underset{(2.6)}{12.3} + \underset{(0.76)}{2.15}X_i + \underset{(0.046)}{0.687}X_i^2 ; \quad \bar{R}^2 = 0.998.$$

We plot $e_i$ against $Y_i$ by the command

DISPLAY 8.2.5

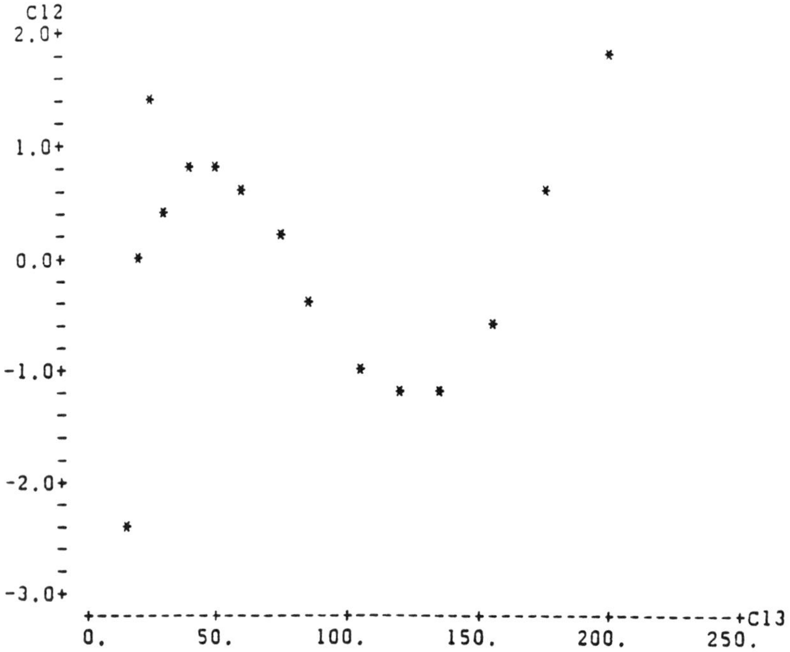


DISPLAY 8.2.6

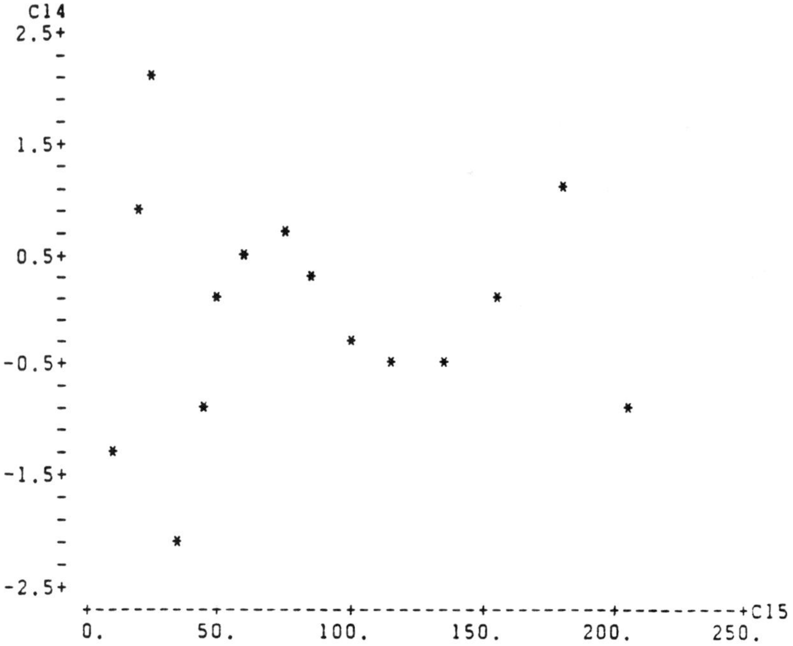

```
PLOT C12 C13
```

and this is shown in Display 8.2.5.

Apart from a couple of points, this plot has definite cubic characteristics. We thus add the variable $X_i^3$ as follows:

```
LET C4 = C2*C3
REGR C1 3 C2 - C4 C50 C15
RESI C14.
```

This regression yields

$$\hat{Y}_i = \underset{(1.37)}{2.43} + \underset{(0.72)}{8.57}X_i - \underset{(0.102)}{0.381}X_i^2 + \underset{(0.0042)}{0.0405}X_i^3\,; \qquad \bar{R}^2 = 0.9997.$$

Again we plot $e_i$ against $\hat{Y}_i$ by the command

```
PLOT C14 C15
```

and this is shown in Display 8.2.6. There is probably still some structure remaining in the residuals, but this will suffice to illustrate the principles involved.

## 8.3 HETEROSCEDASTICITY

A property of the disturbance $u_i$ which has always been assumed so far is that

$$E(u_i^2) = \sigma^2. \tag{8.3.1}$$

This assumption (known as the assumption of <u>homoscedasticity</u>) implies that the conditional variance of $Y_i$ never changes. We refer back to Section 2.1 for a fuller discussion of the assumption. On reading Section 2.3 on the properties of least squares estimators, it is apparent that assumption (8.3.1) plays no part in establishing that OLS estimators are linear and unbiased. However, it is used in proving that these estimators are <u>best</u> in the sense that they are the estimators with minimum variance among all linear, unbiased estimators. We may therefore suspect that if (8.3.1) does not hold, OLS estimators are no longer efficient.

When (8.3.1) is not true, the disturbances are said to be <u>heteroscedastic</u> and (8.3.1) is replaced by

$$E(u_i^2) = \sigma_i^2. \tag{8.3.2}$$

It follows that in this case the conditional variance of $Y_i$ is not constant.

## Occurrences of Heteroscedasticity

Cross-sectional data. Consider the case where cross-sectional data are being used to estimate a regression of savings per household on household income and size of family, in an attempt to derive a marginal propensity to save. For each given family size, one might well expect the variance in savings behavior to be a function of the size of household income. In the cases of households with low incomes, there is little scope to vary savings. As almost all income is required for necessary consumption, only a small proportion is available for savings.

However, in the cases of households with high incomes, there is much more scope for discretionary consumption or savings, depending on the mores of the family concerned. Consequently, it is possible for quite large variations to occur in actual savings between high-income families. Let the regression relationship be

$$S_i = \beta_0 + \beta_1 Y_i + \beta_2 F_i + u_i, \qquad (8.3.3)$$

where $S_i$ is savings per household, $Y_i$ is income per household, and $F_i$ is family size. If we then estimated an ordinary least squares regression equation using data of this type, saved the residuals, and plotted them against household income, we would probably get a graph like Figure 8.3.1, where the residuals tend to fan out as household income increases. This phenomenon of the variance of the disturbance being a function of an exogenous variable (or even several such variables) is a common characteristic of cross-sectional data.

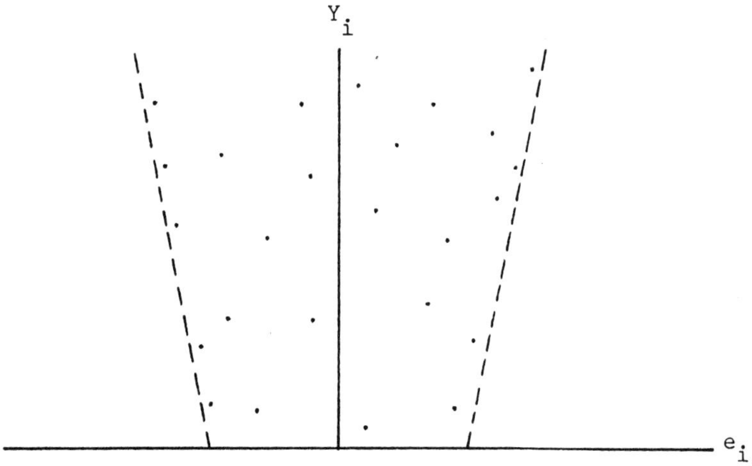


Figure 8.3.1. Variance of residuals positively related to the variable $Y_i$.

Grouped data. Very often the results of large cross-section surveys conducted by government agencies are presented in the form of group averages. This may be done to avoid the problem of publishing large volumes of data, or, more likely, as a means of preserving confidentiality.

Let us suppose, then, that the sample has been divided into g groups, having $N_1, N_2, \ldots, N_g$ observations in the respective groups. We will denote by $X_{ijk}$ the jth observation in the kth group on the variable $X_i$ ($i = 1, 2, \ldots, K$). The same convention will be followed for the dependent variable and the disturbance.

Assume we want to estimate the regression model

$$Y_{jk} = \beta_0 + \beta_1 X_{1jk} + \cdots + \beta_K X_{Kjk} + u_{jk}, \tag{8.3.4}$$

where $E(u_{jk}) = 0$ and $E(u_{jk}^2) = \sigma^2$. If the only data available to us are in the form of group averages, we are unable to estimate (8.3.4) as it stands. Suppose now that we average (8.3.4) over the observations in the kth group. Then

$$\bar{Y}_k = \beta_0 + \beta_1 \bar{X}_{1k} + \cdots + \beta_k \bar{X}_{Kk} + \bar{u}_k, \tag{8.3.5}$$

where

$$\bar{u}_k = \frac{\Sigma_{j=1}^{N_k} u_{jk}}{N_k}. \tag{8.3.6}$$

Because we have observations on the variables in the model (8.3.5), we can regress $\bar{Y}$ on $\bar{X}_1, \ldots, \bar{X}_K$ and estimate the unknown parameters in (8.3.4). However, in view of (8.3.6),

$$E(\bar{u}_k^2) = \frac{\sigma^2}{N_k}, \tag{8.3.7}$$

and the model (8.3.5) thus has a heteroscedastic disturbance unless it happens that $N_1 = N_2 = \cdots = N_g$. More detailed discussions of the problems involved in estimating from grouped data can be found in Haitovsky (1973), Maddala (1977, 268-274), and Kmenta (1971, 322-336).

Aggregated data. It sometimes happens that time-series data are derived from aggregated cross-sectional data, in which the number of elements over which aggregation takes place varies significantly from period to period. With similar notation as for grouped data, we will use $X_{ijt}$ to denote the jth cross-sectional observation on variable $X_i$ in period t. Similarly, $Y_{jt}$ is the jth observation on Y in period t.

Assume that the "true model" is

$$Y_{jt} = \beta_0 + \beta_1 X_{1jt} + \cdots + \beta_K X_{Kjt} + u_{jt}, \tag{8.3.8}$$

where $E(u_{jt}) = 0$ and $E(u_{jt}^2) = \sigma^2$. We will further assume that aggregation takes place over $N_t$ units in period t. Then, in terms of the aggregated time-series data, the model would be

$$Y_t = \beta_0 N_t + \beta_1 X_{1t} + \cdots + \beta_K X_{Kt} + u_t, \tag{8.3.9}$$

where

$$Y_t = \sum_{j=1}^{N_t} Y_{jt},$$

$$X_{kt} = \sum_{j=1}^{N_t} X_{kjt},$$

and

$$u_t = \sum_{j=1}^{N_t} u_{jt}. \tag{8.3.10}$$

The disturbance $u_t$, appropriate to the model (8.3.9) has the following properties, derived from (8.3.10):

$$E(u_t) = \sum_{j=1}^{N_t} E(u_{jt}) = 0,$$

$$E(u_t^2) = N_t \sigma^2. \tag{8.3.11}$$

Thus, by (8.3.11), the time-series model has a heteroscedastic disturbance unless $N_1 = N_2 = \cdots = N_T$, where T is the total number of time periods. Reference is made to Rao and Miller (1971, 101-102 and 118-120) for an example. It is worth noting that when time-series data are derived from national aggregates, heteroscedasticity is not usually a problem because the size of the population varies very little from period to period.

## Effects of Heteroscedasticity

Suppose that the dependent variable $Y_i$ is related to just one regressor variable $X_i$ and is heteroscedastic, as shown in Figure 8.3.2. We will divide the observations into three groups, as shown. Contrast the observations in groups I and III. Clearly, the variation about the true line in group I is much smaller than in group III. That is, there is much more information about the line to be had from group I than group III, and when estimating the line more attention should be paid to the observations having small variances than those with larger variances.

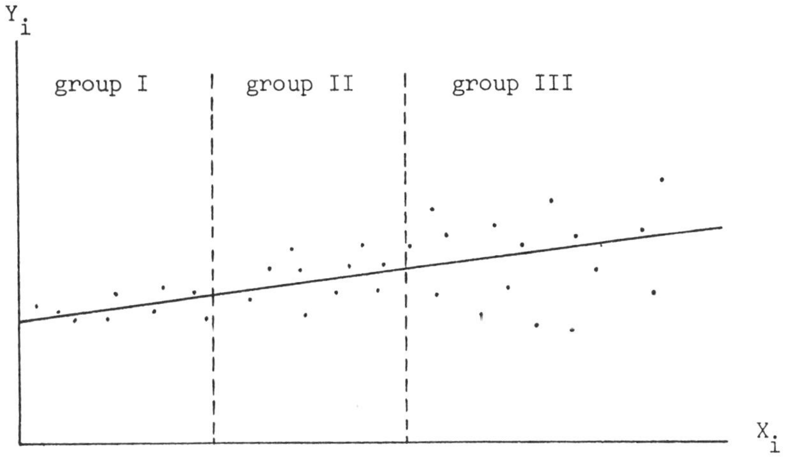


Figure 8.3.2. Heteroscedastic data.

As we have already intimated, when OLS is applied to heteroscedastic data, it is no longer BLUE because it is no longer efficient. The fundamental reason for this is that OLS treats all observations as being of equal importance. In the heteroscedastic situation what is clearly required is a method of weighting the observations according to their relative importance regarding the information they convey about the true line.

Apart from the loss of efficiency, there is another important effect of applying OLS to heteroscedastic data, and that is that the standard errors will be biased, and thus hypothesis tests and confidence intervals based on the t-distribution are likely to be invalid. It can be shown [see, e.g., Kmenta (1986, 278)] that, in the case of the simple regression model, if the variances $\sigma_i^2$ are positively correlated with the regressor $X_i$, the standard errors will be negatively biased. That is, the standard errors will usually be smaller than they should be. This will lead to confidence intervals on the $\beta$'s which are optimistically narrow. Hypothesis tests are also likely to be misleading. Suppose that a test is carried out at the 5 percent level of significance; that means that there is a probability of 0.05 of incorrectly rejecting the null hypothesis. If the standard error is an underestimate, the probability of an incorrect rejection of the null hypothesis will be greater than 0.05. In other words, the level of significance will really be perhaps 10 percent, 20 percent, or even higher, resulting in irrelevant variables appearing to be highly relevant.

It is clearly important to be able to detect the presence of heteroscedasticity and to adjust our estimation procedures accordingly. We will now consider these two aspects but, for convenience, will treat them in reverse order.

Correcting Heteroscedasticity

From an estimation point of view, the problem with heteroscedasticity is that OLS is inefficient—that is, it does not use the sample information to maximum effect. There are at least two ways of approaching this problem. First, we could abandon OLS and search for a new method of estimation that is efficient with heteroscedastic data. Second, we could postulate a model for the heteroscedastic disturbances, and on the basis of this model seek transformations of the data so that with the transformed data OLS is an efficient procedure. We will follow this second strategy.

Let us assume the following model for heteroscedasticity:

$$E(u_i^2) = \sigma^2 Z_i, \tag{8.3.12}$$

where $\sigma^2$ is an unknown parameter and $Z_i$ is a variable on which we have observations. Thus we are assuming that the variance of the disturbance $u_i$ is directly proportional to the variable $Z_i$.[1] This variable is often one of the regressor variables, but need not be.

The full heteroscedastic model may now be written as

$$Y_i = \beta_0 + \beta_1 X_{1i} + \cdots + \beta_K X_{Ki} + u_i, \tag{8.3.13}$$

where

$$E(u_i) = 0, \tag{8.3.14}$$

$$E(u_i^2) = \sigma^2 Z_i, \tag{8.3.15}$$

and

$$E(u_i u_j) = 0 \text{ when } (i \neq j). \tag{8.3.16}$$

Suppose now that we divide equation (8.3.13) by $Z_i^{\frac{1}{2}}$. We obtain

$$\frac{Y_i}{Z_i^{\frac{1}{2}}} = \beta_0 \frac{1}{Z_i^{\frac{1}{2}}} + \beta_1 \frac{X_{1i}}{Z_i^{\frac{1}{2}}} + \cdots + \beta_K \frac{X_{Ki}}{Z_i^{\frac{1}{2}}} + u_i^*,$$

where

$$u_i^* = \frac{u_i}{Z_i^{\frac{1}{2}}}. \tag{8.3.17}$$

If we introduce the notation

$$Y_i^* = \frac{Y_i}{Z_i^{\frac{1}{2}}},$$

$$X_{oi}^* = \frac{1}{Z_i^{\frac{1}{2}}},$$

$$X_{ki}^* = \frac{X_{ki}}{Z_i^{\frac{1}{2}}} \quad (k = 1, 2, \ldots, K),$$

then the transformed model is?

$$Y_i^* = \beta_0 X_{oi}^* + \beta_1 X_{1i}^* + \cdots + \beta_K X_{Ki}^* + u_i^*. \tag{8.3.18}$$

We now claim that if $Y_i^*$ is regressed on the variables $X_{oi}^*$, $X_{1i}^*$, ..., $X_{Ki}^*$ using OLS, we will obtain best, linear, unbiased estimators of the unknown parameters $\beta_0$, $\beta_1$, ..., $\beta_K$. To establish this, all that is necessary is to prove that the disturbance $u_i^*$ of the transformed model, given by (8.3.17), satisfies the classical disturbance assumptions.

(a) $$E(u_i^*) = E\left(\frac{u_i}{Z_i^{\frac{1}{2}}}\right)$$

$$= \frac{1}{Z_i^{\frac{1}{2}}} E(u_i) \quad \text{(because } Z_i \text{ is not stochastic)}$$

$$= 0 \text{ because } E(u_i) = 0.$$

(b) $$E(u_i^{*2}) = E\left(\frac{u_i^2}{Z_i}\right) = \frac{1}{Z_i} E(u_i^2)$$

$$= \frac{1}{Z_i}\sigma^2 Z_i \quad \text{[by (8.3.15)]}$$

$$= \sigma^2.$$

(c) $$E(u_i^* u_j^*) = E\left(\frac{u_i}{Z_i^{\frac{1}{2}}} \frac{u_j}{Z_j^{\frac{1}{2}}}\right)$$

$$= \frac{1}{Z_i^{\frac{1}{2}} Z_j^{\frac{1}{2}}} E(u_i u_j) = 0.$$

From (a), (b), and (c) we have that $u_i^*$ satisfies the classical disturbance assumptions. It follows then that OLS estimates obtained from the transformed model (8.3.18) are BLUE.

It should be noted that when we divide by $Z_i^{\frac{1}{2}}$ we are actually weighting the observations according to their relative importance. Observations corresponding to small values of $Z_i$ have small variance and are therefore relatively important. Weighting inversely by $Z_i^{\frac{1}{2}}$ recognizes this relative importance.

EXAMPLE 1: Grouped Data. In this case

$$\sigma_i^2 = \frac{\sigma^2}{N_i}.$$

For efficient estimation of (8.3.5) we would apply OLS to

$$\sqrt{N_i}\,\bar{Y}_i = \beta_0\sqrt{N_i} + \beta_1\sqrt{N_i}\,\bar{X}_{1i} + \cdots + \beta_K\sqrt{N_i}\,\bar{X}_{Ki} + u_i^*,$$

noting that the transformed model has no intercept.

EXAMPLE 2: Aggregated Data

$$\sigma_t^2 = \sigma^2 N_t.$$

Here, to estimate (8.3.9), we would run the regression

$$\frac{Y_t}{\sqrt{N_t}} = \beta_0\sqrt{N_t} + \frac{\beta_1 X_{1t}}{\sqrt{N_t}} + \cdots + \frac{\beta_K X_{Kt}}{\sqrt{N_t}} + u_t^*.$$

Once again, there is no intercept in the transformed model.

EXAMPLE 3 Suppose that $Z_i = X_{1i}^2$—that is, the variable $Z_i$ is the square of one of the regressor variables, taken without loss of generality to be $X_{1i}$. In this case the transformed model is

$$\frac{Y_i}{X_{1i}} = \beta_0\frac{1}{X_{1i}} + \beta_1 + \beta_2\frac{X_{2i}}{X_{1i}} + \cdots + \beta_K\frac{X_{Ki}}{X_{1i}} + u_i^*$$

and there is an intercept. However, the intercept in this model estimates the coefficient of $X_{1i}$, in the original model, and the coefficient of $(1/X_{1i})$ estimates the original intercept. Beware of this trap!

Finally, it is important to note that when the model is transformed, the dependent variable changes, and it is not valid to assess the effectiveness of the transformation by comparing $R^2$ values from the untransformed and transformed models (see Section 5.2). In order to obtain an $R^2$ value which can be validly compared with that of the untransformed model we may proceed as follows:

(a) Regress $Y_i^*$ on the variables $X_{0i}^*$, $X_{1i}^*$, ..., $X_{Ki}^*$ (with no intercept, except in the case of Example 3 above), and obtain the predictions $\hat{Y}_1^*$, $\hat{Y}_2^*$, ..., $\hat{Y}_N^*$.
(b) Multiply these predictions by the corresponding values of $Z^{\frac{1}{2}}$ to obtain predictions of $Y_i$ denoted by $\hat{\hat{Y}}_i$. That is,

$$\hat{\hat{Y}}_i = \hat{Y}_i^* Z_i^{\frac{1}{2}} \quad (i = 1, 2, \ldots, N).$$

(c) Define $R^{*2} = [\text{corr}(Y_i, \hat{\hat{Y}}_i)]^2$. $R^{*2}$ can be validly compared with $R^2$ from the original model.

## Testing for Heteroscedasticity

The Goldfeld-Quandt test. The only method we have seen so far for detecting heteroscedasticity is an examination of the residuals. This is clearly subjective, and it would be to our advantage to have a statistical test which, as well as detecting heteroscedasticity, would enable us to tell which of a number of transformations does the best job of correcting it. Goldfeld and Quandt (1965) have provided an appropriate procedure.

Let us suppose that we have identified the variable $Z_i$ which possibly is proportional to the variance of $u_i$. That is, we believe that $E(u_i) = \sigma^2 Z_i$. If we were to divide the sample into two equal-sized subsamples according to the values of $Z_i$, we would expect that OLS applied to the two subsamples separately would yield estimates of $\sigma^2$ for which

$$\hat{\sigma}^2(1) < \hat{\sigma}^2(2).$$

Here $\hat{\sigma}^2(1)$ and $\hat{\sigma}^2(2)$ are the estimates from the subsamples corresponding to smaller and larger values of Z, respectively. This intuitively plausible result forms the basis for the following procedure:

(a) Divide the sample into two equal subsamples, each with L observations in such a way that

$$\max\{Z_i : Z_i \in \text{subsample } 1\} \leq \min\{Z_i : Z_i \in \text{subsample } 2\}.$$

If N is odd, the observation corresponding to the median value of $Z_i$ is discarded.
(b) We set up the null and alternative hypotheses as

$$H_0 : \sigma^2(1) = \sigma^2(2) \quad \text{(homoscedasticity)}$$

$$H_1 : \sigma^2(1) < \sigma^2(2) \quad \text{(heteroscedasticity)}$$

(c) Define the F-ratio

Table 8.3.1

| PA | PB | PC | YD | QA |
|---|---|---|---|---|
| 89.20 | 115.32 | 82.61 | 56.14 | 77.17 |
| 91.84 | 100.68 | 72.65 | 61.00 | 75.06 |
| 89.59 | 110.39 | 82.49 | 66.67 | 85.12 |
| 103.29 | 110.23 | 82.68 | 77.46 | 79.01 |
| 94.94 | 106.22 | 85.16 | 67.63 | 76.36 |
| 92.28 | 126.43 | 81.24 | 72.89 | 120.71 |
| 77.35 | 113.29 | 76.79 | 63.08 | 124.56 |
| 87.86 | 100.47 | 89.26 | 96.37 | 153.83 |
| 103.78 | 109.79 | 85.93 | 76.60 | 76.91 |
| 102.66 | 109.77 | 84.97 | 88.68 | 116.37 |
| 92.40 | 100.39 | 96.76 | 93.13 | 99.20 |
| 96.76 | 92.61 | 95.77 | 94.94 | 102.29 |
| 104.66 | 102.79 | 103.63 | 89.71 | 68.77 |
| 102.63 | 96.61 | 90.18 | 108.75 | 118.07 |
| 95.27 | 88.33 | 121.82 | 119.83 | 136.35 |
| 101.79 | 96.13 | 105.43 | 108.34 | 95.16 |
| 109.36 | 68.65 | 112.00 | 124.73 | 63.64 |
| 121.95 | 76.83 | 116.25 | 117.36 | 56.09 |
| 99.80 | 101.13 | 106.70 | 111.02 | 98.09 |
| 108.20 | 105.01 | 116.26 | 104.81 | 89.70 |
| 102.34 | 98.49 | 110.02 | 127.05 | 120.35 |
| 102.31 | 96.87 | 108.37 | 108.97 | 97.10 |
| 108.13 | 88.64 | 116.91 | 124.09 | 89.38 |
| 108.22 | 93.36 | 104.53 | 127.16 | 103.53 |
| 116.67 | 82.41 | 131.65 | 137.58 | 75.21 |
| 105.20 | 92.34 | 121.53 | 131.87 | 116.46 |
| 108.24 | 88.86 | 113.74 | 133.52 | 152.06 |
| 88.20 | 96.23 | 113.66 | 131.90 | 157.91 |
| 126.34 | 83.25 | 135.24 | 137.69 | 75.98 |
| 113.98 | 90.85 | 144.28 | 144.34 | 92.72 |

$$F = \frac{\hat{\sigma}^2(2)}{\hat{\sigma}^2(1)} = \frac{RSS(2)/(L - K - 1)}{RSS(1)/(L - K - 1)} = \frac{RSS(2)}{RSS(1)}.$$

If $H_0$ is true, then F has an F distribution with parameters L - K - 1, L - K - 1.

(d) Reject $H_0$ if F is greater than the critical value of $F_{L-K-1, L-K-1}$. If, using untransformed data, $H_0$ is rejected, the model would be transformed and a second application of the Goldfeld-Quandt test applied. Hopefully, the model using transformed data would show no evidence of heteroscedasticity.

In practice, we may want to try several transforming variables. Other things being equal, we would select that variable which yields the smallest F value after transformation.

EXAMPLE The data in Table 8.3.1 are 30 cross-section observations on the five variables

PA price of commodity A
PB price of commodity B
PC price of commodity C
YD disposable income
QA quantity demanded of commodity A

We will attempt to estimate a constant elasticity of demand model. We define

$X_1$ = ln PA,

$X_2$ = ln PB,

$X_3$ = ln PC,

$X_4$ = ln YD,

Y = ln QA

and therefore estimate a model of the form

$$Y_i = \beta_0 + \beta_1 X_{1i} + \beta_2 X_{2i} + \beta_3 X_{3i} + \beta_4 X_{4i} + u_i.$$

The parameter $\beta_1$ is the own-price elasticity, $\beta_2$ and $\beta_3$ are cross-price elasticities of commodity A with commodities B and C, respectively, and $\beta_4$ is the income elasticity. It is very likely that heteroscedasticity, related to the variable YD, will be a problem with these data.

Let us assume that PA, PB, PC, YD, and QA are stored in columns C1 to C5, respectively. To apply the Goldfeld-Quandt test, we will divide the data into two equal groups of observations according to the value of YD. Thus, as a first step, we order the data using the command

```
SORT C4 C1 - C3 C5 C4 C1 - C3 C5
```

Also, as the variable YD (in C4) will be needed subsequently for transforming the data, we copy it into another column, say C40, by

```
COPY C4 C40
```

The variables $X_1$, $X_2$, $X_3$, $X_4$, and Y are now created:

```
LET C1 = LOGE(C1)
LET C2 = LOGE(C2)
LET C3 = LOGE(C3)
LET C4 = LOGE(C4)
LET C5 = LOGE(C5)
```

The initial estimated model is found by

```
REGR C5 4 C1 - C4
```

The output from this regression is shown in Display 8.3.1.

The data may be divided into two equal groups using the COPY command.

```
COPY C1-C5 C6-C10;
USE 1:15.
COPY C1-C5 C11-C15;
USE 16:30.
```

Observations 1 to 15 are now in C6 to C10 and observations 16 to 30 in C11 to C15. We will also need observations 1 to 15 and 16 to 30 of the variable YD (in C40) to be created as separate columns. Thus

```
COPY C40 C41;
USE 1:15.
COPY C40 C42;
USE 16:30.
```

To perform the Goldfeld-Quandt test on the untransformed data, we carry out the regressions on the two subsets of observations 1 to 15 and 16 to 30, as follows:

```
REGR C10 4 C6 - C9
REGR C15 4 C11 - C14
```

These regressions yield RSS(1) = 0.06647 and RSS(2) = 0.2380. Thus

$$F = \frac{0.02380}{0.006647} = 3.58.$$

DISPLAY 8.3.1

```
THE REGRESSION EQUATION IS
Y =     5.14 -  2.04 X1 +   1.17 X2
    - 0.803 X3 +  1.57 X4
                                                ST. DEV.      T-RATIO =
       COLUMN      COEFFICIENT                  OF COEF.      COEF/S.D.
       --               5.140                    2.202           2.33
X1     C1             -2.0422                   0.3000          -6.81
X2     C2              1.1697                   0.2660           4.40
X3     C3             -0.8034                   0.3050          -2.63
X4     C4              1.5659                   0.1966           7.96

THE ST. DEV. OF Y ABOUT REGRESSION LINE IS
S = 0.1169
WITH (  30- 5) =  25 DEGREES OF FREEDOM
R-SQUARED = 83.4 PERCENT
R-SQUARED = 80.8 PERCENT, ADJUSTED FOR D.F.
ANALYSIS OF VARIANCE
 DUE TO     DF            SS        MS=SS/DF


REGRESSION   4        1.71921        0.42980
RESIDUAL    25        0.34143        0.01366
TOTAL       29        2.06063
```



The critical value for $F_{10,10}$ at the 5 percent level of significance is 2.97, so there appears to be heteroscedasticity, related to the variable YD.

We will investigate two models for heteroscedasticity:

(a) $\sigma_i^2 = kYD_i^2$, and

(b) $\sigma_i^2 = kYD_i$.

To assess whether either of these transformations effectively corrects the heteroscedasticity, we will apply Goldfeld-Quandt tests to the two transformed models,

$$\frac{Y_i}{YD_i} = \beta_0 \frac{1}{YD_i} + \beta_1 \frac{X_{1i}}{YD_i} + \cdots + \beta_4 \frac{X_{4i}}{YD_i} + u_i^*$$

and

$$\frac{Y_i}{YD_i^{\frac{1}{2}}} = \beta_0 \frac{1}{YD_i^{\frac{1}{2}}} + \beta_1 \frac{X_{1i}}{YD_i^{\frac{1}{2}}} + \cdots + \beta_4 \frac{X_{4i}}{YD_i^{\frac{1}{2}}} + u_i^*.$$

The variables associated with the intercept $\beta_0$ will be stored in C48 and C49 by

```
LET C48 = 1/C41
LET C49 = 1/C42
```

The transformed variables for the first of these models are now set up as follows:

```
LET C16 = C6/C41
LET C17 = C7/C41
LET C18 = C8/C41
LET C19 = C9/C41

LET C21 = C11/C42
LET C22 = C12/C42
LET C23 = C13/C42
LET C24 = C14/C42
LET C25 = C15/C42
```

The two regressions on the subsets of data are performed by

```
REGR C20 5 C48 C16-C19;
NOCO.
REGR C25 5 C49 C21-C24;
NOCO.
```

These regressions yield RSS(1) = 0.00000094 and RSS(2) = 0.00000148, so

$$F = \frac{0.00000148}{0.00000094} = 1.57.$$

Thus there is no significant heteroscedasticity when the data are transformed using assumption (a).

The second transformed model is tested in the same way, but first the variable $Y_i^{\frac{1}{2}}$ is created as follows:

```
LET C41 = SQRT(C41)
LET C42 = SQRT(C42)
```

Creation of the appropriate variables and the regression commands are now identical to the preceding set and will not be reproduced here. These regressions yield RSS(1) = 0.000784 and RSS(2) = 0.001870, so that F = 2.38.

Once again there is no significant heteroscedasticity with the transformed model. As the F value for the first transformed model is smaller than for the

DISPLAY 8.3.2

```
THE REGRESSION EQUATION IS
Y = +  5.69 X1 -  2.13 X2 +  1.16 X3
    - 0.819 X4 +  1.56 X5
                                               ST. DEV.     T-RATIO =
       COLUMN      COEFFICIENT                 OF COEF.     COEF/S.D.
NOCONSTANT
X1     C50              5.692                   1.811          3.14
X2     C6             -2.1334                  0.2471         -8.63
X3     C7              1.1620                  0.2288          5.08
X4     C8             -0.8194                  0.2549         -3.21
X5     C9              1.5605                  0.1558         10.01

THE ST. DEV. OF Y ABOUT REGRESSION LINE IS
S = 0.001040
WITH (  30- 5) =  25 DEGREES OF FREEDOM
ANALYSIS OF VARIANCE
 DUE TO       DF           SS          MS=SS/DF
REGRESSION     5    0.07293186      0.01458637
RESIDUAL      25    0.00002702      0.00000108
TOTAL         30    0.07295889
```

Reprinted with permission of Minitab, Inc.

second, we adopt (a) as our model of heteroscedasticity. Finally, the estimated transformed model on the full set of 30 observations is accomplished by the commands

```
LET C1 = C1/C40
LET C2 = C2/C40
LET C3 = C3/C40
LET C4 = C4/C40
LET C5 = C5/C40
LET C40 = 1/C40

REGR C5 5 C40 C1-C4;
NOCO.
```

The output from this regression is shown in Display 8.3.2.

Comparing the results in Display 8.3.1 (prior to correction for heteroscedasticity) and those in Display 8.3.2, we note that the $\beta$ estimates have not changed very much. However, the standard errors are smaller in the transformed model, indicating a gain in efficiency. Finally, the results suggest that the demand for commodity A is very price-elastic ($\hat{\beta}_1$ = -2.13), commodity B is a substitute ($\hat{\beta}_2$ positive, and commodity C is complementary ($\hat{\beta}_3$ negative). As the demand for commodity A is elastic with respect to income ($\hat{\beta}_4$ = 1.56), it is probably a luxury good.

The Breusch-Pagan test. The Goldfeld-Quandt test is based on the assumption that the heteroscedasticity is related to a single, identifiable variable, which gives a means of ordering the data. However, if the heteroscedasticity is related to several variables that do not all move together, it will be impossible to achieve a unique ordering. Breusch and Pagan (1979) have devised a test which does not depend on ordering the data.

This test is an application of the Lagrange multiplier principle, which is beyond the scope of our present discussion, although it is discussed in Chapter 14. Accordingly, we will present the prescription without attempting to give its underlying rationale.

Let us assume that our model for heteroscedasticity is

$$\sigma_i^2 = h(v_i), \tag{8.3.19}$$

where

$$v_i = \alpha_0 + \alpha_1 Z_{1i} + \cdots + \alpha_S Z_{Si}, \tag{8.3.20}$$

and h is any function that does not depend on i. The function $h(\cdot)$ does not have to be specified. The null hypothesis is that there is no heteroscedasticity, which is equivalent to

$$H_0: \alpha_1 = \alpha_2 = \cdots = \alpha_S = 0.$$

If $H_0$ is true, then $v_i = \alpha_0$ (a constant), implying that $\sigma_i^2 = h(\alpha_0) = \text{constant}$ ($= \sigma^2$, say), so by definition, $u_i$ is homoscedastic. The Breusch-Pagan procedure requires two regressions and can be summarized as follows:

(i) Regress Y on $X_1, X_2, \ldots, X_K$ and save the residuals e.
(ii) Define $\tilde{\sigma}^2 = RSS/N = \Sigma e_i^2/N$.
(iii) Regress $e^2/\tilde{\sigma}^2$ on $Z_1, Z_2, \ldots, Z_S$.
(iv) Define the statistic LM by

$$LM = \frac{ESS}{2}, \tag{8.3.21}$$

where ESS is the explained sum of squares from the regression (iii).

THEOREM 8.3.1 If $H_0$ is true, then

$$LM \sim \chi_S^2.$$

The null hypothesis is rejected if LM is greater than the critical value of the $\chi_S^2$ distribution.

EXAMPLE We will once again take the data of Table 8.3.1 and assume a model for heteroscedasticity in which

DISPLAY 8.3.3

```
MTB > LET C1=LOGE(C1)
MTB > LET C2=LOGE(C2)
MTB > LET C3=LOGE(C3)
MTB > LET C4=LOGE(C4)
MTB > LET C5=LOGE(C5)
MTB > REGR C5 4 C1-C4;
SUBC> RESI C6.

The regression equation is
C5 = 5.14 - 2.04 C1 + 1.17 C2 - 0.803 C3 + 1.57 C4

Predictor        Coef          Stdev      t-ratio
Constant        5.140          2.202         2.33
C1            -2.0422         0.3000        -6.81
C2             1.1697         0.2660         4.40
C3            -0.8034         0.3050        -2.63
C4             1.5659         0.1966         7.96

s = 0.1169       R-sq = 83.4%      R-sq(adj) = 80.8%

Analysis of Variance

SOURCE         DF          SS           MS
Regression      4     1.71921      0.42980
Error          25     0.34143      0.01366
Total          29     2.06063

  MTB > LET K1=SSQ(C6)/30
  MTB > LET C7=C6*C6/K1
  MTB > REGR C7 1 C4

  The regression equation is
  C7 = - 8.20 + 2.00 C4

  Predictor        Coef          Stdev      t-ratio
  Constant       -8.198          5.833        -1.41
  C4              1.998          1.265         1.58
 1

  s = 1.900        R-sq = 8.2%       R-sq(adj) = 4.9%

  Analysis of Variance

  SOURCE         DF          SS           MS
  Regression      1       9.012        9.012
  Error          28     101.105        3.611
  Total          29     110.117
```

$$\sigma_i^2 = h(\alpha_0 + \alpha_1 \ln YD_i).$$

Notice that this specification is more general than the models tested earlier with the Goldfeld-Quandt test. If $h(\cdot) = \exp(\cdot)$, then

$$\sigma_i^2 = \exp(\alpha_0 + \alpha_1 \ln YD_i)$$

$$= [\exp(\alpha_0)] \exp(\alpha_1 \ln YD_i)$$

That is,

$$\sigma_i^2 = kYD_i^{\alpha_1},$$

where

$$k = \exp(\alpha_0).$$

The null hypothesis $H_0 : \alpha_1 = 0$ is equivalent to $\sigma_i^2 = k$. The test is now conducted as follows:

(i) Regress ln QA on ln PA, ln PB, ln PC, and ln YD and save the residuals;
(ii) Compute $\tilde{\sigma}^2 = RSS/N$ (here $N = 30$);
(iii) Regress $e^2/\tilde{\sigma}^2$ on ln $YD_i$.

The results are shown in Display 8.3.3. In particular, from regression (iii),

$$ESS = 9.012$$

(denoted by "regression SS" in the output). Therefore

$$LM = \frac{9.012}{2} = 4.51.$$

As the critical value for $\chi_1^2$ at the 5 percent level is 3.84, $H_0$ is rejected. There is evidence that heteroscedasticity is present.

## 8.4 AUTOCORRELATION

The final assumption about the disturbance that has always been made is that

$$E(u_i u_j) = 0 \quad \text{when } j \neq i. \tag{8.4.1}$$

This means that the value of the disturbance at different times (time-series data) or for different elements of the sample (cross-sectional data) are uncorrelated. One disturbance exercises no influence over any other disturbance. This assumption is usually perfectly reasonable when dealing with cross-sectional data. Consider again the example of family savings. If one particular family's saving is "disturbed" by, for example, a car accident, this probably will not affect the savings of other families in the random sample.

When dealing with time-series data,[3] however, the assumption is not nearly as plausible. The disturbance $u_t$ may be regarded as summarizing all the influences on $Y_t$ not accounted for by the regressor variables $X_{1t}$, ..., $X_{Kt}$. As successive values of economic variables are nearly always highly correlated, it is reasonable to assume that the disturbances will also be correlated. Disturbances for which (8.4.1) does not hold are said to be <u>autocorrelated</u> or <u>serially correlated</u>.

When OLS is applied to a model with autocorrelated disturbance, there are two main effects:

(a) There is a loss of efficiency. OLS implicitly assumes that disturbances are independent. If however, disturbances are correlated, then previous values of the disturbance have some information to convey about the current disturbance. If this information is ignored, it is clear that the sample data are not being used with maximum efficiency. Once again, reference to Section 2.3 shows that OLS estimators will still be unbiased.

(b) If there is <u>positive</u> autocorrelation (and this is nearly always the case when dealing with economic data), the standard errors of the $\hat{\beta}$'s will be biased downward. In the results from a particular sample, this is likely to show up in confidence intervals that are much narrower than they should be and values of $R^2$ that give an overly optimistic impression of the worth of the model for prediction. Parameter estimates associated with irrelevant variables may be highly significant. This feature has led Pierce (1977) to remark that probably many nonexistent economic relationships have been "discovered"—purely because the effects of autocorrelation have been ignored. A very useful exposition of these effects is given in Pindyck and Rubinfeld (1976, 106-108).

As in the case of heteroscedasticity, it is therefore important to be able to detect autocorrelation and adjust our procedures to take it into account. In principle, we achieve this in the same way as with heteroscedasticity. A model for autocorrelated disturbances is postulated, and on the basis of this model the variables are transformed in such a way that OLS applied to the transformed variables will yield estimates that are BLUE.

## A Model for Autocorrelated Disturbances: The First-Order Markov Process

We will assume that the disturbance $u_t$ satisfies a relationship of the form

$$u_t = \rho u_{t-1} + v_t, \tag{8.4.2}$$

where

(a) $\rho$ is a constant such that $-1 < \rho < 1$;
(b) $v_t$ is a random variable having the properties[4]

(i) $E(v_t) = 0$,

(ii) $E(v_t^2) = \sigma_v^2$ (constant),

(iii) $E(v_t v_s) = 0 \quad (s \neq t)$.

(c) $E(v_t u_{t-s}) = 0$ for all s and t. That is, $v_t$ is uncorrelated with all previous $u_t$'s.

The model (8.4.2), where $\rho$ and $v_t$ satisfy (a), (b), and (c), is known as the first-order Markov process or the first-order autoregressive process. We will now investigate the properties of this process and see why it can be used as one model for autocorrelated disturbances. From (8.4.2),

$$u_t = \rho u_{t-1} + v_t = \rho(\rho u_{t-2} + v_{t-1}) + v_t$$

$$= \rho^2 u_{t-2} + (v_t + \rho v_{t-1}). \tag{8.4.3}$$

Using (8.4.2) again, to eliminate $u_{t-2}$, we have

$$u_t = \rho^2(\rho u_{t-3} + v_{t-2}) + (v_t + \rho v_{t-1})$$

$$= \rho^3 u_{t-3} + (v_t + \rho v_{t-1} + \rho^2 v_{t-2}). \tag{8.4.4}$$

Comparing (8.4.3) and (8.4.4) the emerging pattern is apparent and we can see that successive substitutions of (8.4.2) would yield

$$u_t = \rho^s u_{t-s} + \sum_{j=0}^{s-1} \rho^j v_{t-j}. \tag{8.4.5}$$

As $s \to \infty$, because $-1 < \rho < 1$ by assumption, $\rho^s \to 0$, and we have

$$u_t = \sum_{j=0}^{\infty} \rho^j v_{t-j}. \tag{8.4.6}$$

Thus the current disturbance $u_t$ is a function of all previous observations on the random variable $v_t$. The relationship (8.4.6) can be used to derive some properties of $u_t$. First,

$$E(u_t) = E\left(\sum_{j=0}^{\infty} \rho^j v_{t-j}\right) = \sum_{j=0}^{\infty} \rho^j E(v_{t-j})$$

$$= 0 \tag{8.4.7}$$

because $E(v_{t-j}) = 0$ for all t and j. Second,[5]

$$\text{var}(u_t) = \text{var}\left(\sum_{j=0}^{\infty} \rho^j v_{t-j}\right)$$

$$= \sum_{j=0}^{\infty} \rho^{2j} \text{var}(v_{t-j})$$

$$= \sigma_v^2 \sum_{j=0}^{\infty} \rho^{2j}$$

because $\text{var}(v_{t-j}) = E(v_{t-j}^2) = \sigma_v^2$ for all t and j. Now $\Sigma_{j=0}^{\infty} \rho^{2j}$ is the infinite sum of a geometrical progression. Because of the assumption that $-1 < \rho < 1$, this sum is finite and

$$\sum_{j=0}^{\infty} \rho^{2j} = \frac{1}{1 - \rho^2}.$$

Thus

$$\text{var}(u_t) = E(u_t^2) = \frac{\sigma_v^2}{1 - \rho^2}. \tag{8.4.8}$$

It follows that because this variance is not a function of t, $u_t$ is homoscedastic. If (8.4.2) is an adequate model for the disturbance, we need not worry about heteroscedasticity.

We have seen how the assumption $-1 < \rho < 1$ has been used above on two occasions. Is this an arbitrary assumption used purely to make the mathematics "come out"? Referring back to (8.4.5), it is clear that if $|\rho| > 1$, the effect on $u_t$ of past observations of both u and v would increase the further back in time we go. This is obviously quite unacceptable, as we would expect these effects to diminish over time. The assumption on $\rho$ implies this diminishing impact of past values.

Let us consider the relationship between different values of the disturbance $u_t$. From (8.4.5)

$$E(u_t u_{t-s}) = E\left[\rho^s u_{t-s}^2 + \left(\sum_{j=0}^{s-1} \rho^j v_{t-j}\right) u_{t-s}\right]$$

$$= \rho^s \operatorname{var}(u_{t-s}) + \sum_{j=0}^{s-1} \rho^j E(v_{t-j}u_{t-s}).$$

In the expression above, because $0 \leq j \leq s - 1$, $u_{t-s}$ occurs <u>before</u> $v_{t-j}$ and so, by assumption (c), $E(v_{t-j}u_{t-s}) = 0$ for $0 \leq j \leq s - 1$. Thus

$$E(u_t u_{t-s}) = \rho^s \operatorname{var}(u_{t-s}). \tag{8.4.9}$$

Also, because $u_t$ is homoscedastic,

$$\operatorname{var}(u_{t-s}) = \operatorname{var}(u_t) = \sigma_u^2$$

and it follows that

$$\operatorname{var}(u_{t-s}) = [\operatorname{var}(u_t) \operatorname{var}(u_{t-s})]^{\frac{1}{2}}.$$

We can now write equation (8.4.9) in the form

$$\rho^s = \frac{E(u_t u_{t-s})}{[\operatorname{var}(u_t) \operatorname{var}(u_{t-s})]^{\frac{1}{2}}}.$$

That is, the correlation between $u_t$ and $u_{t-s}$ is equal to $\rho^s$. By setting $s = 1$, we have

$$\rho = \operatorname{corr}\{u_t, u_{t-1}\}, \tag{8.4.10}$$

and $\rho$ is seen to be simply the correlation between <u>successive</u> disturbances.

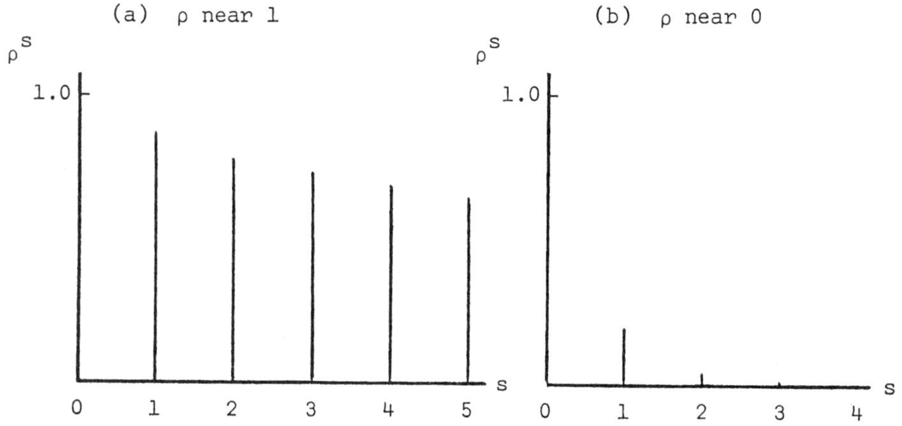


Figure 8.4.1. $\rho^s$ as a function of s when $\rho > 0$.

Table 8.4.1

| s | $\rho = 0.99$ | $\rho = 0.9$ | $\rho - 0.8$ | $\rho = 0.5$ | $\rho = 0.2$ |
|---|---|---|---|---|---|
| 0 | 1 | 1 | 1 | 1 | 1 |
| 1 | 0.99 | 0.9 | 0.8 | 0.5 | 0.2 |
| 2 | 0.98 | 0.81 | 0.64 | 0.25 | 0.04 |
| 3 | 0.97 | 0.73 | 0.51 | 0.13 | 0.01 |
| 4 | 0.96 | 0.66 | 0.41 | 0.06 | 0.002 |
| 5 | 0.95 | 0.59 | 0.33 | 0.03 | 0.000 |

We note that this correlation does not depend on t. It is useful to consider a graph of the correlation between $u_t$ and $u_{t-s}$ separately for the cases of $\rho$ positive and negative.

Case I: $\rho > 0$. See Figure 8.4.1. Table 8.4.1 shows for particular values

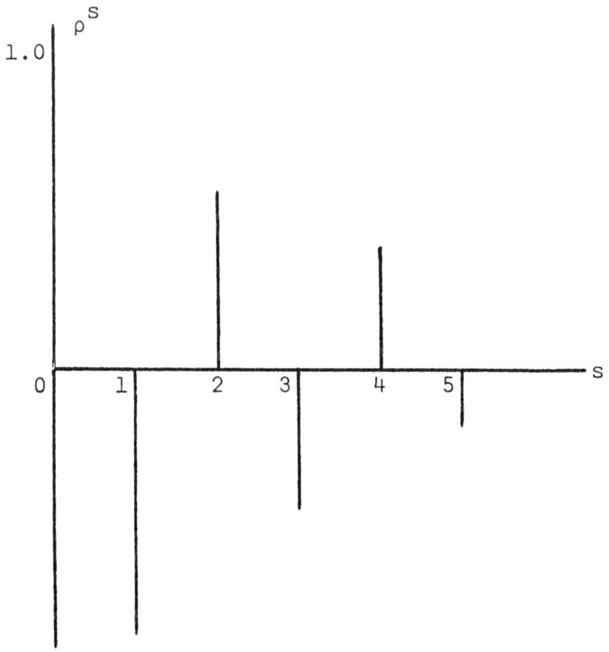


Figure 8.4.2. $\rho^s$ as a function of s when $\rho < 0$.

of $\rho$ how strongly disturbances are related. For example, when disturbances are four time periods apart ($s = 4$), their correlation is 0.96 when $\rho = 0.99$ but only 0.002 when $\rho = 0.2$. Clearly, the higher the value of $\rho$, the longer the effect of a particular disturbance will "linger on."

Case II: $\rho < 0$. See Figure 8.4.2. In this case, the same general remarks apply, except that when s is odd, the correlation is negative, and it is positive when s is even.

## The Effects of Autocorrelation

We have already stated the two main effects of autocorrelation, without providing any justification. Using the first-order Markov process, we will now consider the implications of autocorrelation on OLS. We stress at this point that the first-order Markov process is only one possible model for autocorrelated disturbances, and while our conclusions on the effects of autocorrelation will be based on this model, general conclusions are unlikely to be very different if another model were to be used.

Let us consider just the simple regression model

$$Y_t = \beta_0 + \beta_1 X_t + u_t,$$

where

$$u_t = \rho u_{t-1} + v_t.$$

Let $\tilde{\beta}_1$ be any linear estimator of $\beta_1$. Then

$$\tilde{\beta}_1 = \Sigma a_t Y_t = \Sigma a_t(\beta_0 + \beta_1 X_t + u_t)$$

$$= \beta_0 \Sigma a_t + \beta_1 \Sigma a_t X_t + \Sigma a_t u_t.$$

Thus

$$E(\tilde{\beta}_1) = \beta_0 \Sigma a_t + \beta_1 \Sigma a_t X_t.$$

Thus in order that $\tilde{\beta}_1$ be unbiased, that is, $E(\tilde{\beta}_1) = \beta_1$, it is necessary that the numbers $a_t$ satisfy the two restrictions

$$\Sigma a_t = 0, \quad \Sigma a_t X_t = 1. \tag{8.4.11}$$

Turning now to the variance of $\tilde{\beta}_1$, it can be shown after some algebra that

$$\text{var}(\tilde{\beta}_1) = \sigma_u^2 \sum_t a_t^2 + 2\sigma_u^2 \sum_t \sum_s a_t a_{t-s} \rho^s. \tag{8.4.12}$$

Table 8.4.2 Variance of Efficient Method/Variance OLS

| $\rho$ | Ratio | $\rho$ | Ratio |
|---|---|---|---|
| 0.0 | 1.0 | 0.6 | 0.56 |
| 0.1 | 0.99 | 0.7 | 0.47 |
| 0.2 | 0.94 | 0.8 | 0.35 |
| 0.3 | 0.83 | 0.9 | 0.31 |
| 0.4 | 0.79 | 0.99 | 0.15 |
| 0.5 | 0.74 | | |

To obtain the best linear unbiased estimator of $\beta_1$, we will choose the numbers $a_t$ so that var($\beta_1$) given by (8.4.12) is minimized subject to the constraints (8.4.11).[6] Without actually carrying this out, it is fairly obvious that the optimal $\tilde{\beta}_1$ must involve $\rho$. Now the OLS estimator $\hat{\beta}_1$ is given by

$$\hat{\beta}_1 = \frac{\Sigma xy}{\Sigma x^2}$$

which certainly does not involve $\rho$ and is therefore not BLUE. The larger the absolute value of $\rho$, the greater will be the loss of efficiency incurred in using OLS.

A simulation experiment by Griliches and Rao (1969) yielded the results shown in Table 8.4.2 for samples of size 20. For example, if $\rho = 0.8$, the variance using OLS is almost three times as large as the variance of the efficient method.

A potentially more serious problem than loss of efficiency is that if $\rho$ is positive, the estimates of the variances of the $\hat{\beta}$'s obtained using OLS may be severe underestimates of the true variances. We will again assume the simple regression model together with the first-order Markov process for the disturbances. Also, to facilitate the algebra we will assume that the regressor variable $X_t$ can itself also be described by a first-order Markov scheme.[7] That is,

$$X_t = \lambda X_{t-1} + \epsilon_t.$$

The variance of the OLS estimator $\hat{\beta}_1$ is given by (see Rao and Miller, 1976, 69)

$$\text{var}(\hat{\beta}_1) = \frac{\sigma^2}{\Sigma x^2}\,\frac{1 + \rho\lambda}{1 - \rho\lambda}.$$

If the value of $\lambda$ is close to 1, then

$$\operatorname{var}(\hat{\beta}_1) = \frac{\sigma^2}{\Sigma x^2}\,\frac{1+\rho}{1-\rho}, \text{ approximately.} \tag{8.4.13}$$

When using OLS the variance of $\hat{\beta}_1$ is estimated by $\hat{\sigma}^2/\Sigma x^2$, where $\hat{\sigma}^2 = RSS/(N - 2)$.

This estimator has two possible defects. First, the factor $(1 + \rho)/(1 - \rho)$ in (8.4.13) has not been taken into account. Second, the estimator $\hat{\sigma}^2$ has been shown to be unbiased when $u_t$ is not autocorrelated. There is the possibility that it is biased when there is autocorrelation in the disturbance. We will consider these two effects separately.

(a) If $\rho > 0$, then $(1 + \rho)/(1 - \rho) > 1$. Thus, omission of this factor will lead to an underestimate of $\operatorname{var}(\hat{\beta}_1)$. This underestimate could be very serious. For example, if $\rho = 0.9$, then $(1 + \rho)/(1 - \rho) = 19$, and the OLS estimate of the variance would only be 1/19 as large as it should be.

(b) As pointed out, when OLS is used, $\sigma^2$ is estimated by $\hat{\sigma}^2 = RSS/(N - 2)$. It can be shown that

$$E(\hat{\sigma}^2) = \sigma^2\left[1 - \frac{2\rho}{(N-1)(1-\rho)}\right].$$

If $\rho > 0$, $\hat{\sigma}^2$ also tends to underestimate $\sigma^2$, and the smaller N, the more pronounced will be the underestimate. For example, if $N = 20$ and $\rho = 0.9$,

$$E(\hat{\sigma}^2) = 0.05\sigma^2.$$

On average (for this combination of N and $\rho$), $\hat{\sigma}^2$ would only be 1/20 of $\sigma^2$.

We may show the combined effect of both of these underestimates as follows: Setting $N = 20$ and $\rho = 0.9$ gives us

$$\begin{aligned} E(S^2_{\hat{\beta}_1}) &= \frac{E(\hat{\sigma}^2)}{\Sigma x^2} = 0.05\left(\frac{\sigma^2}{\Sigma x^2}\right) \\ &= 0.05\left(\frac{1-\rho}{1+\rho}\right)\left(\frac{\sigma^2}{\Sigma x^2}\,\frac{1+\rho}{1-\rho}\right) \\ &= \frac{0.05}{19}\left(\frac{\sigma^2}{\Sigma x^2}\,\frac{1+\rho}{1-\rho}\right) = 0.003 \operatorname{var}(\hat{\beta}_1). \end{aligned}$$

With this choice of parameters, the standard error of $\hat{\beta}_1$ is likely to be only $(0.003)^{\frac{1}{2}} = 0.05$ as large as it should be. The general principle that emerges is that if high positive autocorrelation is combined with small sample size, the standard errors of the $\hat{\beta}$'s are likely to be severely underestimated.

## Analysis of Residuals

Given the likely occurrence of autocorrelation in time-series data, it is important to be able to detect its presence. A simple analysis of the residuals may reveal autocorrelation. We outline two procedures at this point.

First, a simple plot of the residuals against time may be useful. If the autocorrelation in the disturbances is reflected in the residuals, the residual plot will not exhibit the completely random character which should normally be expected. Positive autocorrelation would show up in a smooth plot—that is, changes from positive to negative values (and vice-versa) take place less often. Negative autocorrelation, on the other hand, may tend to produce residuals with a sawtooth pattern—positive, negative, positive, and so on.

A more reliable method of observing these characteristics is through a plot of $e_t$ against $e_{t-1}$. In the case of positive autocorrelation, values of $e_{t-1}$ will tend to be associated with values of $e_t$ having the same sign. This will give rise to a scatter diagram with a concentration of points in the first and third quadrants. Negative autocorrelation may be detected by a concentration in the second and fourth quadrants. We should note that while scatter diagrams exhibiting the characteristics listed above strongly indicate autocorrelation, a diagram with the points evenly distributed in the four quadrants does not mean that autocorrelation is not present.

The Durbin-Watson test. The graphical methods outlined above may provide some visual evidence for the existence of autocorrelation. However, this approach falls short of a rigorous testing procedure. We now turn to a test of autocorrelation, proposed by Durbin and Watson (1950, 1951). This test is based on the Durbin-Watson statistic d, defined by

$$d = \frac{\Sigma_{t=2}^{N} (e_t - e_{t-1})^2}{\Sigma_{t=1}^{N} e_t^2}. \tag{8.4.14}$$

Before discussing how this statistic is used to test the hypothesis of zero autocorrelation, it is useful to note two properties.

(a) $0 \leq d \leq 4$. Clearly, from the definition, d cannot be negative. Establishing $d \leq 4$ is not as easy and we will indicate why it cannot exceed 4.

If we multiply (8.4.14) out, we obtain

$$d = \frac{\Sigma_{t=2} e_t^2 + \Sigma_{t=2} e_{t-1}^2 - 2\,\Sigma_{t=2} e_t e_{t-1}}{\Sigma_{t=1} e_t^2}.$$

Except in the case of very small samples, the three sums $\Sigma_{t=2} e_t^2$, $\Sigma_{t=2} e_{t-1}^2$, and $\Sigma_{t=1} e_t^2$ will be approximately equal.

Assuming them to be equal and denoting them by $\Sigma e_t^2$, we have

$$d = \frac{2\,\Sigma e_t^2 - 2\,\Sigma e_t e_{t-1}}{\Sigma e_t^2}$$

$$= 2\left(1 - \frac{\Sigma e_t e_{t-1}}{\Sigma e_t^2}\right).$$

Now if we define $\hat{\rho}$ to be the sample correlation between $e_t$ and $e_{t-1}$, then

$$\hat{\rho} = \frac{\Sigma_{t=2}^{N} e_t e_{t-1}}{\left(\Sigma_{t=2}^{N} e_t^2\, \Sigma_{t=2}^{N} e_{t-1}^2\right)^{\frac{1}{2}}} = \frac{\Sigma e_t e_{t-1}}{\Sigma e_t^2}$$

Thus

$$d = 2(1 - \hat{\rho}) \tag{8.4.15}$$

approximately. Now as $\hat{\rho} \geq -1$, it follows that $d \leq 4$.

(b) The result (8.4.15) enables us to obtain a "feel" for the d-statistic. If there is no autocorrelation, we would expect $\hat{\rho}$ to be close to zero and hence d close to 2. Again, if there is strong positive autocorrelation, $\hat{\rho}$ will probably be near to +1 and thus d will be close to zero. Finally, if there is strong negative autocorrelation, $\hat{\rho}$ should be close to -1 and thus d close to 4. We summarize these results in Table 8.4.3.

When we use the d-statistic to test $H_0 : \rho = 0$ against $H_1 : \rho \neq 0$, we find that it is the first test we have encountered for which the critical values for rejecting the null hypothesis and accepting the alternative hypothesis do not coincide. For a given sample size n and number of regressor variables k' (N and K, respectively, in our notation), Table A1 yields two numbers $d_\ell$ and $d_u$, and these numbers are used to divide the range (0, 4) into five regions. The decision we make with respect to $H_0$ is shown diagrammatically in Figure 8.4.3. If d falls in either of the ranges $(d_\ell, d_u)$ or $(4 - d_u, 4 - d_\ell)$, the test is inconclusive.

Table 8.4.3

| Autocorrelation | d |
|---|---|
| Zero | Near 2 |
| Strong positive | Near 0 |
| Strong negative | Near 4 |

| 0 to $d_\ell$ | $d_\ell$ to $d_u$ | $d_u$ to 2 to $4-d_u$ | $4-d_u$ to $4-d_\ell$ | $4-d_\ell$ to 4 |
|---|---|---|---|---|
| Reject $H_0$<br>Positive auto-correlation | Inconclusive | Accept $H_0$<br>No autocorrelation | Inconclusive | Reject $H_0$<br>Negative auto-correlation |

Figure 8.4.3. Five regions determined by $d_\ell$ and $d_u$.

The Durbin-Watson statistic is computed automatically by almost all regression packages (including MINITAB®). However, it has relevance only when the data are time-series observations. Finally, it is worth noting that there are programs available (e.g., SHAZAM) which compute the exact critical value, and therefore eliminate the inconclusive region. The interested reader is referred to Judge et al. (1982, 453-454).

The Durbin-Watson test when the regressors include lagged dependent variables. The Durbin-Watson test is derived under the assumption that the regressors $X_{1t}$, $X_{2t}$, ..., $X_{Kt}$ are fixed in repeated samples. If some (or all) of the regressors are lagged dependent variables, this assumption is violated. For example, the model

$$Y_t = \beta_0 + \beta_1 Y_{t-1} + \beta_2 Y_{t-2} + \beta_3 X_{3t} + u_t$$

contains two regressor variables ($Y_{t-1}$ and $Y_{t-2}$) which are stochastic and therefore not fixed in repeated samples.

In such cases, Durbin (1970) suggests using the "h-statistic" defined by

$$h = \left(1 - \frac{1}{2}d\right)\left(\frac{N}{1 - NS_{\hat{\alpha}}^2}\right)^{\frac{1}{2}} \tag{8.4.16}$$

where d is the ordinary Durbin-Watson statistic and $S_{\hat{\alpha}}$ is the standard error of $\alpha$, the coefficient of $Y_{t-1}$. In the example above, $\alpha = \beta_1$. Durbin showed that under the null hypothesis $\rho = 0$, h has approximately a standard normal distribution. Thus, at the 5 percent level of significance, we would reject $H_0 : \rho = 0$ if $|h| > 1.96$.[8]

## Overcoming Autocorrelation

Suppose that

$$Y_t = \beta_0 + \beta_1 X_{1t} + \cdots + \beta_K X_{Kt} + u_t, \tag{8.4.17}$$

where $u_t$ is generated by the first-order Markov process (8.4.2). We will assume for the moment that $\rho$ is known, although in practice this is never the case. Our strategy is to transform (8.4.17) so that the disturbance for the transformed model has no autocorrelation. OLS is then used on the new model to produce estimators of the $\beta$'s that are BLUE. We lag (8.4.17) by one period and multiply by the known parameter $\rho$:

$$\rho Y_{t-1} = \rho\beta_0 + \beta_1 \rho X_{1,t-1} + \cdots + \beta_K \rho X_{K,t-1} + \rho u_{t-1}, \qquad t = 2, 3, \ldots, N. \tag{8.4.18}$$

Subtracting (8.4.18) from (8.4.17), we obtain

$$Y_t^* = \beta_0(1 - \rho) + \beta_1 X_{1t}^* + \cdots + \beta_K X_{Kt}^* + (u_t - \rho u_{t-1}), \qquad t = 2, 3, \ldots, N \tag{8.4.19}$$

where $Y_t^* = Y_t - \rho Y_{t-1}$ and $X_{it}^* = X_{it} - \rho X_{i,t-1}$ for $i = 1, 2, \ldots, K$. Now $u_t - \rho u_{t-1} = v_t$, which satisfies the conditions

$$E(v_t) = 0, \quad E(v_t^2) = \sigma_v^2, \quad E(v_t v_s) = 0 \quad (t \neq s).$$

Thus OLS may be applied to (8.4.19). However, even though $v_t$ satisfies the appropriate conditions, OLS estimators are still not fully efficient. The reason is that in lagging (8.4.17) one period, one observation has been lost, and the regression (8.4.19) is carried out on only N - 1 observations.

Provided that N is not too small, it would be reasonable to suppose that the loss in efficiency incurred by losing one observation is probably very slight. However, work by Maeshiro (1979) and Doran (1981) shows that when $\rho$ is positive and close to 1, the loss of efficiency may be surprisingly large. Thus we need to produce a transformation of the first observation. From (8.4.17), on setting $t = 1$, we have

$$Y_1 = \beta_0 + \beta_1 X_{11} + \cdots + \beta_K X_{K1} + u_1, \tag{8.4.20}$$

where $E(u_1^2) = \sigma_v^2/(1 - \rho^2)$. The disturbances $u_1, v_2, v_3, \ldots, v_N$ all have zero mean and are uncorrelated.[9] However, the variances are different, having values of $\sigma_v^2/(1 - \rho^2)$, $\sigma_v^2$, $\sigma_v^2$, $\ldots$, $\sigma_v^2$, respectively. Thus, when the first observation is included, the disturbances are heteroscedastic and may be corrected, as in Section 8.2. We simply weight the first observation appropriately.

If we divide (8.4.20) by $1/(1 - \rho^2)^{\frac{1}{2}}$, we have

$$(1 - \rho^2)^{\frac{1}{2}} Y_1 = (1 - \rho^2)^{\frac{1}{2}}\beta_0 + \beta_1(1 - \rho^2)^{\frac{1}{2}} X_{11} + \cdots + \beta_K(1 - \rho^2)^{\frac{1}{2}} X_{K1} + v_1,$$

where $v_1 = (1 - \rho^2)^{\frac{1}{2}} u_1$. It follows that

$$E(v_1^2) = E(1 - \rho^2)u_1^2 = (1 - \rho^2)E(u_1^2) = \frac{(1 - \rho^2)\sigma_v^2}{1 - \rho^2}$$

$$= \sigma_v^2.$$

Thus, if we define

$$
\begin{aligned}
Y_t^* &= (1 - \rho^2)^{\frac{1}{2}} Y_t && \text{when } t = 1 \\
&= Y_t - \rho Y_{t-1} && \text{otherwise,} \\
X_{it}^* &= (1 - \rho^2)^{\frac{1}{2}} X_{it} && \text{when } t = 1 \\
&= X_{it} - \rho X_{i,t-1} && \text{otherwise;} \\
X_{0t}^* &= (1 - \rho^2)^{\frac{1}{2}} && \text{when } t = 1 \\
&= 1 - \rho && \text{otherwise;}
\end{aligned}
\tag{8.4.21}
$$

then OLS applied to

$$Y_t^* = \beta_0 X_{0t}^* + \beta_1 X_{1t}^* + \cdots + \beta_K X_{Kt}^* + v_t \tag{8.4.22}$$

will produce estimates of $\beta_0$, $\beta_1$, ..., $\beta_K$ that are BLUE.

Estimating $\rho$. The derivation of an appropriate transformation to overcome autocorrelation proceeded on the assumption that $\rho$ was known. In practice, an estimate for $\rho$ would have to be used. We mention three possible estimates.

(a) As $d = 2(1 - \hat{\rho})$, a very quick estimate can be obtained from the OLS output as

$$\hat{\rho} = 1 - \frac{1}{2}d. \tag{8.4.23}$$

(b) The estimate in (a) will be very similar to the sample correlation between $e_t$ and $e_{t-1}$. This in turn will be almost the same as the slope coefficient obtained if $e_t$ were to be regressed on $e_{t-1}$.

(c) Another estimator, suggested by Durbin (1960), is constructed as follows: With the regression model

$$Y_t = \beta_0 + \beta_1 X_{1t} + \cdots + \beta_K X_{Kt} + u_t,$$

where $u_t$ is generated by the Markov process, we have seen how (for known $\rho$) an equation of the form

$$(Y_t - \rho Y_{t-1}) = \beta_0(1-\rho) + \beta_1(X_{1t} - \rho X_{1,t-1}) + \cdots + \beta_K(X_{Kt} - \rho X_{K,t-1}) + v_t$$

is constructed. Rearranging this equation gives us

$$Y_t = \beta_0(1-\rho) + \rho Y_{t-1} + \beta_1 X_{1t} - \beta_1 \rho X_{1,t-1} + \cdots + \beta_K X_{Kt} - \beta_K \rho X_{K,t-1} + v_t.$$

That is,

$$Y_t = \beta_0(1-\rho) + \rho Y_{t-1} + \beta_1 X_{1t} + \beta_1^* X_{1,t-1} + \cdots + \beta_K X_{Kt} + \beta_K^* X_{K,t-1} + v_t. \tag{8.4.24}$$

If we regress $Y_t$ on $Y_{t-1}$, $X_{1t}$, $X_{1,t-1}$, ..., $X_{Kt}$, $X_{K,t-1}$ (2K + 1 regressors), there would be K + 1 possible estimates of $\rho$. These are the coefficient of $Y_{t-1}$ and $-\hat{\beta}_i^*/\hat{\beta}_i$ (i = 1, 2, ..., K). It has been shown that the best estimator of $\rho$ from this model is the coefficient of $Y_{t-1}$. Furthermore, this estimator appears to be the best of the three we have outlined.

## Prediction When Autocorrelation Is Present

So far, the regression model has been used to generate predictions in the following way (see Sections 2.6 and 4.5): If it is desired to predict Y when the regressor variables take the values $X_1^\circ$, $X_2^\circ$, ..., $X_K^\circ$, the optimal predictor $\hat{Y}_0$ is given by

$$\hat{Y}_0 = \hat{\beta}_0 + \hat{\beta}_1 X_1^\circ + \hat{\beta}_2 X_2^\circ + \cdots + \hat{\beta}_K X_K^\circ,$$

where $\hat{\beta}_0$, $\hat{\beta}_1$, $\hat{\beta}_2$, ..., $\hat{\beta}_K$ are the best estimates of the corresponding parameters. We note that implicit in this predictor is the assumption that the disturbance $u_0$ is best predicted by its mean, which is zero. That is,

$$\hat{u}_0 = 0.$$

When we have autocorrelation, information is being passed from disturbance to disturbance through the mechanism

$$u_t = \rho u_{t-1} + v_t,$$

and this enables us to obtain an estimate of $\hat{u}_0$ that is superior to $\hat{u}_0 = 0$.

It can be shown that if we wish to predict n periods ahead, that is, we want $\hat{Y}_{N+n}$, then the optimal predictor of $u_{N+n}$ is

$$\hat{u}_{N+n} = \rho^n e_N, \tag{8.4.25}$$

where $e_N$ is given by

$$e_N = Y_N - (\hat{\beta}_0 + \hat{\beta}_1 X_{1N} + \cdots + \hat{\beta}_K X_{KN})$$

and $\hat{\beta}_0, \hat{\beta}_1, \ldots, \hat{\beta}_K$ have been estimated from the transformed model. Thus

$$\hat{Y}_{N+n} = \hat{\beta}_0 + \hat{\beta}_1 X_{1,N+n} + \hat{\beta}_2 X_{2,N+n} + \cdots + \hat{\beta}_K X_{K,N+n} + \rho^n e_N. \quad (8.4.26)$$

Naturally, in practice $\rho$ is replaced by $\hat{\rho}$.

EXAMPLE Consider the data in Table 8.4.4, which are 30 time-series observations on the five variables

| | |
|---|---|
| LBR1 | unskilled labor |
| LBR2 | skilled labor |
| CAP1 | plant and equipment |
| CAP2 | structures |
| Q | output |

We wish to fit a Cobb-Douglas production function to these data. Accordingly, we define

$$X_1 = \ln \text{LBR1},$$

$$X_2 = \ln \text{LBR2},$$

$$X_3 = \ln \text{CAP1},$$

$$X_4 = \ln \text{CAP2},$$

$$Y = \ln Q,$$

and postulate a model of the form

$$Y_t = \beta_0 + \beta_1 X_{1t} + \beta_2 X_{2t} + \beta_3 X_{3t} + \beta_4 X_{4t} + u_t.$$

The nature of the data leads us to suspect that autocorrelation will be a problem. We will assume that the data have been read into columns C1 to C5 in preparation for analysis using MINITAB.® First, we transform to logarithms using

```
LET C1 = LOGE(C1)
LET C2 = LOGE(C2)
LET C3 = LOGE(C3)
LET C4 = LOGE(C4)
LET C5 = LOGE(C5)
```

The variables $X_1, \ldots, X_4$ and Y are now in columns C1 to C5. To regress Y on $X_1$ to $X_4$, saving the residuals we give the MINITAB command

Table 8.4.4

| Unskilled labor | Skilled labor | Plant and equipment | Structures | Output |
|---|---|---|---|---|
| 92.87 | 99.08 | 107.73 | 91.96 | 99.00 |
| 110.36 | 110.47 | 97.94 | 109.08 | 107.21 |
| 111.09 | 104.98 | 103.19 | 110.02 | 107.12 |
| 102.77 | 110.39 | 102.63 | 125.74 | 108.35 |
| 101.93 | 90.57 | 98.10 | 98.52 | 97.42 |
| 98.37 | 102.11 | 100.70 | 88.90 | 100.54 |
| 113.25 | 92.03 | 83.12 | 79.75 | 94.74 |
| 95.67 | 97.87 | 99.74 | 102.99 | 98.40 |
| 96.56 | 110.52 | 92.54 | 71.08 | 98.20 |
| 100.78 | 95.76 | 101.66 | 100.56 | 99.47 |
| 90.73 | 103.39 | 112.37 | 116.18 | 102.94 |
| 120.29 | 94.29 | 98.63 | 99.69 | 102.35 |
| 110.94 | 97.82 | 94.91 | 94.74 | 101.48 |
| 111.90 | 97.23 | 120.06 | 83.59 | 105.15 |
| 95.30 | 101.50 | 108.81 | 113.47 | 102.61 |
| 96.44 | 96.58 | 106.46 | 101.06 | 100.18 |
| 121.53 | 112.74 | 101.40 | 115.26 | 109.62 |
| 103.68 | 115.59 | 110.84 | 109.72 | 107.86 |
| 98.93 | 106.19 | 87.52 | 87.42 | 96.93 |
| 102.84 | 106.72 | 102.03 | 106.59 | 103.80 |
| 91.77 | 95.24 | 100.14 | 95.20 | 95.16 |
| 105.05 | 111.94 | 94.78 | 122.33 | 107.09 |
| 97.21 | 102.40 | 95.12 | 109.65 | 102.42 |
| 110.02 | 113.16 | 102.40 | 106.17 | 110.41 |
| 109.11 | 106.21 | 107.44 | 102.42 | 106.39 |
| 102.02 | 100.88 | 88.89 | 90.59 | 98.06 |
| 91.51 | 102.45 | 97.24 | 108.41 | 100.36 |
| 114.19 | 109.01 | 93.07 | 95.97 | 104.46 |
| 115.62 | 102.06 | 119.17 | 121.46 | 109.53 |
| 107.05 | 95.32 | 98.13 | 100.14 | 99.26 |

DISPLAY 8.4.1

```
MTB > LET C1=LOGE(C1)
MTB > LET C2=LOGE(C2)
MTB > LET C3=LOGE(C3)
MTB > LET C4=LOGE(C4)
MTB > LET C5=LOGE(C5)
MTB > REGR C5 4 C1-C4;
SUBC> DW.

The regression equation is
C5 = 0.660 + 0.230 C1 + 0.334 C2 + 0.192 C3 + 0.102 C4

Predictor        Coef          Stdev       t ratio
Constant        0.6599        0.1791         3.69
C1             0.22958       0.02196        10.45
C2             0.33424       0.02913        11.47
C3             0.19248       0.02389         8.06
C4             0.10154       0.01621         6.26

s = 0.009658     R-sq = 95.8%      R-sq(adj) = 95.2%

Analysis of Variance

SOURCE         DF          SS           MS
Regression      4     0.053708     0.013427
Error          25     0.002332     0.000093
Total          29     0.056039

SOURCE         DF       SEQ SS
C1              1     0.014711
C2              1     0.022435
C3              1     0.012903
C4              1     0.003658

Unusual Observations
Obs.       C1         C5         Fit  Stdev.Fit   Residual    St.Resid
 24      4.70    4.70420     4.68428    0.00336    0.01992       2.20R

R denotes an obs. with a large st. resid.

Durbin-Watson statistic = 0.93
```

Reprinted with permission of Minitab, Inc.

```
REGR C5 4 C1 - C4;
RESI C10.
```

(The residuals are now stored in C10.) The results of this regression are shown in Display 8.4.1.

The Durbin-Watson statistic has the value

$$d = 0.93.$$

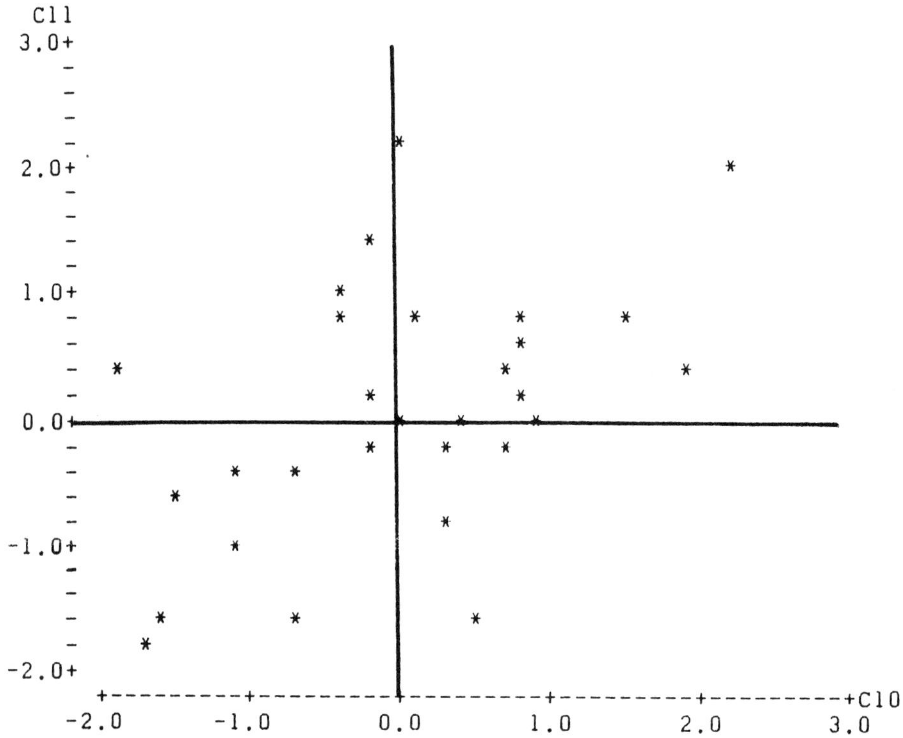


Figure 8.4.4. Scatter diagram of $e_t$ against $e_{t-1}$.

Referring to Table A1, we see that for N = 30 and K = 4,

$$d_\ell = 1.14 \text{ and } d_u = 1.74.$$

Thus, as $d < d_\ell$, there is significant positive autocorrelation. We verify this by plotting the scatter diagram of $e_{t-1}$ against $e_t$. This may be achieved by the two commands

```
LET C11 = LAG(C10)
PLOT C11 C10
```

The result of plotting $e_{t-1}$ against $e_t$ is shown in Figure 8.4.4, with the axes superimposed to clearly delineate the four quadrants. Whereas 18 points fall in quadrants 1 and 3, only 10 fall in the second and fourth quadrants—clear evidence of positive autocorrelation. As there is no evidence of trend in the regressors, we illustrate the estimation of $\rho$ using the first three of the methods given earlier.

(a) $\hat{\rho} = 1 - (1/2)d = 1 - (0.93/2) = 0.535.$
(b) $\hat{\rho} = \text{corr}(e_t, e_{t-1})$. This is found by

```
CORR C10 C11
```

and yields $\hat{\rho} = 0.501$.
(c) The third estimate is obtained by using equation (8.4.24). The appropriate variables are created by

```
LET C6  = LAG(C1)
LET C7  = LAG(C2)
LET C8  = LAG(C3)
LET C9  = LAG(C4)
LET C10 = LAG(C5)
```

The regression is achieved by

```
REGR C5 9 C1 - C4 C6 - C10
```

and $\hat{\rho}$ is the coefficient of C10. In this case we obtain

$$\hat{\rho} = 0.5678.$$

We will use this value for the transformation of the variables.

Noting that $X_1, \ldots, X_4$, Y are in C1-C5 and $X_{1,t-1}, \ldots, X_{4,t-1}, Y_{t-1}$ are in C6-C10, a program to run OLS on the transformed variables is the following:

```
LET K1  = 0.5678

LET C11 = C1-K1*C6
LET C12 = C2-K1*C7
LET C13 = C3-K1*C8
LET C14 = C4-K1*C9
LET C15 = C5-K1*C10

LET K2  = 0.8232
```

$[(1 - \rho^2)^{\frac{1}{2}} = 0.8232.]$

```
LET C11(1) = K2*C1(1)
LET C12(1) = K2*C2(1)
LET C13(1) = K2*C3(1)
LET C14(1) = K2*C4(1)
LET C15(1) = K2*C5(1)
```

```
SET C20
0.8232 29(0.5678)
END
```

(This constructs the intercept variable.) OLS is now applied to the transformed variables, with no intercept.

```
REGR C15 5 C20 C11-C14;
NOCO.
```

The results of this regression are shown in Display 8.4.2.

DISPLAY 8.4.2

```
The regression equation is
C15 = 0.0454 C20 + 0.258 C11 + 0.400 C12 + 0.241 C13 + 0.0893 C14

Predictor          Coef            Stdev       t-ratio
Noconstant
C20             0.04543          0.04395          1.03
C11             0.25758          0.01857         13.87
C12             0.40041          0.02393         16.73
C13             0.24054          0.01932         12.45
C14             0.08925          0.01554          5.74

s = 0.01072
Analysis of Variance

SOURCE          DF           SS             MS
Regression       5      130.537         26.107
Error           25        0.003          0.000
Total           30      130.540

SOURCE          DF       SEQ SS
C20              1      129.747
C11              1        0.597
C12              1        0.156
C13              1        0.034
C14              1        0.004

Unusual Observations
Obs.       C20        C15         Fit  Stdev.Fit   Residual    St.Resid
  1      0.823    3.78270     3.77196    0.01022    0.01074      3.32RX
 17      0.568    2.08118     2.11025    0.00417   -0.02906     -2.94R

R denotes an obs. with a large st. resid.
X denotes an obs. whose X value gives it large influence.

Durbin-Watson statistic = 1.89
```

Reprinted with permission of Minitab, Inc.

Table 8.4.5[a]

| | Before transformation | After transformation |
|---|---|---|
| $\hat{\beta}_0$ | 0.660 | 0.045 |
| | (3.69) | (1.03) |
| $\hat{\beta}_1$ | 0.230 | 0.258 |
| | (10.45) | (13.87) |
| $\hat{\beta}_2$ | 0.334 | 0.400 |
| | (11.47) | (16.73) |
| $\hat{\beta}_3$ | 0.192 | 0.241 |
| | (8.06) | (12.45) |
| $\hat{\beta}_4$ | 0.102 | 0.089 |
| | (6.26) | (5.74) |
| d | 0.93 | 1.89 |

[a]t-ratios in parentheses.

A comparison of the results of the two regressions, before and after transformation, is given in Table 8.4.5. Transforming the variables appears to have resulted in a gain in efficiency, with most of the t-ratios increasing. The Durbin-Watson statistic has risen from 0.93 to 1.89, indicating no significant autocorrelation in the transformed model.

## NOTES

1. We will assume that $Z_i$ is not stochastic. From (8.3.12) it must always be positive.
2. Notice that the transformed model has no intercept.
3. In this context, we will from now on assume that the data are time series, and will use the more conventional suffixes t and s instead of i and j.
4. Notice that $v_t$ has the ideal properties originally postulated for $u_t$, and under which OLS is BLUE. This gives a clue on how to proceed. The model will be transformed so that $v_t$ becomes the disturbance, and OLS will be applied to the transformed model.
5. We have used here the well-known result on the variance of a linear combination of uncorrelated random variables.
6. Most readers will recognize this as a Lagrange multiplier minimization.

7. Granger (1966) shows that most economic variables can be successfully modeled in this way, with $\lambda$ near to 1 in value.

8. It sometimes happens that $NS^2_{\hat{\alpha}} > 1$, and h cannot be computed. In this case an equivalent test of $\rho = 0$ can be performed in two steps. First, the original model is estimated by OLS and the residuals $e_t$ are saved. Second, $e_t$ is regressed on $e_{t-1}$ and the original regressors. If the coefficient of $e_{t-1}$ is significant, $H_0$ is rejected.

9. By assumption, $E(v_t v_s) = 0$ $(t \neq s)$ and v is uncorrelated with previous u disturbances. These assumptions are fundamental to the Markov model.

# 9
# Dynamic Models and Distributed Lags

## 9.1 STATIC AND DYNAMIC MODELS

Most of the models that have been used so far have been static, or, time-independent models. Such models do not often adequately describe economic behavior, which is typically time dependent. To capture this dependency it becomes necessary to develop a class of models known as dynamic models. We will now look at the fundamental differences between static and dynamic models. As we are, for the moment, concerned with the relationships between dependent and regressor variables, we will ignore the disturbance term. This term will be added to our analysis when we consider the estimation aspect of these models.

### Static Models

Consider the simple model

$$Y_t = \beta_0 + \beta_1 X_t,$$

and let us assume that $X_t$ had a constant value $\bar{X}$ until time $t = 0$, and in that instant changed to $\bar{X} + \delta X$ and remained at this new level. What happens to the dependent variable Y? Clearly, it must have also had a constant value, say $\bar{Y}$, given by

$$\bar{Y} = \beta_0 + \beta_1 \bar{X}$$

until $t = 0$. At this instant it changes to $\bar{Y} + \delta Y$, where

$$\bar{Y} + \delta Y = \beta_0 + \beta_1(\bar{X} + \delta X).$$

That is,

$$\delta Y = \beta_1 \delta X. \tag{9.1.1}$$

For as long as X remains at the new level $\bar{X} + \delta X$, Y will remain at the level $\bar{Y} + \delta Y$. The essential feature of such a model is that the total adjustment to Y following a change in the level of X takes place instantaneously. Any model that has this characteristic is known as a static model, and economists talk of the multiplier $\beta_1$—that is, the change in Y that results from a unit change in X.

There are many reasons why static models are inadequate. For example, technical rigidities may prevent producers from adjusting instantaneously to changes in demand, or habit may mean that consumers do not adjust their patterns of consumption immediately to changes in relative prices. There are many other situations that are characterized by a gradual reaction to change.

## Dynamic Models

We will illustrate some fundamental concepts by considering another simple model:

$$Y_t = \beta_0 + \beta_1 Y_{t-1} + \beta_2 X_t. \tag{9.1.2}$$

Let us suppose that prior to time $t = 0$, $X_t$ had the constant value $\bar{X}$. If we assume that $Y_t$ was in equilibrium at some value $\bar{Y}$, we may presume that

$$Y_t = Y_{t-1} = \bar{Y}$$

for all t before $t = 0$.

Substituting this condition into (9.1.2), we see that

$$\bar{Y} = \beta_0 + \beta_1 \bar{Y} + \beta_2 \bar{X} \tag{9.1.3}$$

or

$$\bar{Y} = \frac{\beta_0}{1 - \beta_1} + \frac{\beta_2}{1 - \beta_1} \bar{X}. \tag{9.1.4}$$

Thus, prior to $t = 0$, Y has a constant value $\bar{Y}$, given in terms of $\bar{X}$ by (9.1.4).

Let us now consider what happens to $Y_t$ following a change in $X_t$ from $\bar{X}$ to $\bar{X} + \delta X$ at $t = 0$. From (9.1.2),

$$\begin{aligned} Y_0 &= \beta_0 + \beta_1 Y_{-1} + \beta_2 (\bar{X} + \delta X) \\ &= (\beta_0 + \beta_1 \bar{Y} + \beta_2 \bar{X}) + \beta_2 \delta X. \end{aligned}$$

Now, by (9.1.3), $\beta_0 + \beta_1 \bar{Y} + \beta_2 \bar{X}$ is simply the old equilibrium value $\bar{Y}$. Thus

$$Y_0 = \bar{Y} + \beta_2 \delta X.$$

Setting t = 1 in (9.1.2),

$$\begin{aligned} Y_1 &= \beta_0 + \beta_1 Y_0 + \beta_2(\bar{X} + \delta X) \\ &= \beta_0 + \beta_1(\bar{Y} + \beta_2 \delta X) + \beta_2(\bar{X} + \delta X) \\ &= (\beta_0 + \beta_1 \bar{Y} + \beta_2 \bar{X}) + \beta_2(1 + \beta_1)\delta X. \end{aligned}$$

That is,

$$Y_1 = \bar{Y} + \beta_2(1 + \beta_1)\delta X.$$

Following this procedure for t = 2, we obtain

$$Y_2 = \bar{Y} + \beta_2(1 + \beta_1 + \beta_1^2)\delta X.$$

In general,

$$Y_t = \bar{Y} + \beta_2(1 + \beta_1 + \beta_1^2 + \cdots + \beta_1^t)\delta X, \tag{9.1.5}$$

which, on using the expression for sum of a geometrical progression, becomes

$$Y_t = \bar{Y} + \frac{\beta_2(1 - \beta_1^{t+1})}{1 - \beta_1} \delta X. \tag{9.1.6}$$

The analysis shows us that this model is quite different in its response to a once-and-for-all change in X from the static model considered earlier. The adjustment does not take place immediately, but over a period (in this case, infinite) of time. Such a model is known as a dynamic model. There are three questions that arise when dealing with dynamic models:

(i) Will Y ever achieve a new equilibrium value? Proceeding formally as before, except that $\bar{X} + \delta X$ replaces X, we derive a new equilibrium, say $\bar{Y}^*$, as

$$\bar{Y}^* = \frac{\beta_0}{1 - \beta_1} + \frac{\beta_2}{1 - \beta_1}(\bar{X} + \delta X) = \bar{Y} + \frac{\beta_2}{1 - \beta_1} \delta X.$$

From this we deduce that the new equilibrium is $[\beta_2/(1 - \beta_1)]\delta X$ above the previous equilibrium.

This comparison between old and new equilibriums is called comparative statics. In deriving the expression for $\bar{Y}^*$, we have assumed that it does exist. However, if we return to (9.1.5), we have an alternative definition of $\bar{Y}^*$ as $\bar{Y}_\infty$, given by

$$\bar{Y}^* = \bar{Y} + \beta_2(1 + \beta_1 + \beta_1^2 + \cdots)\ \delta X.$$

This infinite sum will exist only if $|\beta_1| < 1$. If $|\beta_1| < 1$, the expression for the sum of an infinite geometrical progression gives

$$1 + \beta_1 + \beta_1^2 + \cdots = \frac{1}{1 - \beta_1}.$$

Then

$$\bar{Y}^* = \bar{Y} + \frac{\beta_2}{1 - \beta_1} \delta X$$

as before. Thus comparative statics is a meaningless exercise unless changes in X lead eventually to a new equilibrium. If a new equilibrium does exist, the model is said to be stable—otherwise, it is unstable.

If the model is stable, we see that there is not just one multiplier, but an infinite number. The response to a unit change in X is $\beta_2$ immediately, $\beta_2(1 + \beta_1)$ after one period, $\beta_2(1 + \beta_1 + \beta_1^2)$ after two periods, and so on. After an infinite time the response is $\beta_2(1 + \beta_1 + \beta_1^2 + \cdots) = \beta_2/(1 - \beta_1)$. Economists often talk about short-run and long-run multipliers. For this model, we could define the short-run (or impact) multiplier as $\beta_2$, and the long-run multiplier as $\beta_2/(1 - \beta_1)$.

(ii) What is the speed of adjustment? We have seen that it takes an infinite number of time periods for adjustment to the new equilibrium to be complete. However, for all practical purposes, adjustment may be complete in quite a short period. For example, if $\beta_1 = 0.1$, 99 percent of the adjustment has taken place after two periods. Again, if $\beta_1 = 0.5$, 99 percent adjustment takes about seven periods; for $\beta_1 = 0.9$ it takes 44 periods. Clearly, in cases of rapid adjustment comparative statics is very relevant, whereas when adjustment is slow, intermediate effects are likely to be more important.

(iii) What path does Y follow as it moves toward the new equilibrium? We can illustrate this as shown in Figure 9.1.1. If the model is stable and if adjustment is fairly slow, the path Y follows may be of interest. We have shown just two possible adjustment paths.

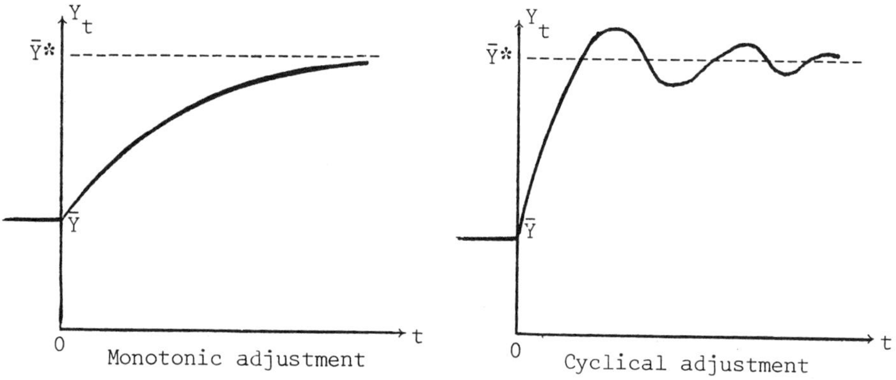


Figure 9.1.1. Two examples of adjustment paths.

The foregoing discussion is intended to introduce some fundamental notions concerning models of dynamic behavior. It is worth noting that merely by including a lagged value of the dependent variable among the regressors, we have changed a simple static model into a far richer model, which has the potential to represent economic behavior far more adequately. It is true in general that whenever lagged dependent variables occur among the regressors, dynamic behavior patterns are implied. As we will now see, this is not the only way of modeling dynamic behavior. The inclusion of lagged independent variables can also achieve this.

## 9.2 DISTRIBUTED LAG MODELS

Consider a model of the form

$$Y_t = \alpha + \sum_{s=0}^{\infty} \beta_s X_{t-s}. \tag{9.2.1}$$

(a) A disturbance should, of course, be included in the model. We have omitted it for the moment in order to concentrate on the relationship between Y and the lagged X's.

(b) We have denoted the intercept by $\alpha$ instead of the usual $\beta_0$, which has been used as the coefficient of $X_t$. This convention makes for a neater formulation of the model.

(c) In (9.2.1) we have an infinite number of regressor variables $X_t$, $X_{t-1}$, $X_{t-2}$, .... The case of K regressors

$$Y_t = \alpha + \sum_{s=0}^{K-1} \beta_s X_{t-s} \tag{9.2.2}$$

is simply the special case in which

$$\beta_K = \beta_{K+1} = \beta_{K+2} = \cdots = 0.$$

(d) We will further assume that the coefficients $\beta_s$ (s = 0, 1, 2, ...) have the property

$$\sum_{s=0}^{\infty} |\beta_s| < \infty. \tag{9.2.3}$$

From the theory of infinite series, this is sufficient to ensure that the sum $\Sigma_{s=0}^{\infty} \beta_s$ is a finite number. This is equivalent to assuming that the model is stable. We will denote the sum of coefficients by H. That is,

$$H = \sum_{s=0}^{\infty} \beta_s. \tag{9.2.4}$$

Models of the form (9.2.1) have had a long history in empirical economics [dating back to Fisher (1937)], as a means of modeling inertia induced by technological and institutional rigidities or psychological factors.

We will interpret the model by using the same device as in the preceding section. We will assume that $X_t$ has the constant value $\bar{X}$ until $t = 0$, when it changed to $\bar{X} + \delta X$ and remained at this new level. Then, prior to $t = 0$, $Y_t$ would also have been constant, with value $\bar{Y}$, say, given by

$$\bar{Y} = \alpha + \sum_{s=0}^{\infty} \beta_s \bar{X} = \alpha + \Big(\sum_{s=0}^{\infty} \beta_s\Big)\bar{X}.$$

That is,

$$\bar{Y} = \alpha + H\bar{X}. \tag{9.2.5}$$

Setting $t = 0$, we have

$$\begin{aligned} Y_0 &= \alpha + [\beta_0(\bar{X} + \delta X) + \beta_1 \bar{X} + \beta_2 \bar{X} + \cdots] \\ &= \alpha + H\bar{X} + \beta_0 \delta X. \end{aligned}$$

Thus

$$Y_0 = \bar{Y} + \beta_0 \delta X.$$

Similarly,

$$\begin{aligned} Y_1 &= \bar{Y} + (\beta_0 + \beta_1)\,\delta X, \\ Y_2 &= \bar{Y} + (\beta_0 + \beta_1 + \beta_2)\,\delta X, \end{aligned}$$

and so on. The complete impact on Y of a change in X is not felt immediately but distributed over time. The model (9.2.1) is thus called a distributed lag model and is clearly dynamic. The multipliers are quite obvious in this case being $\beta_0$, $\beta_0 + \beta_1$, $\beta_0 + \beta_1 + \beta_2$, .... The total (or final multiplier) is $H = \Sigma_{s=0}^{\infty} \beta_s$. It is worth noting at this stage that long-run behavior is determined by the sum of all the coefficients.

## 9.3 LAG OPERATORS

We will pause at this point in our discussion of distributed lag models to introduce a technique that can greatly facilitate this study. We define the lag operator L by

$$L^r X_t \equiv X_{t-r}. \tag{9.3.1}$$

In particular,

$$LX_t \equiv L^1X_t = X_{t-1},$$

$$IX_t \equiv L^oX_t = X_t.$$

I, known as the unit operator, leaves a variable unchanged in the same way that multiplying a number by 1 leaves it unchanged. The principal fact to realize about lag operators is that they can be treated as ordinary algebraic symbols. For example,

$$L^r(L^sX_t) = L^r(X_{t-s}) = X_{t-r-s} = L^{r+s}X_t.$$

Thus

$$L^rL^s = L^{r+s},$$

just as with algebraic symbols.

We define a lag operator polynomial by

$$\Phi(L) = \alpha_0 I + \alpha_1 L + \alpha_2 L^2 + \cdots + \alpha_n L^n. \tag{9.3.2}$$

Then, using the definition of the lag operator, we have

$$\begin{aligned}\Phi(L)X_t &= (\alpha_0 I + \alpha_1 L + \alpha_2 L^2 + \cdots + \alpha_n L^n)X_t \\ &= \alpha_0 X_t + \alpha_1 X_{t-1} + \alpha_2 X_{t-2} + \cdots + \alpha_n X_{t-n} \\ &= \sum_{s=0}^{n} \alpha_s X_{t-s}.\end{aligned}$$

This illustrates how a finite distributed lag model could be reformulated in terms of a polynomial lag operator.

We define the <u>product</u> of two polynomial lag operators $\Phi(L)$ and $\Psi(L)$, given by

$$\Phi(L) = \sum_{s=0}^{n} \alpha_s L^s,$$

$$\Psi(L) = \sum_{s=0}^{m} \alpha_s L^s,$$

as

$$\Theta(L) = \Phi(L)\Psi(L),$$

where

$$\Theta(L) = \sum_{s=0}^{n} \sum_{t=0}^{m} \alpha_s \beta_t L^{s+t}.$$

This is precisely the way ordinary polynomials are multiplied. Finally, we can define the inverse operator to $\Phi(L)$ as that operator which when multiplied by $\Phi(L)$ results in the unit operator I. That is, if

$$\Phi(L)\Psi(L) = I,$$

then $\Psi(L)$ is the inverse of $\Phi(L)$ and can be denoted by $\Phi^{-1}(L)$ or $I/\Phi(L)$. We will now give some examples of the use of lag operators.

## Operation on a Constant

Let c be a constant. Then, clearly, Lc is the value of c in the preceding period, which is c (it is a constant!). Similarly, $L^2c = c$, and so on. Thus

$$\begin{aligned}\Phi(L)c &= (\alpha_0 I + \alpha_1 L + \alpha_2 L^2 + \cdots)c \\ &= \alpha_0 c + \alpha_1 c + \alpha_2 c + \cdots \\ &= (\alpha_0 + \alpha_1 + \alpha_2 + \cdots)c.\end{aligned}$$

That is,

$$\Phi(L)c = \Phi(1)c. \tag{9.3.3}$$

Thus, when a lag operator polynomial acts on a constant, the symbols I and L are replaced by the number 1. For example,

$$\frac{I}{I - \lambda L}\beta_0 = \frac{1\cdot\beta_0}{1 - \lambda} = \frac{\beta_0}{1 - \lambda}.$$

## A Very Important Special Case

We remind the reader of the formula for the sum of an infinite geometrical progression

$$\sum_{s=0}^{\infty} \lambda^s = \frac{1}{1 - \lambda}, \tag{9.3.4}$$

provided that $|\lambda| < 1$. This result enables us to write down an expression of fundamental importance in lag operator manipulations, namely

$$\sum_{s=0}^{\infty} (\lambda L)^{s} = \frac{I}{I - \lambda L}. \tag{9.3.5}$$

Consider now the first-order Markov process studied in Chapter 8. We had

$$u_t = \rho u_{t-1} + v_t \tag{9.3.6}$$

with $|\rho| < 1$. We may express this equation in lag operator form as

$$Iu_t = \rho L u_t + I v_t$$

or

$$u_t = \frac{I}{I - \rho L} v_t. \tag{9.3.7}$$

Equations (9.3.6) and (9.3.7) give two entirely equivalent formulations of the Markov process. Now, referring to the geometrical progression result (9.3.5), we may rewrite (9.3.7) as

$$u_t = \left[ \sum_{s=0}^{\infty} (\rho L)^{s} \right] v_t$$

$$= \sum_{s=0}^{\infty} \rho^{s} L^{s} v_t.$$

That is,

$$u_t = \sum_{s=0}^{\infty} \rho^{s} v_{t-s},$$

which is precisely the result derived (with much greater effort!) in (8.4.6).

## The Connection Between the Two Types of Dynamic Models

In our introduction to dynamic models, we looked at models that had lagged *dependent* variables. We have since seen that lagged *regressor* variables also imply dynamic behavior. Is there a connection?

The distributed lag model (9.2.1) may be written in the form

$$Y_t - \alpha = \left( \sum_{s=0}^{\infty} \beta_s L^{s} \right) X_t = \Phi(L) X_t, \tag{9.3.8}$$

where

$$\Phi(L) = \sum_{s=0}^{\infty} \beta_s L^s.$$

Suppose now that it is possible to find the inverse operator to $\Phi(L)$, which we denote by $\Phi^{-1}(L)$. Then multiplying (9.3.8) by $\Phi^{-1}(L)$, we have

$$\Phi^{-1}(L)Y_t - \Phi^{-1}(L)\alpha = \Phi^{-1}(L)\Phi(L)X_t = IX_t = X_t.$$

Thus

$$\Phi^{-1}(L)Y_t = \gamma + X_t, \tag{9.3.9}$$

where $\gamma = \Phi^{-1}(L)\alpha = \Phi^{-1}(1)\alpha$, and is just a constant.

The model (9.3.9), which is completely equivalent to (9.2.1), is characterized by lagged dependent variables. This shows the very close connection between the two types of dynamic models. As an example of this, consider the consumption function

$$C_t = \beta_0 + \beta_1 C_{t-1} + \beta_2 Y_t,$$

which we may write as

$$(I - \beta_1 L)C_t = \beta_0 + \beta_2 IY_t.$$

Multiplying through by $I/(I - \beta_1 L)$, we have

$$C_t = \frac{I}{I - \beta_1 L}\beta_0 + \beta_2 \frac{I}{I - \beta_1 L} Y_t.$$

Using the rule for operating on the constant $\beta_0$ and the geometrical progression result, we have

$$C_t = \frac{\beta_0}{1 - \beta_1} + \beta_2 \left[\sum_{s=0}^{\infty} (\beta_1 L)^s\right] Y_t.$$

That is,

$$C_t = \frac{\beta_0}{1 - \beta_1} + \beta_2 \sum_{s=0}^{\infty} \beta_1^s Y_{t-s}.$$

This is a completely equivalent formulation of the original consumption function. We note that a single lagged dependent variable is equivalent to an infinite lag distribution of the regressor variable $Y_t$.

## 9.4 RESTRICTIONS ON THE DISTRIBUTED LAG MODEL

In our discussion we are now leading up to a consideration of the estimation problems associated with distributed lag models and therefore must include a disturbance in the model specification. We thus take

$$Y_t = \alpha + \sum_{s=0}^{\infty} \beta_s X_{t-s} + u_t \tag{9.4.1}$$

as our basic model, where the disturbance $u_t$ satisfies the usual assumptions:

(a) $E(u_t) = 0$;

(b) $E(u_t^2) = \sigma^2$;

(c) $E(u_t u_s) = 0$ when $t \neq s$; and

(d) $u_t$ is normally distributed.

The fundamental problem is that (9.4.1) involves an infinite number of parameters $\alpha$, $\beta_0$, $\beta_1$, ... and can therefore never be estimated. Estimation can proceed only if restrictions are placed on the number of parameters. We will now consider some of the ways that restrictions have been imposed.

### Finite Lag Models

In this case the model becomes

$$Y_t = \alpha + \sum_{s=0}^{K} \beta_s X_{t-s} + u_t, \tag{9.4.2}$$

where K is known and $K + 2 < N - K$. That is, the total number of parameters is less than the number of available observations. From our earlier discussion, specifying this model implies that it is known in advance that all adjustments to a change in $X_t$ will be worked out in <u>exactly</u> K time periods. This model is just an ordinary linear regression model, and the properties of OLS estimates can be outlined as follows:

(i) If $X_t$, $X_{t-1}$, ..., $X_{t-K}$ are fixed regressors, then, provided that K is correctly specified, OLS estimators are BLUE. If, on the other hand, the X's are stochastic but independent of the disturbance, OLS estimators still have good properties when the sample is large. These properties are discussed in Chapter 11.

(ii) In most practical situations it is not possible to specify the value of K with certainty. If K is chosen <u>too large</u>, irrelevant regressors have

Table 9.4.1

| K | $\hat{\beta}_0$ | $\hat{\beta}_1$ | $\hat{\beta}_2$ | $\hat{\beta}_3$ | $\hat{\beta}_4$ | $\Sigma\hat{\beta}_s$ |
|---|---|---|---|---|---|---|
| 0 | 0.7932<br>(0.0088) | — | — | — | — | 0.7932<br>(0.0088) |
| 1 | 0.4773<br>(0.0757) | 0.3262<br>(0.0778) | — | — | — | 0.8035<br>(0.0077) |
| 2 | 0.4783<br>(0.0771) | 0.2428<br>(0.1147) | 0.0866<br>(0.0852) | — | — | 0.8077<br>(0.0089) |
| 3 | 0.4954<br>(0.0781) | 0.2270<br>(0.1121) | -0.0523<br>(0.1123) | 0.1416<br>(0.0822) | — | 0.8117<br>(0.0098) |
| 4 | 0.4834<br>(0.0926) | 0.2288<br>(0.1472) | -0.0447<br>(0.1228) | 0.0590<br>(0.1215) | 0.0912<br>(0.0995) | 0.8176<br>(0.0131) |

been included and OLS becomes inefficient. If K is chosen too small, relevant regressors have been omitted and OLS estimators are biased.

(iii) For most economic variables $X_t$, $X_{t-1}$, ..., $X_{t-K}$ will be highly correlated, and multicollinearity will be a severe problem in obtaining precise estimates of the $\beta_s$. However, it is an interesting feature of distributed lag models that even though multicollinearity may prevent reasonable estimation of the individual $\beta_s$, the long-run effects (which are determined by $\Sigma\beta_s$) may be estimated with considerable precision. We illustrate this using the consumption and income data of Table 2.10.1 to estimate the model

$$C_t = \alpha + \sum_{s=0}^{K} \beta_s Y_{t-s} + u_t,$$

where K is assigned the values zero (static model), 1, 2, 3, and 4. The results of these regressions are shown in Table 9.4.1. The last column of this table shows that in varying K from zero to 4, a 95% confidence interval for the long-run multiplier varies from (0.776, 0.811) to (0.791, 0.844). This small change contrasts with the changes in the individual $\hat{\beta}$'s as K is changed.

## Polynomial (Almon) Lags

Before we discuss Almon lags, we will given an interpretation of the lag coefficients $\beta_0$, $\beta_1$, $\beta_2$, .... Let us suppose that up to t = 0, $X_t$ had the value zero. At $t = 0$ it increases to $X_t = 1$ and then returns to zero thereafter. We say that $X_t$ has had a unit impulse at $t = 0$. How does the dependent variable $Y_t$ respond to this impulse? Substituting $\cdots X_{-2} = X_{-1} = 0$, $X_0 = 1$, $X_1 = X_2 = \cdots = 0$ into (9.2.1), it is easily seen that

$$Y_0 - \alpha = \beta_0,$$

$$Y_1 - \alpha = \beta_1,$$

$$Y_2 - \alpha = \beta_2,$$

and so on. Thus the lag distribution $\beta_0, \beta_1, \beta_2, \ldots$ may be thought of as describing the way the dependent variable $Y_t$ (adjusted for the mean $\alpha$) reacts through time to a unit impulse in the exogenous variable.

Bearing in mind that the distributed response is due to frictions caused by psychological or institutional rigidities, it seems reasonable to assume that the lag distribution would be smooth, as shown in Figure 9.4.1. That is, if $\beta_s$ were plotted against s for $s = 0, 1, 2, \ldots$, the graph would not fluctuate wildly.

This assumption that the lag distribution is smooth underlies the approach pioneered by Almon (1965). Her idea was that if the lag distribution were smooth, it should be possible to approximate it by a low-order polynomial. That is,

$$\beta_s = \gamma_0 + \gamma_1 s + \gamma_2 s^2 + \cdots + \gamma_p s^p$$

or

$$\beta_s = \sum_{j=0}^{p} \gamma_j s^j. \tag{9.4.3}$$

It is assumed that $p < K$ (otherwise, there is no point in approximating the lag distribution by a pth-order polynomial). Furthermore, in practice,

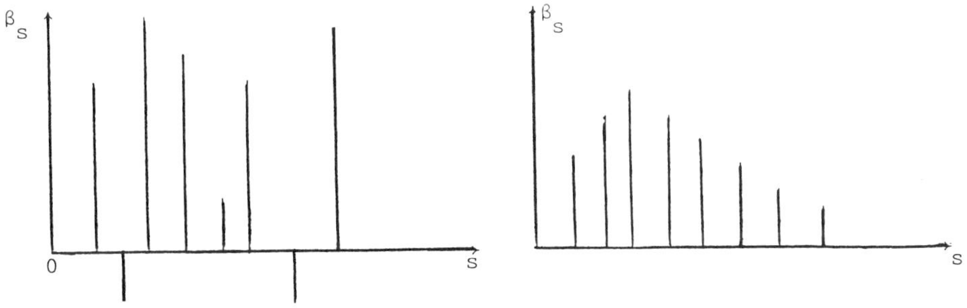


Figure 9.4.1. Contrast between nonsmooth and smooth lag distributions.

p is usually no larger than 3 or 4. If we now substitute (9.4.3) into (9.4.2), we have

$$Y_t = \alpha + \sum_{s=0}^{K} \Big(\sum_{j=0}^{p} \gamma_j s^j\Big) X_{t-s} + u_t$$

$$= \alpha + \sum_{j=0}^{p} \gamma_j \Big(\sum_{s=0}^{K} s^j X_{t-s}\Big) + u_t.$$

That is,

$$Y_t = \alpha + \sum_{j=0}^{p} \gamma_j Z_{tj} + u_t, \tag{9.4.4}$$

where

$$Z_{tj} = \sum_{s=0}^{K} s^j X_{t-s}. \tag{9.4.5}$$

Thus estimation is carried out by regressing $Y_t$ on the Almon variables $Z_{tj}$ to obtain $\hat{\alpha}$ and $\hat{\gamma}_j$ ($j = 0, 1, \ldots, p$). The $\beta_s$ are then estimated by using (9.4.3), namely

$$\hat{\beta}_s = \sum_{j=0}^{p} \hat{\gamma}_j s^j. \tag{9.4.6}$$

The advantage in this method is that the number of parameters to be estimated is reduced from K + 1 to p + 1, and if p is substantially less than K, there will be a significant gain in degrees of freedom and hence efficiency.

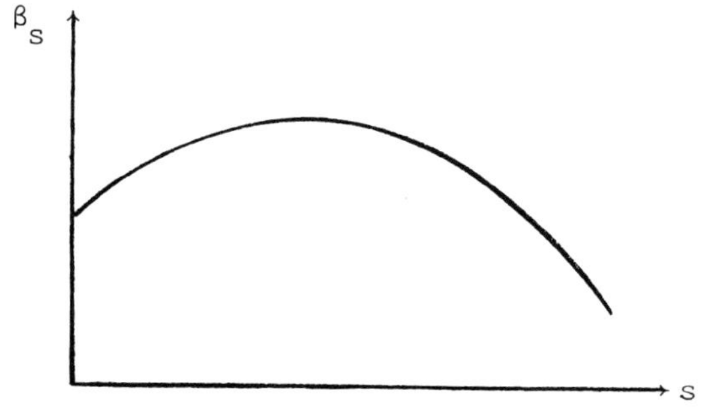


Figure 9.4.2. Parabolic lag distribution.

To illustrate the use of the Almon procedure, let us assume K = 6 (i.e., all adjustments are completed in six time periods) and p = 2. This assumption about p implies a belief that the lag distribution has a parabolic shape, as shown in Figure 9.4.2. Then, by (9.4.4),

$$Y_t = \alpha + \gamma_0 Z_{t0} + \gamma_1 Z_{t1} + \gamma_2 Z_{t2} + u_t. \tag{9.4.7}$$

The Almon variables $Z_{t0}$, $Z_{t1}$, and $Z_{t2}$ are defined by (9.4.5); namely,

$$Z_{t0} = X_t + X_{t-1} + X_{t-2} + \cdots + X_{t-6},$$

$$Z_{t1} = 1X_{t-1} + 2X_{t-2} + \cdots + 6X_{t-6},$$

$$Z_{t2} = 1^2 X_{t-1} + 2^2 X_{t-2} + \cdots + 6^2 X_{t-6}.$$

The first step is to construct the variables $Z_{t0}$, $Z_{t1}$, $Z_{t2}$ for t = 7, 8, ..., N. Then $Y_t$ is regressed on these variables to obtain $\hat{\alpha}$, $\hat{\gamma}_0$, $\hat{\gamma}_1$, $\hat{\gamma}_2$. Finally, the $\beta_s$ (s = 0, 1, ..., 6) are obtained from (9.4.6). Once the value of K is known (and as has already been pointed out, this will always be a problem area), determination of the best choice of p becomes a simple exercise in hypothesis testing—we start by assuming that p = 0 and then successively increase p, ceasing when the last $\hat{\gamma}$ is not significantly different from zero.

A further restriction suggested by Almon is to impose endpoint restrictions. Returning to our example, let us suppose that we impose the constraint $\beta_7 = 0$. Then assuming as before that p = 2, by (9.4.3)

$$\gamma_0 + 7\gamma_1 + 7^2\gamma_2 = 0.$$

This is an ordinary linear restriction on the $\gamma$'s and enables us to eliminate one of them. Suppose that we choose to eliminate $\gamma_0$. Then

$$\gamma_0 = -7\gamma_1 - 7^2\gamma_2,$$

and substituting into (9.4.7) gives us

$$Y_t = \alpha + (-7\gamma_1 - 7^2\gamma_2)Z_{t0} + \gamma_1 Z_{t1} + \gamma_2 Z_{t2} + u_t$$

$$= \alpha + \gamma_1(Z_{t1} - 7Z_{t0}) + \gamma_2(Z_{t2} - 7^2 Z_{t0}) + u_t.$$

That is,

$$Y_t = \alpha + \gamma_1 Z^*_{t1} + \gamma_2 Z^*_{t2} + u_t,$$

where

$$Z^*_{t1} = Z_{t1} - 7Z_{t0},$$

$$Z^*_{t2} = Z_{t2} - 7^2 Z_{t0}.$$

Estimation then proceeds as follows:

(i) $Y_t$ is regressed on $Z^*_{t1}$ and $Z^*_{t2}$ to obtain $\hat{\alpha}$, $\hat{\gamma}_1$, and $\hat{\gamma}_2$;

(ii) $\hat{\gamma}_0$ is obtained as $\hat{\gamma}_0 = -7\hat{\gamma}_1 - 7^2\hat{\gamma}_2$;

(iii) The $\beta_s$ are obtained from $\hat{\gamma}_0$, $\hat{\gamma}_1$, and $\hat{\gamma}_2$, as before.

Thus the imposition of endpoint restrictions introduces no new technical problems. Each restriction will imply a linear restriction on the $\gamma$'s which enables parameters to be eliminated. However, imposing endpoint restrictions always has implications for the shape of the lag distribution—that is, all $\beta$'s are affected. As Dhrymes (1971, 232) points out, "its [endpoint restriction] imposition may have the unintended effect of predetermining the shape of the lag structure." Schmidt and Waud (1973) have also strongly argued against the validity and usefulness of endpoint restrictions.

## Geometric (Koyck) Lags

In the previous two subsections we have looked at parameter restrictions to limit the number of $\beta$'s to be less than the number of available observations. In a well-known investment study, Koyck (1954) suggested a different approach—allowing the number of $\beta$'s to be infinite, but restricting them to obey a relationship involving only a finite number of parameters. In particular, Koyck suggested the geometric lag scheme defined by

$$\beta_s = \gamma\beta^s, \qquad (9.4.8)$$

where $|\beta| < 1$. The characteristics of this scheme can be seen in Figure 9.4.3.

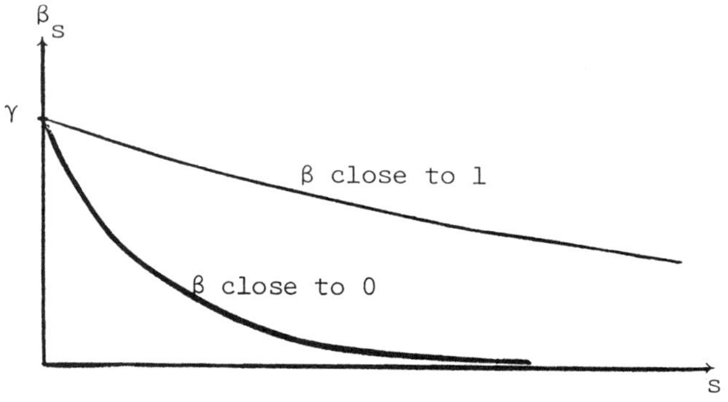


Figure 9.4.3. Geometric lag scheme.

If $\beta$ is close to zero, the lag distribution will die out very quickly, whereas if $\beta$ is close to 1, the distribution dies out much more slowly. Thus, any change in $X_t$ will give rise to an adjustment time in $Y_t$ which is infinite. However, if $\beta$ is close to zero, the total adjustment will, for all intents and purposes, take place rapidly, whereas if $\beta$ is close to 1, the adjustment process will be much slower.

Substituting (9.4.8) into (9.4.1), we have

$$Y_t = \alpha + \gamma \sum_{s=0}^{\infty} \beta^s X_{t-s} + u_t. \tag{9.4.9}$$

In lag operator notation,

$$Y_t = \alpha + \gamma\Big(\sum_{s=0}^{\infty} \beta^s L^s\Big) X_t + u_t.$$

We now use the geometrical sum formula (9.3.5) and obtain

$$Y_t = \alpha + \frac{\gamma I X_t}{I - \beta L} + u_t,$$

which on multiplying through by $(I - \beta L)$ gives

$$(I - \beta L) Y_t = \alpha(1 - \beta) + \gamma I X_t + (I - \beta L) u_t.$$

Thus, finally,

$$Y_t = \alpha(1 - \beta) + \beta Y_{t-1} + \gamma X_t + v_t, \tag{9.4.10}$$

where

$$v_t = u_t - \beta u_{t-1}. \tag{9.4.11}$$

There are several important features to notice about this scheme.

(i) The infinite lag model gives rise to the apparently much simpler model (9.4.10), which involves only three unknown parameters $\alpha$, $\beta$, and $\gamma$.

(ii) The disturbance $v_t$ is generated by what is known as a first-order moving-average process. Although it appears superficially similar to the Markov process, it is in fact quite different. For our purposes, the important properties are

$$E(v_t^2) = (1 + \beta^2)\sigma^2,$$

$$E(v_t v_{t-1}) = -\beta\sigma^2,$$

$$E(v_t v_{t-s}) = 0 \quad (s > 1).$$

Thus, there is correlation between values of $v_t$, which are one time period apart.

(iii) As $Y_t$ is correlated with $v_t$ in (9.4.10), it follows that $Y_{t-1}$ is correlated with $v_{t-1}$. Then, because $v_{t-1}$ is correlated with $v_t$, $Y_{t-1}$ will also be correlated with $v_t$. Thus the regressor $Y_{t-1}$ is correlated with the disturbance in (9.4.10). As we shall see in Chapter 11, OLS estimates of $\alpha$, $\beta$, and $\gamma$ can no longer be relied on. They are inconsistent. When we discuss the adaptive expectations hypothesis, we will give an efficient method of estimation. A general statement is that whenever lagged <u>dependent</u> variables occur in a model involving a <u>moving-average disturbance</u>, efficient estimation is problematical. If, however, the disturbance is Markov (autoregressive), simple transformations of the variables will usually be sufficient to enable efficient estimation.

## A Modified Almon Scheme

There are two fundamental problems with the Almon approach.

(i) Estimation requires the maximum lag length K to be specified in advance. Rarely, if ever, will our a priori knowledge be sufficient to enable us to do that.

(ii) It is not possible to <u>test</u> the existence of a lagged response within the Almon framework because the purely static model is not a special case. Thus, when the Almon scheme is used, a lag distribution will always result, even if in fact there are no lags at all.

To overcome these two problems, Schmidt (1974) suggested a modified Almon scheme, defined as follows:

$$Y_t = \alpha + \sum_{s=0}^{\infty} \beta_s X_{t-s} + u_t,$$

where

$$\beta_s = \delta^s \sum_{j=0}^{p} \alpha_j s^j, \qquad 0 \le \delta < 1. \tag{9.4.12}$$

The scheme combines the geometric lag and Almon polynomial lag ideas. First, a finite lag (specification of K) is not involved. Second, as $s \to \infty$, $\beta_s \to 0$ as in the Koyck scheme, but for small s it is the polynomial part $\Sigma\gamma_j s^j$ that dominates. Finally, the special case $\delta = 0$ is the case of no lag response (static model). The hypothesis $\delta = 0$ can always be tested within this framework. Estimation of this model is a little difficult and will not be pursued here. The interested reader is referred to Schmidt (1974).

## 9.5 UNOBSERVABLE VARIABLES

A very common way in which dynamic models have been generated is by recognizing that economic agents often react to unobservable variables such as expected price, permanent income, desired production, or desired level of inventory. Modeling the way these unobservables are generated from observable variables gives rise to dynamic models. As we shall see, some justification is found for the Koyck form of a distributed lag.

### Partial Adjustment

In very simple models of agricultural supply[1] it may be assumed that the amount of production of some commodity (beef, wheat, etc.) is determined by the current price. That is,

$$Q_t = \alpha + \beta P_t.$$

A more realistic assumption, however, is probably that price determines the desired level of production $Q_t^*$. That is,

$$Q_t^* = \alpha + \beta P_t. \tag{9.5.1}$$

As this equation stands, $\alpha$ and $\beta$ cannot be estimated, because $Q_t^*$ is not observable. To estimate the parameters, we need to express $Q_t^*$ in terms of observable variables. A common hypothesis, known as the partial adjustment hypothesis, is that because of cost, capital, or other constraints, a farmer is unable to adjust his enterprise completely in response to a price change. For example, if the price of wheat changes, a farmer may desire to increase his wheat acreage by 100 hectares. However, because of the demands of his sheep enterprise, he may be able to increase by only 25 hectares in the next year. Thus, we assume that he is able to adjust only a proportion $\gamma$ of the way to his desired level.

As $Q_{t-1}$ is production in the preceding period, the desired adjustment is $Q_t^* - Q_{t-1}$. The actual adjustment is $Q_t - Q_{t-1}$. Thus we hypothesize that

$$Q_t - Q_{t-1} = \gamma(Q_t^* - Q_{t-1}). \tag{9.5.2}$$

In neither (9.5.1) nor (9.5.2) has a disturbance been included. It is probably reasonable to assume that desired production $Q_t^*$ is deterministic. That is, $Q_t^*$ is completely determined by $P_t$. However, actual production, being influenced by uncontrollable effects such as weather, will be stochastic. Hence a disturbance $u_t$ should be included in (9.5.2) to give

$$Q_t - Q_{t-1} = \gamma(Q_t^* - Q_{t-1}) + u_t, \tag{9.5.3}$$

and we assume that $u_t$ obeys the usual assumptions. We may now use (9.5.3) to eliminate the unobservable $Q_t^*$ from (9.5.1). Writing (9.5.3) in lag operator form,

$$[I - (1 - \gamma) L]Q_t = \gamma Q_t^* + u_t,$$

we have from equation (9.5.1),

$$[I - (1 - \gamma) L]Q_t = \alpha\gamma + \beta\gamma P_t + u_t. \tag{9.5.4}$$

This equation may be written in two forms. First, setting $IQ_t = Q_t$ and $LQ_t = Q_{t-1}$,

$$Q_t = \alpha\gamma + (1 - \gamma)Q_{t-1} + \beta\gamma P_t + u_t. \tag{9.5.5}$$

This is the form we would use for estimation. Although the lagged dependent variable $Q_{t-1}$ appears on the right-hand side, the disturbance $u_t$ is not autocorrelated (by assumption) and OLS estimators are consistent and asymptotically efficient.[2] We thus estimate the equation

$$\hat{Q}_t = \hat{\pi}_0 + \hat{\pi}_1 Q_{t-1} + \hat{\pi}_2 P_t, \tag{9.5.6}$$

and estimators of $\gamma$, $\alpha$, and $\beta$ are given by

$$\begin{aligned} \hat{\gamma} &= 1 - \hat{\pi}_1, \\ \hat{\alpha} &= \frac{\hat{\pi}_0}{\hat{\gamma}}, \\ \beta &= \frac{\hat{\pi}_2}{\hat{\gamma}}. \end{aligned} \tag{9.5.7}$$

The second form in which (9.5.4) may be written is obtained by dividing it by the operator $[I - (1 - \gamma) L]$. Then[3]

$$Q_t = \alpha + \beta\gamma \frac{I}{I - (1 - \gamma) L} P_t + v_t$$

where $v_t = [I/(I - (1 - \gamma) L)]u_t$. Now from (9.3.5),

$$\frac{I}{I - (1 - \gamma) L} P_t = \sum_{s=0}^{\infty} (1 - \gamma)^s P_{t-s}.$$

Thus

$$Q_t = \alpha + \beta\gamma \sum_{s=0}^{\infty} (1 - \gamma)^s P_{t-s} + v_t. \tag{9.5.8}$$

The infinite lag distribution on $P_t$ is therefore seen to be geometric, where

$$\beta_s = \beta\gamma(1 - \gamma)^s.$$

Thus the partial adjustment hypothesis gives rise to the geometric (Koyck) form of distributed lag.

### Permanent Income Hypothesis

The familiar Keynesian absolute income hypothesis states that consumption is determined by income level. That is,

$$C_t = \alpha + \beta Y_t + u_t.$$

Friedman's (1957) permanent income hypothesis states that consumption is not determined by absolute income but by "permanent" income. That is,

$$C_t = \alpha + \beta Y_t^* + u_t, \tag{9.5.9}$$

where $Y_t^*$ is permanent income and is unobservable. One approach to the definition of permanent income is to formulate assumptions about how a consumer changes his or her concept of permanent income in relation to changes in actual income $Y_t$. We hypothesize that

$$Y_t^* - Y_{t-1}^* = \gamma(Y_{t-1} - Y_{t-1}^*). \tag{9.5.10}$$

That is, a consumer considers that difference between actual and permanent income is the last period and adjusts his permanent income by a constant proportion of this amount. As this model is identical to the adaptive expectations model, we leave a detailed consideration to the next section.

### Adaptive Expectations Hypothesis

This hypothesis was originally put forward by Cagan (1956) in a study of hyperinflation. A more familiar context is in agricultural supply response, pioneered by Nerlove (1958). Let us assume that the quantity $Q_t$ of a product supplied depends on the producer's expected price $P_t^*$. Thus

$$Q_t = \alpha + \beta P_t^* + u_t. \tag{9.5.11}$$

Furthermore, we assume that expected price is updated in response to actual prices in the same way as permanent income. That is,

$$P_t^* - P_{t-1}^* = \gamma(P_{t-1} - P_{t-1}^*). \tag{9.5.12}$$

Another way of writing this is

$$P_t^* = \gamma P_{t-1} + (1 - \gamma)P_{t-1}^*. \tag{9.5.13}$$

As the coefficients attached to $P_{t-1}$ and $P^*_{t-1}$ sum to unity, the right-hand side of (9.5.13) is a weighted average of $P_{t-1}$ and $P^*_{t-1}$. Expressed in this form, current expected price is a weighted average of actual and expected prices in the preceding period. If $\gamma$ is close to 1, then producers attach most importance to past actual price in forming current expectations. In this case producers will react relatively rapidly to changes in price. On the other hand, if $\gamma$ is close to zero, past actual price will not change their expected price very much (they are very fixed in their ideas!). In such cases reaction to price changes will be much slower.

Writing (9.5.12) or (9.5.13) in lag operator form, we have

$$[I - (1 - \gamma)L]P^*_t = \gamma P_{t-1}.$$

Thus, eliminating $P^*_t$ in (9.5.11), we obtain

$$Q_t = \alpha + \frac{\beta\gamma I P_{t-1}}{I - (1 - \gamma)L} + u_t. \tag{9.5.14}$$

This equation, as in the partial adjustment case, may be written in two forms.

First, using the geometrical series formula (9.3.5), we have

$$Q_t = \alpha + \beta\gamma \sum_{s=0}^{\infty} (1 - \gamma)^s P_{t-1-s} + u_t, \tag{9.5.15}$$

showing that the adaptive expectations hypothesis also gives rise to a Koyck distributed lag (in this case on $P_{t-1}$ rather than $P_t$, but this is an unimportant detail).

Second, if we multiply (9.5.14) through by the operator $[I - (1 - \gamma)L]$, we have

$$[I - (1 - \gamma)L]Q_t = \alpha\gamma + \beta\gamma P_{t-1} + [I - (1 - \gamma)L]u_t,$$

which gives

$$Q_t = \alpha\gamma + (1 - \gamma)Q_{t-1} + \beta\gamma P_{t-1} + v_t, \tag{9.5.16}$$

where

$$v_t = u_t - (1 - \gamma)u_{t-1}.$$

A comparison between this equation and the corresponding equation (9.5.5) for the partial adjustment model shows that they are virtually identical. However, these models differ in one crucial respect. The disturbance in (9.5.5)—partial adjustment—obeys the classical assumptions, which enabled us to conclude that the OLS estimators were consistent and asymptotically efficient. In this case, however, the disturbance $v_t$ is a first-order moving-average disturbance, which, from the discussion on Koyck lags, leads to OLS being inconsistent. We now turn to devising an efficient estimator.

In our earlier studies on heteroscedasticity and autocorrelation, our strategy for obtaining efficient estimates was to transform the equations to obtain a disturbance satisfying the classical assumptions. We then used OLS on the transformed equations. Using this guiding principle, estimation should be carried out on equation (9.5.15), which has the disturbance $u_t$—satisfying the usual assumptions. The problem, of course, is that there are an infinite number of regressors, and somehow we have to constrain the parameters $\beta_s$ to have the form $\beta\gamma(1-\gamma)^s$. That is, $\beta_{s+1} = (1-\gamma)\beta_s$.

If we write out the infinite summation, we have

$$\begin{aligned}\sum_{s=0}^{\infty}(1-\gamma)^s P_{t-1-s} &= [P_{t-1} + (1-\gamma)P_{t-2} + (1-\gamma)^2 P_{t-3} + \cdots + (1-\gamma)^{t-2}P_1] \\ &\quad + [(1-\gamma)^{t-1}P_0 + (1-\gamma)^t P_{-1} + (1-\gamma)^{t+1}P_{-2} + \cdots] \\ &= [P_{t-1} + (1-\gamma)P_{t-2} + (1-\gamma)^2 P_{t-3} + \cdots + (1-\gamma)^{t-2}P_1] \\ &\quad + (1-\gamma)^{t-1}[P_0 + (1-\gamma)P_{-1} + (1-\gamma)^2 P_{-2} + \cdots].\end{aligned}$$

We have included the two sets of brackets to indicate that we have two fundamentally different types of terms. Provided that $\gamma$ is known, those in the first set of brackets involve observable terms $P_{t-1}, \ldots, P_1$. Those in the second set of brackets involve presample values $P_0, P_{-1}, P_{-2}, \ldots$. By definition they can never be observed.

Rewriting in summation notation yields

$$\sum_{s=0}^{\infty}(1-\gamma)^s P_{t-1-s} = \sum_{s=0}^{t-2}(1-\gamma)^s P_{t-1-s} + (1-\gamma)^{t-1}\sum_{s=0}^{\infty}(1-\gamma)^s P_{-s}. \qquad (9.5.17)$$

The first summation is observable (if $\gamma$ is known), the second is not. As the second summation is a summary of all observations before the sample period, we may regard it as being fixed and therefore estimate it as a parameter.

Thus we may rewrite (9.5.15) in the form

$$Q_t = \alpha + \beta X_{1t}(\gamma) + a_0^*(\gamma)X_{2t}(\gamma) + u_t, \qquad (9.5.18)$$

where

$$X_{1t}(\gamma) = \gamma\sum_{s=0}^{t-2}(1-\gamma)^s P_{t-1-s}, \qquad (9.5.19)$$

$$X_{2t}(\gamma) = (1 - \gamma)^{t-1}, \tag{9.5.20}$$

and

$$a_0^*(\gamma) = \beta\gamma \sum_{s=0}^{\infty} (1 - \gamma)^s P_{-s}. \tag{9.5.21}$$

We now note the important features of (9.5.18) to (9.5.21):

(i) From the definition of (9.5.19) it is clear that $X_{1t}(\gamma)$ cannot be constructed for $t = 1$. Hence we regress using observations $t = 2, 3, \ldots, N$.
(ii) If $\gamma$ is known, the variables $X_{1t}(\gamma)$ and $X_{2t}(\gamma)$ can be constructed.
(iii) The parameter $a_0^*(\gamma)$ is a nuisance parameter. Its value is of no interest, but it must be included in the model to avoid misspecification bias.

Thus, if $\gamma$ were known, our procedure would be to regress $Q_t$ on $X_{1t}(\gamma)$ and $X_{2t}(\gamma)$ to estimate $\alpha$ and $\beta$. But of course, $\gamma$ is never known, so how do we proceed? We know that $\gamma$ lies in the range (0, 1), and we search over this range to find the best $\hat{\gamma}$, in the sense that this $\hat{\gamma}$ yields the smallest residual sum of squares. The method, then, is to give $\gamma$ a range of values, say 0(0.1)1, and for each of these $\gamma$'s, construct $X_{1t}(\gamma)$ and $X_{2t}(\gamma)$ and regress $Q_t$ on these variables. The value of $\gamma$ giving smallest RSS is taken as the first approximation. A range of $\gamma$'s around this value is then searched and the $\gamma$ giving the smallest RSS is selected as second approximation, and so on. The process is stopped when the required accuracy for $\gamma$ is achieved. The values of $\hat{\alpha}$ and $\hat{\beta}$ for the optimal $\gamma$ are taken as the appropriate estimators for $\alpha$ and $\beta$.

This process sounds very involved but can be done very easily on a computer. We should note that the appropriate standard errors are not those coming from the OLS regression, but have to be found somewhat differently. We need not go into this aspect of the estimation here. It can be shown that this procedure yields estimators of $\alpha$, $\beta$, and $\gamma$ which are consistent and asymptotically efficient.

It should be clear from the preceding two sections that there would be many situations when a combination of both the partial adjustment and the adaptive expectations hypotheses would be appropriate. In our agricultural example, it may be reasonable to postulate that it is desired level of production (rather than actual level) that is determined by expected price. Then we would have

$$Q_t^* = \alpha + \beta P_t^*,$$

$$Q_t - Q_{t-1} = \gamma(Q_t^* - Q_{t-1}) + u_t,$$

$$P_t^* - P_{t-1}^* = \delta(P_{t-1} - P_{t-1}^*).$$

It is an easy exercise in lag operator calculus to show that

$$Q_t = \alpha\gamma\delta + [(1-\gamma) + (1-\delta)]Q_{t-1} - (1-\gamma)(1-\delta)Q_{t-2} + \beta\gamma\delta P_{t-1} + v_t, \quad (9.5.22)$$

where

$$v_t = u_t - (1 - \delta)u_{t-1}.$$

Estimation of this model is basically an elaboration of the method we have just discussed. The reader is referred to Doran and Griffiths (1978) for a full discussion.

## 9.6 AN EXAMPLE

We conclude this chapter with an example of the application of the adaptive expectations hypothesis. Following Anderson (1974), we will examine the relationship between Australian barley acreages and price. Our results differ from Anderson's, as he appears simply to have used OLS on (9.5.16). The relevant data are shown in Table 9.6.1.

The regression (9.5.18) was run for $\gamma$ taking the values 0.1(0.1)0.9, with the following results:

| $\gamma$ | RSS $\times 10^{-6}$ |
|---|---|
| 0.1 | 1.30 |
| 0.2 | 1.51 |
| 0.3 | 2.25 |
| 0.4 | 3.07 |
| 0.5 | 3.78 |
| 0.6 | 4.36 |
| 0.7 | 4.83 |
| 0.8 | 5.23 |
| 0.9 | 5.57 |

A progressively finer division of $\gamma$ finally yielded the minimum RSS at $\hat{\gamma} = 0.1239$. At this value for $\hat{\gamma}$, the estimates for $\alpha$ and $\beta$ were $\hat{\alpha} = -800.37$ and $\hat{\beta} = 50.256$.

From (9.5.13) we have

$$P^*_t = 0.1239P_{t-1} + 0.8761P^*_{t-1}.$$

Thus, barley farmers appear to place very little weight on the actual previous

Table 9.6.1

| Year | $Q_t$ [a] (thousands of acres) | $P_t$ [b] (cents/bushel) |
|---|---|---|
| 1946-47 | 748 | 40 |
| 1947-48 | 839 | 40 |
| 1948-49 | 1012 | 48 |
| 1949-50 | 1040 | 50 |
| 1950-51 | 1079 | 60 |
| 1951-52 | 1118 | 60 |
| 1952-53 | 1377 | 80 |
| 1953-54 | 1803 | 108 |
| 1954-55 | 1691 | 68 |
| 1955-56 | 1894 | 70 |
| 1956-57 | 2093 | 60 |
| 1957-58 | 2121 | 65 |
| 1958-59 | 2381 | 65 |
| 1959-60 | 2380 | 65 |
| 1960-61 | 2830 | 65 |
| 1961-62 | 2383 | 65 |
| 1962-63 | 2027 | 68 |
| 1963-64 | 2013 | 80 |
| 1964-65 | 2064 | 77 |
| 1965-66 | 2298 | 79 |
| 1966-67 | 2498 | 80 |
| 1967-68 | 2611 | 85 |
| 1968-69 | 3314 | 85 |

[a] Area sown to barley in Australia. Source: Commonwealth Bureau of Census and Statistics.
[b] First advance paid on delivery by the Australian barley board, net of freight. Source: Australian Barley Board Annual Reports.

price in forming their price expectations. We would therefore expect that they would react slowly to changes in price. The long-run multiplier for this model is $\beta$. We can therefore deduce that the proportion of the total adjustment that has taken place after t periods (call it $h_t$) is given by

$$h_t = \frac{[\beta\gamma \Sigma_{s=0}^{t} (1 - \gamma)^s]}{\beta}$$

$$= \frac{\gamma[1 - (1 - \gamma)^{t+1}]}{1 - (1 - \gamma)}$$

That is,

$$h_t = 1 - (1 - \gamma)^{t+1}.$$

Thus, on setting $\gamma = 0.1239$, we find that after 5 years ($t = 5$), $h_t = 0.548$, and after 10 years ($t = 10$), $h_t = 0.767$. That is, about 55 percent of total adjustment has taken place after 5 years and about 77 percent after 10 years. Our estimate of $\beta$, $\hat{\beta} = 50.256$, implies that a 1 cent/bushel increase in the price of barley would eventually lead to an increase of about 50,000 acres being sown to barley, with the increase being about 27,500 acres after 5 years and about 38,500 acres after 10 years.

## NOTES

1. These models have much wider applicability than agricultural supply. For the sake of clarity we will describe the principles in this context.

2. The terms consistent and asymptotically efficient are defined and discussed in Chapter 11. For the moment it is sufficient to understand that such estimators give good results, at least in large samples.

3. Note that $[I/(I - (1 - \gamma)L]\alpha\gamma = [1/(1 - (1 - \gamma))]\alpha\gamma = \alpha$, by using the rule that lag operators are replaced by 1 when operating on a constant.

# 10

# Modeling and Estimation of Economic Relationships: An Example

## 10.1 ESTIMATION OF CONSUMPTION FUNCTIONS

In this chapter our attention is concentrated on the problems involved in moving from the relationships hypothesized in economic theory as existing between idealized concepts, to relationships that involve actually measured variables and which are amenable to econometric estimation. To a large degree this is an area where one learns by doing, so that at most this chapter and the associated readings will provide some troubleshooting guidelines together with an indication of how some econometricians have tackled the estimation of a few of the possible types of relationships in which economists may be interested.

Since they represent relationships involving relatively simple macroeconomic concepts, our example will involve the estimation of a number of alternative consumption functions for Australia. Readers not familiar with Australian National Accounts definitions and data are referred to an Australian Bureau of Statistics (ABS) publication (1981), so that they may familiarize themselves with the differences that exist between such abstract concepts as current consumption and the actual measured concept. The problems of interpretation of econometric relationships involving measured variables which differ from their conceptual definitions will become even more apparent as one widens the class of variables affecting the level of consumption to include others suggested by various theories of consumption behavior.

Assuming that readers are familiar with the main theories of consumption from previous or concurrent work in economics, this chapter will provide only incidental exposition of theoretical material. Should a review of the economic theory be needed, reference should be made to any recent intermediate macroeconomics textbook. In addition, reference is made to a number of the original sources [see particularly Deaton and Muellbauer

---

*This chapter was written by Professor J. W. B. Guise.

(1980)] and several review papers [see particularly Ferber (1973)] where particular approaches to the explanation of consumption behavior have been outlined.

The four main theories of the consumption function that we will attempt to model econometrically, using annual postwar Australian data, are

(a) the Keynesian absolute income hypothesis (Keynes, 1936);
(b) Duesenberry's relative income hypothesis (Duesenberry, 1949);
(c) Friedman's permanent income hypothesis (Friedman, 1957); and
(d) the Modigliani-Ando life cycle hypothesis (Modigliani and Ando, 1957).

## 10.2 THE ABSOLUTE INCOME HYPOTHESIS

In its simplest form the Keynesian absolute income hypothesis may be represented econometrically as

$$C_t = \beta_0 + \beta_1 Y_t + u_t \tag{10.2.1}$$

since the level of aggregate disposable income is the most obvious determinant of aggregate consumption. In this form $\beta_1$ is the marginal propensity to consume and a positive value of $\beta_0$ is assumed to account for autonomous consumption at zero income level. The average propensity to consume exceeds the marginal propensity to consume in this context.

Additional factors considered to have lesser effects on the level of consumption are (a) the degree of inequality in income distribution—with greater equality implying a higher marginal propensity to consume in aggregate—since cross-sectional studies have found higher marginal propensities to consume among low-income earners than among high-income earners; (b) the level of wealth of the community—since wealthy individuals, and hence by implication communities, do not need to save to the same extent for precautionary motives from current income as do less wealthy individuals or communities; and (c) the rate of interest prevailing.

The precise impact of this last factor is regarded as uncertain. An increase in interest rates would positively encourage saving from current income, and the same increase would reduce the value of existing assets yielding fixed-income streams, and hence people's conception of their wealth, so further tending to reduce consumption. However, it would also be the case that smaller capital sums would need to be saved from current and expected future incomes to provide for given levels of income at retirement, and this could lead to an increase in current consumption as a result.

When one attempts to estimate a model such as (10.2.1) using time-series data, a variety of measurement problems require attention. The basic definitional problem of the congruence between the concept and the measure of a variable has already been mentioned. Naturally, one must

Table 10.2.1 Macroeconomic Series Used in Modeling the Consumption Function: Financial Years 1947-48 to 1982-83

-- PRINT C2
COLUMN C2 Private Consumption Expenditure—1979-80 Base
COUNT 36 ($000 million)

| | | | | | |
|---|---|---|---|---|---|
| 19.9814 | 21.1814 | 22.4738 | 24.1522 | 24.0179 | 23.4557 |
| 25.1131 | 26.7693 | 27.6149 | 27.8641 | 28.7904 | 29.6817 |
| 31.7519 | 32.3429 | 33.1506 | 35.1878 | 37.3587 | 39.0633 |
| 40.2244 | 42.1904 | 44.5542 | 47.0131 | 49.9650 | 51.9650 |
| 54.0940 | 57.0480 | 60.1270 | 61.9680 | 63.9360 | 65.5450 |
| 66.4600 | 68.5850 | 69.8340 | 72.2110 | 74.8320 | 75.3940 |

-- PRINT C3
COLUMN C3 Household Disposable Income—1979-80 Base
COUNT 36 ($000 million)

| | | | | | |
|---|---|---|---|---|---|
| 22.9185 | 24.9625 | 27.1924 | 31.6635 | 27.4567 | 28.1624 |
| 29.0062 | 30.6061 | 31.7704 | 32.1620 | 31.6964 | 33.6056 |
| 35.5862 | 36.5793 | 37.7433 | 40.0290 | 43.2895 | 45.0983 |
| 45.8353 | 49.1285 | 49.5470 | 54.1568 | 56.6599 | 59.4825 |
| 62.3951 | 67.5097 | 72.2937 | 73.9188 | 74.1711 | 75.3053 |
| 75.8668 | 79.4352 | 79.3240 | 81.7051 | 84.3279 | 84.2429 |

-- PRINT C4
COLUMN C4 Implicit Deflator Index for GNE—1979-80 Base
COUNT 36

| | | | | | |
|---|---|---|---|---|---|
| 12.280 | 13.420 | 14.721 | 16.634 | 20.115 | 21.987 |
| 22.537 | 23.123 | 24.217 | 25.428 | 25.861 | 26.299 |
| 27.185 | 28.180 | 28.371 | 28.649 | 29.227 | 30.312 |
| 31.188 | 32.248 | 33.259 | 34.343 | 35.823 | 38.258 |
| 40.990 | 43.814 | 49.577 | 59.308 | 68.816 | 76.835 |
| 84.104 | 90.581 | 100.000 | 110.085 | 121.266 | 135.267 |

-- PRINT C5
COLUMN C5 Measure of Household Cash Balances—1979-80 Base
COUNT 36 ($000 million)

| | | | | | |
|---|---|---|---|---|---|
| 12.3927 | 11.8330 | 11.3586 | 10.8932 | 9.5087 | 9.1764 |
| 9.4908 | 9.7782 | 9.8986 | 10.1135 | 10.5019 | 11.0600 |
| 11.6900 | 11.7122 | 12.7830 | 14.3704 | 16.0502 | 16.9882 |
| 17.8650 | 19.1044 | 20.2998 | 21.7396 | 22.8499 | 23.8359 |
| 25.6467 | 30.2469 | 30.2150 | 29.3239 | 29.4027 | 30.3444 |
| 31.2720 | 32.9243 | 33.2903 | 33.7070 | 33.4055 | 35.2492 |

-- PRINT C7
COLUMN C7 Mean Population for Financial Year
COUNT 36

| | | | | | |
|---|---|---|---|---|---|
| 7.6395 | 7.7965 | 8.0443 | 8.3030 | 8.5286 | 8.7342 |
| 8.9003 | 9.0897 | 9.3142 | 9.5325 | 9.7424 | 9.9483 |
| 10.1639 | 10.3909 | 10.6448 | 10.8466 | 11.0593 | 11.2786 |
| 11.5006 | 11.7293 | 11.9972 | 12.2900 | 12.6054 | 12.9071 |
| 13.1922 | 13.4066 | 13.6120 | 13.8182 | 13.9658 | 14.1110 |
| 14.2802 | 14.4364 | 14.6018 | 14.8098 | 15.0515 | 15.2761 |

Source: Data derived from Australian Bureau of Statistics and Reserve Bank sources; publications of various dates. Reprinted with permission of Minitab, Inc.

allow for changes in the nominal price level over time to obtain comparable values for consumption and disposable income, with the attendant index number problems that this entails. Further, one should allow for changes in population so as to distinguish income and consumption increases arising as a result of there being more people in the economy from those attributable to rising per capita incomes and consumption. This is of particular importance if average and marginal propensities to consume differ, as suggested by the Keynesian hypothesis.

The Australian data set out in Table 10.2.1 are based on Australian Bureau of Statistics and Reserve Bank sources, but have been transformed so that all of the monetary variables are consistently related in real terms at 1979-80 price levels. With these data, equation (10.2.1) can be estimated using MINITAB® for the period 1948-49 to 1979-80 as outlined in Display 10.2.1. The final three years of data are omitted to allow one to test the forecast accuracy of fitted regression models using Hendry's chi-square

DISPLAY 10.2.1

```
-- DIVI C2 C7 C2
-- DIVI C3 C7 C3
-- REGR C2 1 C3

THE REGRESSION EQUATION IS
Y =   .0510 + 0.852 X1

                                        ST. DEV.     T-RATIO =
        COLUMN      COEFFICIENT         OF COEF.     COEF/S.D.
        --             0.05104          0.09670         0.53
X1      C3             0.85189          0.02298        37.07

THE ST. DEV. OF Y ABOUT REGRESSION LINE IS
S = 0.1044
WITH (  32- 2) =  30 DEGREES OF FREEDOM

R-SQUARED = 97.9 PERCENT
R-SQUARED = 97.8 PERCENT, ADJUSTED FOR D.F.

ANALYSIS OF VARIANCE

 DUE TO      DF           SS       MS=SS/DF
REGRESSION    1     14.98994       14.98994
RESIDUAL     30      0.32722        0.01091
TOTAL        31     15.31716

            X1           Y     PRED. Y    ST.DEV.
ROW         C3          C2      VALUE     PRED. Y    RESIDUAL       ST.RES.
  3       3.81      2.9089     3.2997     0.0199     -0.3909        -3.81R

R DENOTES AN OBS. WITH A LARGE ST. RES.

DURBIN-WATSON STATISTIC = 1.25
-- STOP
```

test and to test for possible structural change in the post estimation period using Chow's F-test. These tests will only be applied to models that are satisfactory on other counts and will be explained at that stage.

Interestingly, Display 10.2.1 indicates that equation (10.2.1) does not possess a significant intercept term; consequently, the marginal propensity to consume ($\hat{\beta}_1 = 0.852$) does not differ significantly from the average propensity to consume. Although the equation explains 97.8 percent of the variation in per capita consumption over the period, it is clearly unsatisfactory from an econometric viewpoint. The Durbin-Watson statistic ($d = 1.25$) is significant at the 5 percent level, indicating that the disturbances in the model (10.2.1) do not satisfy the independence assumption [i.e., $E(u_t u_{t-1}) = 0$]. In addition, the observation for 1951-52 is clearly an outlier since its residual in this set of estimates falls 3.81 standard deviations below the expected value. These factors, together with the economic theory supporting additional explanatory variables, indicate that model (10.2.1) is inadequate.

If one attempts to take into account the other factors hypothesized to affect consumption under this hypothesis—the distribution of income, the level of wealth of the community, and the rate of interest—one runs into considerable additional difficulties with measurement. Measures of the degree of inequality in the distribution of incomes may be constructed from income taxation data, but taxable incomes do not correspond to disposable incomes. Further, the degree of inequality in income distributions usually tends to change only slowly over time, although this is not always the case—witness the years 1972 to 1975 in Australia.

Adequate measures of community wealth are even more difficult to obtain, particularly as time series, since they depend on a somewhat arbitrary valuation of wealth at a particular date and annual adjustments for price-level changes, new investments, and depreciation. Fortunately, the level of wealth of a community also tends to change only slowly over time, and it also tends to be strongly correlated with the level of disposable income. As a consequence, estimates of the coefficient $\beta_1$ in equation (10.2.1) are likely to be biased to some extent by the misspecification of omitting wealth from the equation.

The rate of interest is less difficult to measure than the other variables, but even for it there is a wide choice of possible rates for different types of investment from which to choose. Typically, the choice falls on the long-term rate for government loans, but movements in this are often not closely correlated with movements in rates in areas where discretionary personal savings are likely to be made, such as savings banks, building societies, and credit unions. Overall, the available empirical evidence does not suggest that interest rate changes have an important effect on the overall level of consumption expenditure, so this variable will be omitted from further consideration.

Recent evidence relating to the measurement of consumption relationships in the United Kingdom (Davidson et al., 1978) indicates that the rate of inflation prevailing in the economy is likely to be a significant factor in explaining changes in per capita consumption expenditure. This is supported by Australian evidence presented by Davy (1975), Freebairn (1977), and Bladen-Hovell and Richards (1983), as well as Italian evidence (Rossi and Schiantarelli, 1982). A number of explanations may be put forward for this phenomenon, but within the context of the Keynesian absolute income hypothesis probably the most acceptable explanation is that consumption will fall when inflation increases (and vice versa) because of the need for increased savings to maintain the real value of financial assets required to facilitate transactions and provide a hedge against uncertainty.

Extending the model (10.2.1) to take account of this variable, one obtains the econometric representation

$$C_t = \beta_0 + \beta_1 Y_t + \beta_2 I_t + u_t, \tag{10.2.2}$$

where $\beta_0 > 0$, $0 < \beta_1 < 1$, and $\beta_2 < 0$. As a measure of inflation the implicit price deflator for gross national expenditure is utilized, the variable used being $(P_t - P_{t-1})/P_{t-1}$. Estimates for the model (10.2.2) are presented in Display 10.2.2. These clearly support the relevance of the inflation variable in explaining per capita consumption. While the Durbin-Watson statistic now supports the hypothesis of independent disturbances, the residual display indicates that there are now two very large residuals of opposite sign which relate to the consecutive years 1950-51 and 1951-52.

Considering these anomalous years, the first is the year of the "wool boom" when sheep farmers' incomes suddenly skyrocketed before dropping back to more normal levels in the following year. A natural reaction to the receipt of such windfall income would be to save a large part of it in the short run and then to spend it on both consumption and investment over a longer planning period. This is consistent with a level of consumption well above that predicted by the usual explanatory variables in 1951-52, following a level of consumption well below that predicted by $Y_t$ and $I_t$ in 1950-51. These effects could be conveniently modeled by a dummy variable taking the value +1.0 in 1950-51 and -1.0 in 1951-52 and zero in other years, this being identified as an effect of windfall income carry forward.

Adding this variable to the model (10.2.2), one obtains the econometric representation

$$C_t = \beta_0 + \beta_1 Y_t + \beta_2 I_t + \beta_3 D_t + u_t, \tag{10.2.3}$$

where $\beta_0 > 0$, $0 < \beta_1 < 1$, $\beta_2 < 0$, and $\beta_3 < 0$. This equation was estimated as in Display 10.2.3. Now 99.3 percent of the sample variation in per capita consumption is explained by the model, with all three regressors highly significant. The marginal propensity to consume has now increased to $\hat{\beta}_1 =$ 0.899, which is substantially <u>higher</u> than the average propensity to consume, which is $\overline{C/Y} = 0.865$ over the same period. In this situation the average propensity to consume is <u>reduced</u> by the effects of inflation.

DISPLAY 10.2.2

```
-- REGR C2 2 C3 C6 C5

   32 CASES USED
    4 CASES CONTAINED MISSING VALUES

THE REGRESSION EQUATION IS
Y =   .0053 + 0.881 X1 - 1.07 X2

                                         ST. DEV.      T-RATIO =
     COLUMN      COEFFICIENT             OF COEF.      COEF/S.D.
     --              0.00532             0.08636         0.06
X1   C3              0.88077             0.02224        39.60
X2   C6             -1.0671               0.3421        -3.12

THE ST. DEV. OF Y ABOUT REGRESSION LINE IS
S = 0.09192
WITH (  32- 3) = 29 DEGREES OF FREEDOM

R-SQUARED = 98.4 PERCENT
R-SQUARED = 98.3 PERCENT, ADJUSTED FOR D.F.

ANALYSIS OF VARIANCE

 DUE TO        DF          SS        MS=SS/DF
REGRESSION      2     15.07214        7.53607
RESIDUAL       29      0.24502        0.00845
TOTAL          31     15.31716

FURTHER ANALYSIS OF VARIANCE
SS EXPLAINED BY EACH VARIABLE WHEN ENTERED IN THE ORDER GIVEN

 DUE TO        DF          SS
REGRESSION      2     15.07214
C3              1     14.98994
C6              1      0.08220

            X1          Y     PRED. Y    ST.DEV.
ROW         C3         C2       VALUE    PRED. Y    RESIDUAL    ST.RES.
  4       3.81     2.9089      3.2255     0.0295     -0.3166    -3.64R
  5       3.22     2.8162      2.6176     0.0616      0.1986     2.91RX
 28       5.35     4.4845      4.5075     0.0437     -0.0229    -0.28 X

R DENOTES AN OBS. WITH A LARGE ST. RES.
X DENOTES AN OBS. WHOSE X VALUE GIVES IT LARGE INFLUENCE.

DURBIN-WATSON STATISTIC = 2.18
```

As a check on the validity of this equation, Chow's F-test was applied to determine whether or not structural change became apparent when the equation was estimated over a longer time period to include the observations for 1980-81 to 1982-83. When the number of additional observations used is less than or equal to the number of regressors in the equation, Chow (1960) has shown that the statistic [see equation (7.3.2)]

$$\frac{(RSS - RSS_1)/N_2}{RSS_1/(N_1 - K - 1)} \sim F_{N_2, (N_1-K-1)},$$

where $N_2$ is the number of additional observations ($N_2 \leq K + 1$), $RSS_1$ the residual sum of squares from the original equation, and RSS the residual sum of squares from the augmented equation.

Applying the test to equation (10.2.3), the residual sum of squares when the extra three observations were included was RSS = 0.14824. Consequently, the estimated F value obtained is

$$F = \frac{(0.14824 - 0.10207)/3}{0.10207/28} = 4.22.$$

As the value of $F_{3,28}$ exceeded in 5 percent of cases under the null hypothesis of no structural change is $F_{3,28} = 2.95$, the value obtained indicates that one would accept the alternative hypothesis of structural change in the postsample period. This throws serious doubt on the validity and adequacy of the model (10.2.3) since its parameter estimates do not seem to remain constant, apart from simple sampling variations, when the estimation period is changed. This suggests that one should reassess the specification and perhaps attempt to add variables measuring wealth and/or income distribution.

Rather than attempting to allow for wealth and income distribution directly, by extending equation (10.2.2) to include them, an alternative procedure that is economical in data requirements would be to use a proxy variable to represent the effects of changes in these variables. A suitable proxy variable, through which changes in wealth and income distribution over time could operate, would be the series of actual <u>lagged</u> consumption values. If these variables affect consumption levels out of a given income, one would expect different levels of consumption to be associated with different levels of lagged consumption even when current disposable income was constant between the relevant periods. Of course, lagged consumption may also be interpreted as an "inertia" variable—consumption will be much the same as last year because people's habits in aggregate change only slowly.

The model estimated including the lagged consumption variable was specified as

$$C_t = \beta_0 + \beta_1 Y_t + \beta_2 I_t + \beta_3 C_{t-1} + u_t, \qquad (10.2.4)$$

since inclusion of a lagged dependent variable was likely to affect the impact of the dummy variable, $D_t$. From Chapter 9 it is clear that equation (10.2.4)

DISPLAY 10.2.3

```
-- REGR C2 3 C3 C6 C14 C5

   32 CASES USED
    4 CASES CONTAINED MISSING VALUES

THE REGRESSION EQUATION IS
Y = - .0434 + 0.899 X1 -  1.43 X2
    - 0.277 X3

                                           ST. DEV.       T-RATIO =
      COLUMN      COEFFICIENT              OF COEF.       COEF/S.D.
      --            -0.04336               0.05726          -0.76
X1    C3             0.89868               0.01489          60.36
X2    C6            -1.4337                 0.2322          -6.17
X3    C14           -0.27746               0.04431          -6.26

THE ST. DEV. OF Y ABOUT REGRESSION LINE IS
S = 0.06038
WITH (  32- 4) =  28 DEGREES OF FREEDOM

R-SQUARED = 99.3 PERCENT
R-SQUARED = 99.3 PERCENT, ADJUSTED FOR D.F.

ANALYSIS OF VARIANCE

 DUE TO        DF          SS       MS=SS/DF
REGRESSION      3     15.21509       5.07170
RESIDUAL       28      0.10207       0.00365
TOTAL          31     15.31716

FURTHER ANALYSIS OF VARIANCE
SS EXPLAINED BY EACH VARIABLE WHEN ENTERED IN THE ORDER GIVEN

 DUE TO        DF          SS
REGRESSION      3     15.21509
C3              1     14.98994
C6              1      0.08220
C14             1      0.14295

          X1          Y     PRED. Y    ST.DEV.
ROW       C3         C2       VALUE    PRED. Y     RESIDUAL     ST.RES.
  4     3.81     2.9089      2.9200     0.0525      -0.0111     -0.37 X
  5     3.22     2.8162      2.8273     0.0525      -0.0111     -0.37 X
 26     5.04     4.2552      4.3832     0.0172      -0.1280     -2.21R
 27     5.31     4.4172      4.5410     0.0202      -0.1238     -2.18R

R DENOTES AN OBS. WITH A LARGE ST. RES.
X DENOTES AN OBS. WHOSE X VALUE GIVES IT LARGE INFLUENCE.

DURBIN-WATSON STATISTIC = 1.58
```

DISPLAY 10.2.4

```
-- REGR C2 3 C4 C12 C5

   32 CASES USED
   11 CASES CONTAINED MISSING VALUES

THE REGRESSION EQUATION IS
Y = - .0372 + 0.353 X1 - 0.736 X2
    + 0.628 X3

                                         ST. DEV.      T-RATIO =
       COLUMN      COEFFICIENT           OF COEF.      COEF/S.D.
       --            -0.03718            0.03923          -0.95
X1     C3             0.35298            0.05045           7.00
X2     C4            -0.7357              0.1577          -4.67
X3     C12            0.62811            0.05883          10.68

THE ST. DEV. OF Y ABOUT REGRESSION LINE IS
S = 0.04154
WITH (  32- 4) =  28 DEGREES OF FREEDOM

R-SQUARED = 99.7 PERCENT
R-SQUARED = 99.7 PERCENT, ADJUSTED FOR D.F.

ANALYSIS OF VARIANCE

 DUE TO        DF           SS       MS=SS/DF
REGRESSION      3     15.26884        5.08961
RESIDUAL       28      0.04832        0.00173
TOTAL          31     15.31716

FURTHER ANALYSIS OF VARIANCE
SS EXPLAINED BY EACH VARIABLE WHEN ENTERED IN THE ORDER GIVEN

 DUE TO        DF           SS
REGRESSION      3     15.26884
C3              1     14.98994
C4              1      0.08220
C12             1      0.19670

             X1         Y     PRED. Y    ST.DEV.
ROW          C3        C2       VALUE    PRED. Y    RESIDUAL     ST.RES.
 11        3.81   2.90885     2.96810    0.02755    -0.05924     -1.91 X
 12        3.22   2.81616     2.77233    0.03137     0.04382      1.61 X
 13        3.22   2.68550     2.80139    0.01435    -0.11589     -2.97R

R DENOTES AN OBS. WITH A LARGE ST. RES.
X DENOTES AN OBS. WHOSE X VALUE GIVES IT LARGE INFLUENCE.

DURBIN-WATSON STATISTIC = 2.55
```

Reprinted with permission of Minitab, Inc.

is a dynamic model, including a distributed lag, unlike the previous equations, which were static specifications. If (10.2.4) is regarded as a partial adjustment model, the disturbance specification is unaffected by the introduction of the lagged dependent variable. If, however, equation (10.2.4) is considered to arise from a Koyck distributed lag formulation, the disturbance is likely to take the form of a first-order moving average. This would imply that OLS estimates would not be consistent, and consequently an instrumental variable estimator might be required.

Estimates of equation (10.2.4) are set out in Display 10.2.4. The results appear to be significantly better than earlier estimates with 99.7 percent of the sampling variation in per capita consumption being accounted for. All partial regression coefficients are highly significant, with the short-run marginal propensity to consume being 0.353 and the long-run marginal propensity to consume, $\hat{\beta}_1/(1 - \hat{\beta}_3) = 0.949$. Of the residuals, those for 1950-51 and 1951-52 have a large influence in determining the estimated parameters, but it is now the residual for 1952-53 which is unusually large. The Durbin-Watson statistic falls in the indeterminate zone for negative autocorrelation, while the more appropriate Durbin's h-statistic takes the value $h = -1.65$. This is suggestive of a moving-average disturbance, but not conclusive. When Chow's F-test was conducted with this model, it provided a test statistic of $F_{3,28} = 0.55$, thus supporting the null hypothesis of no structural change. Overall these estimates indicate that lagged consumption is a very important determinant of current consumption, but whether this is as a result of inertia, or of changes in income distribution and wealth over time, or of other as yet unconsidered factors is not clear from the statistical estimates. This equation will be reconsidered following an examination of other hypotheses relating to the determination of consumption.

## 10.3 THE RELATIVE INCOME HYPOTHESIS

Empirical work undertaken in the 1940s by Kuznets and others to test the Keynesian hypothesis found that it conformed with the evidence from cross-section household data and from short time periods of aggregate data. However, the long-term implication of a decline in the average propensity to consume as a community became richer was not upheld, and in fact the evidence suggested that in the long term, consumption tended to be proportional to the level of income, as suggested by Display 10.2.1. This implies a consumption function of the form

$$C_t = \beta Y_t + u_t.$$

In an attempt to reconcile the short-term and long-term evidence, Smithies (1945) proposed a consumption drift hypothesis, which suggested that for a variety of plausible reasons the observed short-run relationship tended to drift upward over time and so maintain the observed proportionality between consumption and disposable income. The main deficiency with this hypothesis was that it did not provide any reasons why the upward drift should be such as to maintain the long-term proportionality that had been observed.

It was in this context that Duesenberry put forward his relative income hypothesis, which maintains that the consumption of any household will depend on its income relative to that of other households, so that in aggregate if the distribution of income is constant but all incomes rise, consumption will tend to rise proportionately. This invokes a "keeping up with the Joneses" mentality among consumers, where consumer satisfaction depends on the level of consumption of the peer group (interdependent utility functions), together with continual activity on the part of entrepreneurs to create and exploit new wants among consumers and so obtain a share of the discretionary income that arises as incomes increase.

Given this basis for the long-term proportionality of consumption and income, Duesenberry proceeded to explain the short-term nonproportionality in terms consistent with it. Noting a remark of Keynes concerning consumption behavior of households at different stages of the business cycle, Duesenberry suggested that while consumption out of established levels of income might be proportional to income, consumption during prosperous periods out of increases in income above previously established levels was likely to be less than proportional, while consumption out of reduced levels of income during recession periods was likely to be more than proportional.

A possible model of consumption behavior consistent with both the long- and short-term characteristics suggested by Duesenberry is

$$C_t = \beta_0 + \beta_1 Y_t^* + \beta_2 (Y_t - Y_t^*) + u_t, \tag{10.3.1}$$

where $Y_t^*$ represents the highest level of income achieved prior to year t. In this model $\beta_1$ represents the long-run marginal propensity to consume, so that if this is to be close to the long-run average propensity to consume, one would expect $\beta_0$ to be close to zero. The parameter $\beta_2$ represents the short-run marginal propensity to consume out of income additional to established levels when $(Y_t - Y_t^*) > 0$ and the short-run marginal propensity to reduce consumption below established levels when $(Y_t - Y_t^*) < 0$.

Within the context of the model, one would expect $\beta_2$ to be smaller than $\beta_1$, this being consistent with the assumption that the short-run marginal propensity to consume is less than the long-run marginal propensity. If one thought that consumer behavior in booms and recessions was likely to be different, one could postulate the revised model

$$C_t = \beta_0 + \beta_1 Y_t^* + \beta_2 Z_{1t} + \beta_3 Z_{2t} + u_t, \tag{10.3.2}$$

where

DISPLAY 10.3.1

```
-- REGR C2 2 C22 C21 C23

   32 CASES USED
   11 CASES CONTAINED MISSING VALUES

THE REGRESSION EQUATION IS
Y = - 0.174 + 0.901 X1 + 0.670 X2

                                         ST. DEV.      T-RATIO =
      COLUMN      COEFFICIENT            OF COEF.      COEF/S.D.
      --             -0.1744              0.1068          -1.63
X1    C22             0.90148             0.02466         36.55
X2    C21             0.66952             0.05757         11.63

THE ST. DEV. OF Y ABOUT REGRESSION LINE IS
S = 0.09002
WITH (  32- 3) = 29 DEGREES OF FREEDOM

R-SQUARED = 98.5 PERCENT
R-SQUARED = 98.4 PERCENT, ADJUSTED FOR D.F.

ANALYSIS OF VARIANCE

 DUE TO        DF          SS        MS=SS/DF
REGRESSION      2     15.08218        7.54109
RESIDUAL       29      0.23498        0.00810
TOTAL          31     15.31716

FURTHER ANALYSIS OF VARIANCE
SS EXPLAINED BY EACH VARIABLE WHEN ENTERED IN THE ORDER GIVEN

 DUE TO        DF          SS
REGRESSION      2     15.08218
C22             1     13.98615
C21             1      1.09603

            X1          Y     PRED. Y    ST.DEV.
ROW        C22         C2       VALUE    PRED. Y     RESIDUAL      ST.RES.
 11       3.38     2.9089      3.1630     0.0440      -0.2541      -3.24RX
 13       3.81     2.6855      2.8690     0.0319      -0.1835      -2.18R

R DENOTES AN OBS. WITH A LARGE ST. RES.
X DENOTES AN OBS. WHOSE X VALUE GIVES IT LARGE INFLUENCE.

DURBIN-WATSON STATISTIC = 0.87
```

$$Z_{1t} = (Y_t - Y_t^*) \text{ when } Y_t > Y_t^* \text{ and zero otherwise, and}$$

$$Z_{2t} = (Y_t - Y_t^*) \text{ when } Y_t < Y_t^* \text{ and zero otherwise,}$$

so that $\beta_2$ would represent the short-run marginal propensity to consume out of incremental income and $\beta_3$ would represent the short-run marginal propensity to reduce consumption below established levels when income falls.

When one estimates (10.3.1) from the Australian data of Table 10.2.1 for the period 1948-49 to 1979-80, one obtains the results set out in Display 10.3.1, which suggest a long-run marginal propensity to consume of 0.9015 compared with 0.8520 obtained in Display 10.2.1, which did not distinguish current changes in income from established levels of income. With a t-ratio of -1.63, one could not decisively reject the hypothesis that $\beta_0 = 0$, and hence there is some support for long-run equality of the average and marginal propensities to consume. As the Duesenberry theory suggests, we find that the short-run marginal propensity to consume measured by $\hat{\beta}_2$ is substantially lower than the long-run propensity measured by $\hat{\beta}_1$. With a Durbin-Watson statistic of $d = 0.87$, the equation is clearly not satisfactory, even though it explains 98.5 percent of the sample variation in consumption per capita.

When one attempts to distinguish between the propensity to consume out of incremental income and the propensity to reduce consumption below established levels when income falls as per model (10.3.2), the econometric estimates presented in Display 10.3.2 turn out rather strangely, as the estimated value for $\hat{\beta}_3$ exceeds the estimate for $\hat{\beta}_1$ while the estimated value $\hat{\beta}_2$ is not significantly different from zero. In this case the long-run estimate of the marginal propensity to consume is slightly lower at 0.8611, a value comparable to that obtained in Display 10.2.1. In this instance one would not reject the hypothesis that the marginal propensity to consume was equal to the average propensity to consume, since $\hat{\beta}_0$ is not significantly different from zero. In this regression model the marginal propensity to consume out of incremental income is effectively zero, whereas the propensity to reduce consumption when income falls is unexpectedly high. This suggests an attitude of caution on the part of consumers and/or high levels of contractual savings obligations. However, the value obtained for $\hat{\beta}_3$ is not consistent with the substance of the relative income hypothesis, which suggests that $\beta_3 < \beta_1$. The equation again seems unsatisfactory because of the value of the Durbin-Watson statistic obtained ($d = 1.05$) despite the fact that it explains 99.1 percent of the sample variation in per capita consumption. It also throws doubt on the previous specification (10.3.2) since $(Y_t - Y_t^*)$ is simply the sum of $Z_{1t}$ and $Z_{2t}$, which have very different estimated regression coefficients in Display 10.3.2.

Given the previous finding that the rate of inflation seems to be a relevant variable in the explanation of per capita consumption, this variable was added to each of the two models under consideration, thus giving

DISPLAY 10.3.2

```
-- REGR C2 3 C22 C11 C20 C23

   32 CASES USED
   11 CASES CONTAINED MISSING VALUES

THE REGRESSION EQUATION IS
Y =    .0879 + 0.861 X1 + .0705 X2
    + 0.954 X3
```

| | COLUMN | COEFFICIENT | ST. DEV. OF COEF. | T-RATIO = COEF/S.D. |
|---|---|---|---|---|
| | -- | 0.0879 | 0.1013 | 0.87 |
| X1 | C22 | 0.86110 | 0.02112 | 40.76 |
| X2 | C11 | 0.0705 | 0.1404 | 0.50 |
| X3 | C20 | 0.95425 | 0.07744 | 12.32 |

```
THE ST. DEV. OF Y ABOUT REGRESSION LINE IS
S = 0.06979
WITH (  32- 4) =  28 DEGREES OF FREEDOM

R-SQUARED = 99.1 PERCENT
R-SQUARED = 99.0 PERCENT, ADJUSTED FOR D.F.

ANALYSIS OF VARIANCE
```

| DUE TO | DF | SS | MS=SS/DF |
|---|---|---|---|
| REGRESSION | 3 | 15.18077 | 5.06026 |
| RESIDUAL | 28 | 0.13639 | 0.00487 |
| TOTAL | 31 | 15.31716 | |

```
FURTHER ANALYSIS OF VARIANCE
SS EXPLAINED BY EACH VARIABLE WHEN ENTERED IN THE ORDER GIVEN
```

| DUE TO | DF | SS |
|---|---|---|
| REGRESSION | 3 | 15.18077 |
| C22 | 1 | 13.98615 |
| C11 | 1 | 0.45508 |
| C20 | 1 | 0.73954 |

| ROW | X1 C22 | Y C2 | PRED. Y VALUE | ST.DEV. PRED. Y | RESIDUAL | ST.RES. |
|---|---|---|---|---|---|---|
| 11 | 3.38 | 2.9089 | 3.0292 | 0.0452 | -0.1204 | -2.27RX |
| 35 | 5.31 | 4.4845 | 4.6639 | 0.0242 | -0.1794 | -2.74R |

```
R DENOTES AN OBS. WITH A LARGE ST. RES.
X DENOTES AN OBS. WHOSE X VALUE GIVES IT LARGE INFLUENCE.

DURBIN-WATSON STATISTIC = 1.05
```

Reprinted with permission of Minitab, Inc.

$$C_t = \beta_0 + \beta_1 Y_t^* + \beta_2(Y_t - Y_t^*) + \beta_3 I_t + u_t \tag{10.3.3}$$

and

$$C_t = \beta_0 + \beta_1 Y_t^* + \beta_2 Z_{1t} + \beta_3 Z_{2t} + \beta_4 I_t + u_t \tag{10.3.4}$$

as the two models being entertained. The first of these was estimated and is presented in Display 10.3.3. It is apparent that the inflation variable provides a significant incremental explanation of per capita consumption, while $\hat{\beta}_1 > \hat{\beta}_2$ as required by theory, but now $\hat{\beta}_0$ is significantly different from zero while $\hat{\beta}_1$ is again significantly larger than the average propensity to consume, as found with Display 10.2.4. Although the Durbin-Watson statistic is satisfactory, there are several very large residuals during the early fifties from 1950-51 to 1952-53.

Equation (10.3.4) was then estimated and is presented in Display 10.3.4. The inflation variable again clearly provided a significant incremental explanation, and interestingly, the value of $\hat{\beta}_2$ was now significant at the 5 percent level on a one-tail test. Further, the values of $\hat{\beta}_1$ and $\hat{\beta}_3$ obtained are virtually identical, this suggesting that when all the community becomes poorer in recession periods, consumption is reduced proportionately, even though this action may not be observed in cross-sectional data for individuals whose income drops below that of their reference group. Given this empirical regularity, the assumption that $\beta_1 = \beta_3$ was introduced into equation (10.3.4), to give

$$C_t = \beta_0 + \beta_1(Y_t^* + Z_{2t}) + \beta_2 Z_{1t} + \beta_4 I_t + u_t, \tag{10.3.5}$$

where $(Y_t^* + Z_{2t})$ represents either the highest previous level of per capita disposable income (if current per capita income is above the previous level) or the current level of per capita disposable income (where this is lower than the highest previous level attained).

The estimates using specification (10.3.5) are presented in Display 10.3.5. The restriction $\beta_1 = \beta_3$ is obviously accepted by the sample data since to four decimal places there is no change in the residual sum of squares between Displays 10.3.4 and 10.3.5. Further, specification (10.3.5) dominates specification (10.3.3), so this specification will be used as a basis for further exploration of the factors affecting per capita consumption under this hypothesis. Since lagged consumption was an important explana- variable in the context of the absolute income hypothesis, it is again introduced here with some questions about the nature of its influence—as an indicator of habit persistence or as a proxy for changes in wealth and income distribution—and about the nature of its effects on the disturbances of the model.

The specified equation becomes

$$C_t = \beta_0 + \beta_1 Y_t^{**} + \beta_2 Z_{1t} + \beta_3 I_t + \beta_4 C_{t-1} + u_t, \tag{10.3.6}$$

**DISPLAY 10.3.3**

```
-- REGR C2 3 C22 C21 C6

   32 CASES USED
   11 CASES CONTAINED MISSING VALUES

THE REGRESSION EQUATION IS
Y = - 0.246 + 0.938 X1 + 0.684 X2
    -  1.18 X3

                                      ST. DEV.      T-RATIO =
      COLUMN     COEFFICIENT          OF COEF.      COEF/S.D.
      --           -0.24624           0.08402          -2.93
X1    C22           0.93810           0.02069          45.34
X2    C21           0.68417           0.04459          15.34
X3    C6           -1.1794             0.2599          -4.54

THE ST. DEV. OF Y ABOUT REGRESSION LINE IS
S = 0.06954
WITH (  32- 4) =  28 DEGREES OF FREEDOM

R-SQUARED = 99.1 PERCENT
R-SQUARED = 99.0 PERCENT, ADJUSTED FOR D.F.

ANALYSIS OF VARIANCE

 DUE TO        DF          SS       MS=SS/DF
REGRESSION      3     15.18177       5.06059
RESIDUAL       28      0.13539       0.00484
TOTAL          31     15.31716

FURTHER ANALYSIS OF VARIANCE
SS EXPLAINED BY EACH VARIABLE WHEN ENTERED IN THE ORDER GIVEN

 DUE TO        DF          SS
REGRESSION      3     15.18177
C22             1     13.98615
C21             1      1.09603
C6              1      0.09959

           X1          Y     PRED. Y    ST.DEV.
ROW       C22         C2       VALUE    PRED. Y     RESIDUAL      ST.RES.
 11       3.38    2.9089      3.0679     0.0399      -0.1591      -2.79RX
 12       3.81    2.8162      2.6779     0.0483       0.1383       2.76RX
 13       3.81    2.6855      2.8184     0.0271      -0.1329      -2.08R

R DENOTES AN OBS. WITH A LARGE ST. RES.
X DENOTES AN OBS. WHOSE X VALUE GIVES IT LARGE INFLUENCE.

DURBIN-WATSON STATISTIC = 1.99
```

DISPLAY 10.3.4

```
-- REGR C2 4 C22 C11 C20 C6 C23

   32 CASES USED
   11 CASES CONTAINED MISSING VALUES

THE REGRESSION EQUATION IS
Y = - .0259 + 0.899 X1 + 0.213 X2
    + 0.903 X3 - 0.925 X4

                                          ST. DEV.        T-RATIO =
      COLUMN      COEFFICIENT             OF COEF.        COEF/S.D.
      --             -0.02594             0.08313            -0.31
X1    C22             0.89865             0.01858            48.38
X2    C11              0.2133              0.1142             1.87
X3    C20             0.90336             0.06148            14.69
X4    C6              -0.9246              0.2117            -4.37

THE ST. DEV. OF Y ABOUT REGRESSION LINE IS
S = 0.05440
WITH (  32- 5) =  27 DEGREES OF FREEDOM

R-SQUARED = 99.5 PERCENT
R-SQUARED = 99.4 PERCENT, ADJUSTED FOR D.F.

ANALYSIS OF VARIANCE

 DUE TO        DF           SS       MS=SS/DF
REGRESSION      4     15.23724        3.80931
RESIDUAL       27      0.07991        0.00296
TOTAL          31     15.31716

FURTHER ANALYSIS OF VARIANCE
SS EXPLAINED BY EACH VARIABLE WHEN ENTERED IN THE ORDER GIVEN

 DUE TO        DF           SS
REGRESSION      4     15.23724
C22             1     13.98615
C11             1      0.45508
C20             1      0.73954
C6              1      0.05647

            X1          Y     PRED. Y    ST.DEV.
ROW        C22         C2       VALUE    PRED. Y     RESIDUAL      ST.RES.
 11       3.38    2.90885     2.98402    0.03676     -0.07516      -1.87 X
 12       3.81    2.81616     2.67088    0.03779      0.14528       3.71RX

R DENOTES AN OBS. WITH A LARGE ST. RES.
X DENOTES AN OBS. WHOSE X VALUE GIVES IT LARGE INFLUENCE.

DURBIN-WATSON STATISTIC = 1.99
```

DISPLAY 10.3.5

```
-- REGR C2 3 C16 C11 C6

   32 CASES USED
   11 CASES CONTAINED MISSING VALUES

THE REGRESSION EQUATION IS
Y = - .0302 + 0.899 X1 + 0.218 X2
    - 0.928 X3

                                          ST. DEV.       T-RATIO =
       COLUMN     COEFFICIENT             OF COEF.       COEF/S.D.
       --           -0.03016              0.05042          -0.60
X1     C16           0.89949              0.01316          68.34
X2     C11           0.21782              0.08813           2.47
X3     C6           -0.9284                0.1997          -4.65

THE ST. DEV. OF Y ABOUT REGRESSION LINE IS
S = 0.05343
WITH (  32- 4) =  28 DEGREES OF FREEDOM

R-SQUARED = 99.5 PERCENT
R-SQUARED = 99.4 PERCENT, ADJUSTED FOR D.F.

ANALYSIS OF VARIANCE

 DUE TO      DF          SS        MS=SS/DF
REGRESSION    3    15.23723         5.07908
RESIDUAL     28     0.07993         0.00285
TOTAL        31    15.31716

FURTHER ANALYSIS OF VARIANCE
SS EXPLAINED BY EACH VARIABLE WHEN ENTERED IN THE ORDER GIVEN

 DUE TO      DF          SS
REGRESSION    3    15.23723
C16           1    15.16639
C11           1     0.00914
C6            1     0.06170

           X1          Y     PRED. Y    ST.DEV.
ROW        C1         C2       VALUE    PRED. Y     RESIDUAL      ST.RES.
 11      3.38    2.90885     2.98409    0.03608     -0.07523      -1.91 X
 12      3.22    2.81616     2.67133    0.03647      0.14483       3.71RX
 35      5.31    4.48452     4.57315    0.02685     -0.08863      -1.92 X

R DENOTES AN OBS. WITH A LARGE ST. RES.
X DENOTES AN OBS. WHOSE X VALUE GIVES IT LARGE INFLUENCE.

DURBIN-WATSON STATISTIC = 1.98
```

DISPLAY 10.3.6

-- REGR C2 4 C16 C11 C6 C12 C23

32 CASES USED
11 CASES CONTAINED MISSING VALUES

THE REGRESSION EQUATION IS
Y = - .0383 + 0.485 X1 + 0.268 X2
- 0.771 X3 + 0.478 X4

| | COLUMN | COEFFICIENT | ST. DEV. OF COEF. | T-RATIO = COEF/S.D. |
|---|---|---|---|---|
| | -- | -0.03829 | 0.03764 | -1.02 |
| X1 | C16 | 0.48532 | 0.08630 | 5.62 |
| X2 | C11 | 0.26836 | 0.06656 | 4.03 |
| X3 | C6 | -0.7709 | 0.1525 | -5.06 |
| X4 | C12 | 0.47771 | 0.09890 | 4.83 |

THE ST. DEV. OF Y ABOUT REGRESSION LINE IS
S = 0.03985
WITH ( 32- 5) = 27 DEGREES OF FREEDOM

R-SQUARED = 99.7 PERCENT
R-SQUARED = 99.7 PERCENT, ADJUSTED FOR D.F.

ANALYSIS OF VARIANCE

| DUE TO | DF | SS | MS=SS/DF |
|---|---|---|---|
| REGRESSION | 4 | 15.27428 | 3.81857 |
| RESIDUAL | 27 | 0.04288 | 0.00159 |
| TOTAL | 31 | 15.31716 | |

FURTHER ANALYSIS OF VARIANCE
SS EXPLAINED BY EACH VARIABLE WHEN ENTERED IN THE ORDER GIVEN

| DUE TO | DF | SS |
|---|---|---|
| REGRESSION | 4 | 15.27428 |
| C16 | 1 | 15.16639 |
| C11 | 1 | 0.00914 |
| C6 | 1 | 0.06170 |
| C12 | 1 | 0.03705 |

| ROW | X1 C16 | Y C2 | PRED. Y VALUE | ST.DEV. PRED. Y | RESIDUAL | ST.RES. |
|---|---|---|---|---|---|---|
| 11 | 3.38 | 2.90885 | 2.95290 | 0.02768 | -0.04404 | -1.54 X |
| 12 | 3.22 | 2.81616 | 2.75239 | 0.03196 | 0.06377 | 2.68RX |
| 13 | 3.22 | 2.68550 | 2.80014 | 0.01378 | -0.11464 | -3.07R |
| 35 | 5.31 | 4.48452 | 4.50836 | 0.02410 | -0.02384 | -0.75 X |

R DENOTES AN OBS. WITH A LARGE ST. RES.
X DENOTES AN OBS. WHOSE X VALUE GIVES IT LARGE INFLUENCE.

DURBIN-WATSON STATISTIC = 2.50

where $Y_t^{**} = Y_t^* + Z_{2t}$, and this is presented in Display 10.3.6. This apparently provides a very good explanation of the sample data, with all estimated partial regression coefficients highly significant and of correct sign. The value of $\hat{\beta}_1$ is substantially larger than that of $\hat{\beta}_2$, as predicted by the relative income hypothesis, with the long-run marginal propensity to consume out of established income levels being 0.929 and from incremental income only 0.514. The elasticity of consumption with respect to inflation is $\epsilon$ = -0.0285 at the sample means of the variables, with $\hat{\beta}_3$ and $\hat{\beta}_4$ suggesting a decline in consumption of $14.76 per capita at 1979-80 prices per 1 percent increase in the inflation rate. While the model seems to have stable structure—Chow's F-test gave a value of $F_{3,27} = 0.71$—where are some large residuals in the early 1950s and Durbin's h-test for autocorrelation gives a value of h = -1.71, which is suggestive of negative autocorrelation in the residuals. This equation will be reconsidered following an examination of the remaining hypotheses that attempt to explain variations in consumption.

## 10.4 THE PERMANENT INCOME HYPOTHESIS

Both the permanent income hypothesis and the life cycle hypothesis differ from the absolute and relative income hypotheses in placing primary emphasis on the wealth rather than the current income of the consumer. The concept of wealth is a broad one encompassing both current nonhuman wealth and expected inheritances, as well as the human wealth represented by the expected value of sales of labor services over a person's lifetime. The key feature of the permanent income hypothesis relative to consumption behavior is that it postulates that the reaction of current consumption to an increase or decrease in current income depends on the individual's expectations about whether the increase or decrease in income is likely to continue in future periods or is simply a windfall gain or loss. Permanent changes in income levels are assumed to directly affect expected consumption, whereas transitory changes are assumed to have no effect on expected consumption (see pages 000-000).

If one proceeds to define the variables:

$C_t = C_t^* + u_t$, where $C_t^*$ is expected or permanent consumption, and

$Y_t = Y_t^* + v_t$, where $Y_t^*$ is permanent income and $v_t$ transitory income,

then the basic permanent income relationship may be specified as

$$C_t^* = \beta Y_t^* \quad \text{or} \quad C_t = \beta Y_t^* + u_t,$$

where $u_t$ represents transitory consumption. Given the need to ensure that the sample mean of the residuals is zero so that the central t-test and F-test can be obtained correctly, this model should include an intercept term (which should not be significantly different from zero). Further, noting the important role that the rate of inflation seems to play in models estimated under alternative hypotheses, it would seem appropriate to extend this model to include an expected inflation variable, $I_t^*$, as a consumption determinant, thus giving the model

$$C_t = \beta_0 + \beta_1 Y_t^* + \beta_2 I_t^* + u_t \tag{10.4.1}$$

as a basis for further consideration.

The main problem with model (10.4.1) is to find suitable measures for the permanent income and expected inflation variables since only measures for actual income and actual inflation are available. One possible way of linking these hypothetical concepts with actual measurements, given the theoretical importance of expectations, both with respect to inflation and also in changing the evaluation of permanent income by individuals, is by way of the adaptive expectations hypothesis proposed by Nerlove (1958), which was outlined in Chapter 9. In this case extensions of the Nerlove hypothesis which incorporate something of the flavor of pseudorational expectations are proposed, these being

$$Y_t^* - Y_{t-1}^* = (1 - \gamma_1)(Y_t - Y_{t-1}) + (1 - \gamma_2)(Y_{t-1} - Y_{t-1}^*) \tag{10.4.2}$$

and

$$I_t^* - I_{t-1}^* = (1 - \lambda_1)(I_t - I_{t-1}) + (1 - \lambda_2)(I_{t-1} - I_{t-1}^*), \tag{10.4.3}$$

where $0 < \gamma_1, \gamma_2 < 1$ and $0 < \lambda_1, \lambda_2 < 1$. [This notation differs slightly from that used in Chapter 9, but it is entirely arbitrary whether one defines the parameter being measured as $\gamma_1$ or $(1 - \gamma_1)$ since $(\gamma_1 + 1 - \gamma_1) = 1$.]

The standard Nerlovian hypothesis omits the first terms on the right-hand side, thus making the change in expectations a simple function of the difference between actual and expected values in the most recent past period. If the rational expectations hypothesis (Bilson, 1980) is appropriate, the second set of terms on the right-hand side should be relevant only insofar as the most recent past actual value contains unanticipated innovations, and similarly, the first pair of terms also becomes relevant, again insofar as the current change in actual value has not been anticipated. Many past applications of the Nerlovian hypothesis have not discriminated between the two portions of the right-hand side in (10.4.2), but have confused the issues involved by assuming $\gamma_1 = \gamma_2 = \gamma$, thus reducing the equation to

$$Y_t^* - Y_{t-1}^* = (1 - \gamma)(Y_t - Y_{t-1}^*).$$

Rearranging (10.4.2) and (10.4.3), one obtains, using lag operator notation,

$$(1 - \gamma_2 L)Y_t^* = (1 - \gamma_1)Y_t + (\gamma_1 - \gamma_2)LY_t \tag{10.4.4}$$

DISPLAY 10.4.1

```
-- REGR C2 8 C4 C13 C14 C5-C7 C12 C11 C15

NOTE - THE FOLLOWING VARIABLES ARE HIGHLY
CORRELATED WITH OTHER PREDICTOR VARIABLES
C12
C11

   32 CASES USED
   11 CASES CONTAINED MISSING VALUES

THE REGRESSION EQUATION IS
Y = - .0531 + 0.326 X1 + 0.277 X2
    - 0.115 X3 -  1.32 X4 + .0143 X5
    + 0.247 X6 + 0.599 X7 - 0.121 X8

                                       ST. DEV.      T-RATIO =
      COLUMN      COEFFICIENT          OF COEF.      COEF/S.D.
      --            -0.05307           0.03451          -1.54
X1    C4             0.32625           0.04289           7.61
X2    C13             0.2765            0.1104           2.50
X3    C14            -0.1150            0.1061          -1.08
X4    C5             -1.3203            0.4025          -3.28
X5    C6              0.0143            0.4246           0.03
X6    C7              0.2469            0.2033           1.21
X7    C12             0.5988            0.1757           3.41
X8    C11            -0.1206            0.1471          -0.82

THE ST. DEV. OF Y ABOUT REGRESSION LINE IS
S = 0.03369
WITH (  32- 9) =  23 DEGREES OF FREEDOM

R-SQUARED = 99.8 PERCENT
R-SQUARED = 99.8 PERCENT, ADJUSTED FOR D.F.

ANALYSIS OF VARIANCE

 DUE TO         DF           SS        MS=SS/DF
REGRESSION       8     15.29105         1.91138
RESIDUAL        23      0.02611         0.00114
TOTAL           31     15.31716

           X1          Y      PRED. Y    ST.DEV.
ROW        C4         C2       VALUE     PRED. Y     RESIDUAL     ST.RES.
 12      3.22    2.81616     2.81746     0.02944     -0.00130     -0.08 X
 13      3.22    2.68550     2.69837     0.02920     -0.01287     -0.77 X
 14      3.26    2.82160     2.82004     0.02782      0.00157      0.08 X
 20      3.50    3.12399     3.03937     0.01613      0.08461      2.86R

R DENOTES AN OBS. WITH A LARGE ST. RES.
X DENOTES AN OBS. WHOSE X VALUE GIVES IT LARGE INFLUENCE.

DURBIN-WATSON STATISTIC = 1.86
```

and

$$(1 - \lambda_2 L)I_t^* = (1 - \lambda_1)I_t + (\lambda_1 - \lambda_2) LI_t, \tag{10.4.5}$$

so that on substitution in (10.4.1) the resulting equation is

$$C_t = \beta_0 + \frac{\beta_1(1 - \gamma_1)Y_t + \beta_1(\gamma_1 - \gamma_2)Y_{t-1}}{1 - \gamma_2 L} + \frac{\beta_2(1 - \lambda_1)I_t + \beta_2(\lambda_1 - \lambda_2)I_{t-1}}{1 - \lambda_2 L} + u_t.$$

On multiplying this result through by the two denominator lag operators, canceling, and rearranging, we obtain

$$\begin{aligned} C_t &= \beta_0(1 - \gamma_2)(1 - \lambda_2) + \beta_1(1 - \gamma_1)Y_t + \beta_1[(\gamma_1 - \gamma_2) - \lambda_2(1 - \gamma_1)]Y_{t-1} \\ &\quad - \beta_1\lambda_2(\gamma_1 - \gamma_2)Y_{t-2} + \beta_2(1 - \lambda_1)I_t + \beta_2[(\lambda_1 - \lambda_2) - \gamma_2(1 - \lambda_1)]I_{t-1} \\ &\quad - \beta_2\gamma_2(\lambda_1 - \lambda_2)I_{t-2} + (\gamma_2 + \lambda_2)C_{t-1} - \gamma_2\lambda_2 C_{t-2} + u_t - (\gamma_2 + \lambda_2)u_{t-1} \\ &\quad + \gamma_2\lambda_2 u_{t-2}, \end{aligned} \tag{10.4.6}$$

which is a model highly nonlinear in the parameters and with an explicit second-order moving-average form for the disturbance, provided that $u_t$ has the usually assumed properties.

This equation can readily be estimated using nonlinear least squares methods, but since these fall outside the scope of the present book, a linearized version of this model

$$\begin{aligned} C_t &= \alpha_0 + \alpha_1 Y_t + \alpha_2 Y_{t-1} + \alpha_3 Y_{t-2} + \alpha_4 I_t + \alpha_5 I_{t-1} + \alpha_6 I_{t-2} \\ &\quad + \alpha_7 C_{t-1} + \alpha_8 C_{t-2} + \epsilon_t \end{aligned} \tag{10.4.7}$$

will be estimated. The results of this estimation are presented in Display 10.4.1. It is clear that one would accept the hypothesis that $\alpha_5 = 0$ in this instance. Dropping the variable $I_{t-1}$ from the model and reestimating, one obtains the results presented in Display 10.4.2. In this case it seems that the hypothesis $\alpha_8 = 0$ would be accepted with a fairly large probability, so dropping the variable $C_{t-2}$ from the model, one obtains the results presented in Display 10.4.3.

The significance of the variable $Y_{t-2}$ has improved considerably in the latter display as a result of the reduced levels of multicollinearity between the variables in the equation. Finally, the significance of $\alpha_6$ is questionable in this set of estimates, and given the rejection of $I_{t-1}$ as an explanatory variable, there is little theoretical justification remaining for the inclusion of $I_{t-2}$. Therefore, dropping this variable one finishes up with the results presented in Display 10.4.4, where all of the partial regression coefficients are significantly nonzero at the 5 percent level of significance. Comparing these results with Display 10.2.4, the two additional explanatory variables

DISPLAY 10.4.2

```
--REGR C2 7 C4 C13 C14 C5 C7 C12 C11

NOTE - THE FOLLOWING VARIABLES ARE HIGHLY
CORRELATED WITH OTHER PREDICTOR VARIABLES
C12
C11

   32 CASES USED
   11 CASES CONTAINED MISSING VALUES

THE REGRESSION EQUATION IS
Y = - .0531 + 0.326 X1 + 0.276 X2
    - 0.113 X3 -  1.31 X4 + 0.250 X5
    + 0.597 X6 - 0.121 X7

                                      ST. DEV.     T-RATIO =
         COLUMN      COEFFICIENT      OF COEF.     COEF/S.D.
         --           -0.05309        0.03378        -1.57
X1       C4            0.32608        0.04169         7.82
X2       C13            0.2758         0.1061         2.60
X3       C14          -0.11261        0.07849        -1.43
X4       C5            -1.3119         0.3095        -4.24
X5       C7             0.2503         0.1725         1.45
X6       C12            0.5973         0.1671         3.58
X7       C11           -0.1208         0.1440        -0.84

THE ST. DEV. OF Y ABOUT REGRESSION LINE IS
S = 0.03298
WITH (  32- 8) =  24 DEGREES OF FREEDOM

R-SQUARED = 99.8 PERCENT
R-SQUARED = 99.8 PERCENT, ADJUSTED FOR D.F.

ANALYSIS OF VARIANCE

 DUE TO       DF            SS       MS=SS/DF
REGRESSION     7      15.29105        2.18444
RESIDUAL      24       0.02611        0.00109
TOTAL         31      15.31716

             X1          Y     PRED. Y     ST.DEV.
ROW          C4         C2       VALUE     PRED. Y     RESIDUAL      ST.RES.
 12        3.22    2.81616     2.81774     0.02765     -0.00158     -0.09 X
 13        3.22    2.68550     2.69820     0.02814     -0.01270     -0.74 X
 14        3.26    2.82160     2.82003     0.02723      0.00158      0.08 X
 20        3.50    3.12399     3.03935     0.01577      0.08464      2.92R

R DENOTES AN OBS. WITH A LARGE ST. RES.
X DENOTES AN OBS. WHOSE X VALUE GIVES IT LARGE INFLUENCE.

DURBIN-WATSON STATISTIC = 1.85
```

Reprinted with permission of Minitab, Inc.

DISPLAY 10.4.3

```
-- REGR C2 6 C4 C13 C14 C5 C7 C12 C15


NOTE - THE FOLLOWING VARIABLES ARE HIGHLY
CORRELATED WITH OTHER PREDICTOR VARIABLES
C12

   32 CASES USED
   11 CASES CONTAINED MISSING VALUES

THE REGRESSION EQUATION IS
Y = - .0466 + 0.326 X1 + 0.247 X2
    - 0.157 X3 -  1.16 X4 + 0.234 X5
    + 0.558 X6

                                      ST. DEV.     T-RATIO =
        COLUMN      COEFFICIENT       OF COEF.     COEF/S.D.
        --            -0.04662        0.03269        -1.43
X1      C4             0.32620        0.04144         7.87
X2      C13            0.24719        0.09989         2.47
X3      C14           -0.15681        0.05783        -2.71
X4      C5            -1.1567          0.2466        -4.69
X5      C7             0.2341          0.1704         1.37
X6      C12            0.5577          0.1593         3.50

THE ST. DEV. OF Y ABOUT REGRESSION LINE IS
S = 0.03279
WITH (  32- 7) =  25 DEGREES OF FREEDOM

R-SQUARED = 99.8 PERCENT
R-SQUARED = 99.8 PERCENT, ADJUSTED FOR D.F.

ANALYSIS OF VARIANCE

 DUE TO      DF          SS        MS=SS/DF
REGRESSION    6     15.29028        2.54838
RESIDUAL     25      0.02688        0.00108
TOTAL        31     15.31716

            X1          Y      PRED. Y     ST.DEV.
ROW         C4         C2        VALUE     PRED. Y     RESIDUAL       ST.RES.
 11       3.81    2.90885      2.96030     0.02185     -0.05145       -2.10RX
 12       3.22    2.81616      2.81899     0.02745     -0.00283       -0.16 X
 13       3.22    2.68550      2.69631     0.02788     -0.01081       -0.63 X
 14       3.26    2.82160      2.82641     0.02599     -0.00480       -0.24 X
 20       3.50    3.12399      3.04928     0.01037      0.07471        2.40R

R DENOTES AN OBS. WITH A LARGE ST. RES.
X DENOTES AN OBS. WHOSE X VALUE GIVES IT LARGE INFLUENCE.

DURBIN-WATSON STATISTIC = 1.82
```

DISPLAY 10.4.4

```
-- REGR C2 5 C4 C13 C14 C5 C12 C15


NOTE - THE FOLLOWING VARIABLES ARE HIGHLY
CORRELATED WITH OTHER PREDICTOR VARIABLES
C12

   32 CASES USED
   11 CASES CONTAINED MISSING VALUES

THE REGRESSION EQUATION IS
Y = - .0520 + 0.337 X1 + 0.289 X2
     - 0.104 X3 -  1.21 X4 + 0.443 X5

                                          ST. DEV.     T-RATIO =
        COLUMN      COEFFICIENT           OF COEF.     COEF/S.D.
        --           -0.05198             0.03301         -1.57
X1      C4            0.33703             0.04137          8.15
X2      C13           0.28901             0.09674          2.99
X3      C14          -0.10361             0.04367         -2.37
X4      C5            -1.2084              0.2478         -4.88
X5      C12            0.4432              0.1380          3.21

THE ST. DEV. OF Y ABOUT REGRESSION LINE IS
S = 0.03334
WITH (  32- 6) =  26 DEGREES OF FREEDOM

R-SQUARED = 99.8 PERCENT
R-SQUARED = 99.8 PERCENT, ADJUSTED FOR D.F.

ANALYSIS OF VARIANCE

 DUE TO       DF           SS       MS=SS/DF
REGRESSION     5     15.28825        3.05765
RESIDUAL      26      0.02891        0.00111
TOTAL         31     15.31716


              X1          Y     PRED. Y     ST.DEV.
ROW           C4         C2       VALUE     PRED. Y     RESIDUAL       ST.RES.
 11         3.81    2.90885     2.95965     0.02222     -0.05080      -2.04RX
 12         3.22    2.81616     2.82129     0.02786     -0.00513      -0.28 X
 13         3.22    2.68550     2.70576     0.02748     -0.02026      -1.07 X
 14         3.26    2.82160     2.80474     0.02101      0.01686       0.65 X
 20         3.50    3.12399     3.04889     0.01054      0.07510       2.37R

R DENOTES AN OBS. WITH A LARGE ST. RES.
X DENOTES AN OBS. WHOSE X VALUE GIVES IT LARGE INFLUENCE.

DURBIN-WATSON STATISTIC = 1.67
```

have substantially reduced the residual mean square, $\hat{\sigma}^2$, although they have changed the $R^2$ value only slightly. The d-statistic has moved from the indeterminate negative autocorrelation range to the indeterminate positive autocorrelation range, while the h-statistic is now $h = 1.12$. Although the observations from the early 1950s are most influential in determining the regression coefficients, there is no indication of a structural change when the equation is estimated up to 1982-83 as Chow's F-test takes a value of $F_{3,26} = 1.89$.

A very interesting feature of Display 10.4.4 is the significance of the variables $Y_{t-1}$ and $Y_{t-2}$, since if the rational expectations hypothesis is being considered fully in conjunction with the permanent income hypothesis, neither of these variables should be relevant. Under rational expectations current consumption should be fully in equilibrium with the known level of permanent income, so that a change in the level of current consumption should only arise whenever new information comes to hand which affects the level of permanent income. By definition, new information not already taken fully into account in determining permanent income at time t should only arise from innovations in variables of date t, so the influence of $Y_{t-1}$ and $Y_{t-2}$ tends to throw doubt on the validity of the rational expectations hypothesis in this context, despite its popularity among economists and econometricians in recent years. (The situation is even worse than indicated in Display 10.4.4, since if one models first differences of consumption, then first differences of household disposable income at times t, t - 1, and t - 2 are all significant explanatory variables.)

Considering other possible explanatory variables compatible with the permanent income hypothesis, Zellner et al. (1965) have emphasized the importance of liquid assets of households as a component of wealth having a definite impact on consumption [see also Hendry and Ungern-Sternberg (1981)]. The measurement of a suitable variable is, however, a problem. In the present study the variable utilized was the sum of depositors' balances at savings banks, building societies, and credit unions on June 30 of the previous year, which was adjusted to 1979-80 money values using the GNE deflator and then put on a per capita basis. On the basis of the Zellner et al. study, this variable should have a positive effect which is lagged, since it is probably tied up with the purchase of consumer durables, with cash balances being built up prior to purchase. For this reason the model incorporating this variable was specified using the results of Display 10.4.4 as

$$C_t = \alpha_0 + \alpha_1 Y_t + \alpha_2 Y_{t-1} + \alpha_3 Y_{t-2} + \alpha_4 I_t + \alpha_5 L_t + \alpha_6 L_{t-1} + \alpha_7 L_{t-2} + \alpha_8 C_{t-1} + \epsilon_t. \qquad (10.4.8)$$

On estimation of this model (see Display 10.4.5), it was apparent that $\alpha_5 = 0$ and $\alpha_6 = 0$ would be an acceptable null hypothesis. Therefore, the equation was reestimated omitting $L_t$ and $L_{t-1}$, the results appearing in Display 10.4.6. In this case $\hat{\alpha}_7$ is not quite significant at the 5 percent level

DISPLAY 10.4.5

```
-- REGR C2 8 C4 C13 C14 C5 C8 - C10 C12

NOTE - THE FOLLOWING VARIABLES ARE HIGHLY
CORRELATED WITH OTHER PREDICTOR VARIABLES
C12

   32 CASES USED
   11 CASES CONTAINED MISSING VALUES

THE REGRESSION EQUATION IS
Y = - .0625 + 0.319 X1 + 0.288 X2
    - 0.124 X3 -  1.31 X4 + .0093 X5
    - .0671 X6 + .0962 X7 + 0.478 X8
```

| | COLUMN | COEFFICIENT | ST. DEV. OF COEF. | T-RATIO = COEF/S.D. |
|---|---|---|---|---|
| | -- | -0.06248 | 0.05204 | -1.20 |
| X1 | C4 | 0.31855 | 0.05194 | 6.13 |
| X2 | C13 | 0.2875 | 0.1022 | 2.81 |
| X3 | C14 | -0.12384 | 0.05768 | -2.15 |
| X4 | C5 | -1.3107 | 0.3050 | -4.30 |
| X5 | C8 | 0.0093 | 0.1352 | 0.07 |
| X6 | C9 | -0.0671 | 0.1632 | -0.41 |
| X7 | C10 | 0.09623 | 0.08746 | 1.10 |
| X8 | C12 | 0.4784 | 0.1482 | 3.23 |

```
THE ST. DEV. OF Y ABOUT REGRESSION LINE IS
S = 0.03349
WITH (  32- 9) =  23 DEGREES OF FREEDOM

R-SQUARED = 99.8 PERCENT
R-SQUARED = 99.8 PERCENT, ADJUSTED FOR D.F.

ANALYSIS OF VARIANCE
```

| DUE TO | DF | SS | MS=SS/DF |
|---|---|---|---|
| REGRESSION | 8 | 15.29136 | 1.91142 |
| RESIDUAL | 23 | 0.02580 | 0.00112 |
| TOTAL | 31 | 15.31716 | |

| ROW | X1 C4 | Y C2 | PRED. Y VALUE | ST.DEV. PRED. Y | RESIDUAL | ST.RES. |
|---|---|---|---|---|---|---|
| 9 | 3.20 | 2.71678 | 2.70258 | 0.02675 | 0.01420 | 0.70 X |
| 11 | 3.81 | 2.90885 | 2.96079 | 0.02567 | -0.05194 | -2.41RX |
| 12 | 3.22 | 2.81616 | 2.81973 | 0.02895 | -0.00357 | -0.21 X |
| 13 | 3.22 | 2.68550 | 2.69928 | 0.02919 | -0.01378 | -0.84 X |
| 20 | 3.50 | 3.12399 | 3.04346 | 0.01312 | 0.08053 | 2.61R |
| 34 | 5.31 | 4.41721 | 4.42079 | 0.02973 | -0.00359 | -0.23 X |
| 35 | 5.35 | 4.48452 | 4.45609 | 0.02884 | 0.02843 | 1.67 X |

```
R DENOTES AN OBS. WITH A LARGE ST. RES.
X DENOTES AN OBS. WHOSE X VALUE GIVES IT LARGE INFLUENCE.

DURBIN-WATSON STATISTIC = 1.68
```

DISPLAY 10.4.6

```
-- REGR C2 6 C4 C13 C14 C5 C10 C12 C15

NOTE - THE FOLLOWING VARIABLES ARE HIGHLY
CORRELATED WITH OTHER PREDICTOR VARIABLES
C12

   32 CASES USED
   11 CASES CONTAINED MISSING VALUES

THE REGRESSION EQUATION IS
Y = - .0528 + 0.314 X1 + 0.300 X2
    - 0.117 X3 -  1.38 X4 + .0549 X5
    + 0.453 X6

                                          ST. DEV.      T-RATIO =
      COLUMN     COEFFICIENT              OF COEF.      COEF/S.D.
      --           -0.05276               0.03199         -1.65
X1    C4            0.31396               0.04248          7.39
X2    C13           0.30036               0.09399          3.20
X3    C14          -0.11705               0.04310         -2.72
X4    C5           -1.3785                 0.2615         -5.27
X5    C10           0.05494               0.03347          1.64
X6    C12            0.4526                0.1339          3.38

THE ST. DEV. OF Y ABOUT REGRESSION LINE IS
S = 0.03231
WITH (  32- 7) =  25 DEGREES OF FREEDOM

R-SQUARED = 99.8 PERCENT
R-SQUARED = 99.8 PERCENT, ADJUSTED FOR D.F.

ANALYSIS OF VARIANCE

 DUE TO        DF          SS       MS=SS/DF
REGRESSION      6     15.29106       2.54851
RESIDUAL       25      0.02609       0.00104
TOTAL          31     15.31716

             X1          Y    PRED. Y    ST.DEV.
ROW          C4         C2      VALUE    PRED. Y    RESIDUAL      ST.RES.
  9        3.20    2.71678    2.70533    0.02456     0.01145       0.55 X
 11        3.81    2.90885    2.95761    0.02156    -0.04876      -2.03RX
 12        3.22    2.81616    2.81656    0.02715    -0.00040      -0.02 X
 13        3.22    2.68550    2.70158    0.02675    -0.01608      -0.89 X
 20        3.50    3.12399    3.04120    0.01124     0.08279       2.73R

R DENOTES AN OBS. WITH A LARGE ST. RES.
X DENOTES AN OBS. WHOSE X VALUE GIVES IT LARGE INFLUENCE.

DURBIN-WATSON STATISTIC = 1.69
```

in a one-tail test, but it does have the correct sign. When estimation was extended to 1982-83, the Chow F-test took the value $F_{3,25} = 1.58$, while the t-statistic associated with $\hat{\alpha}_7$ increased to $t = 1.94$, thus supporting the inclusion of this variable in the model. The long-run MPC for the model is estimated at 0.908, which seems to be an acceptable value. Overall, then, this final model seems fairly satisfactory in that residual variance is 40 percent smaller than for the best models of the previous sections, while the Durbin h-statistic takes a value of $h = 1.04$. Of course, given the additional variables that seem to be relevant, the models of the previous section require extension and reevaluation. Before this is done, the final hypothesis, the life cycle hypothesis, will be introduced and examined.

## 10.5 THE LIFE CYCLE HYPOTHESIS

This is similar to the permanent income hypothesis in stressing the importance of wealth for consumption but differs in its analysis of the consumption behavior of individuals over their life cycles. Beginning from the midteens when full-time work becomes a legal option, a large proportion of young people will save at significant levels in preparation for establishing a home of their own. The remaining young people (an increasing proportion) will consume more than they earn, while investing in further education (which is treated as consumption in the national accounts), but by their midtwenties virtually all young people will be saving a high proportion of their income as they accumulate such assets as a home and its associated consumer durables (which are treated as consumption in the national accounts).

As the children of the household grow, consumption expenditure will necessarily rise to meet their increasing requirements, then decline again as they become independent and leave home. Finally, as the parents reach retirement age, income will fall and consumption probably exceed income by drawing upon accumulated savings. These descriptions relate to the behavior of individuals, but they have demographic implications for aggregate behavior and so for the functional form used to model consumption behavior.

For example, an increase in the proportion of dependents in a population (infants plus full-time students plus retired people) is likely to be associated with an increase in consumption as a proportion of disposable income. On the other hand, an increase in the proportion of married women in paid employment, although perhaps motivated primarily by consumption needs initially, is likely to lead to a decrease in consumption as a proportion of disposable income, as it raises the income of the affected households relative to comparable one-income households, and so according to the relative income hypothesis, should raise their savings ratios.

Unfortunately, neither of these series are readily available for the full number of years over which we have been considering the Australian consumption function, so they cannot be utilized in an econometric model. However, as both are likely to change only slowly over time, one could use lagged consumption as a proxy variable to pick up any such demographic effects. This would take us once again back to the estimates obtained in Display 10.2.4 or 10.4.6 and provide one more possible justification for using lagged consumption as an explanatory variable. Note that in both Section 10.2 and here, $C_{t-1}$ is introduced in a rather ad hoc fashion and not through an explicit model as in the developments of Section 10.4. For this reason, because these sections do not provide us an a priori basis to assume anything different about the disturbance structure, this has been assumed to satisfy the usual least squares assumptions. Thus one avoids the logical impasse reached with equation (10.4.6), but since (10.4.6) may be true, a very careful investigation of residual autocorrelation is warranted in all models.

One further demographic variable that it may be of some interest to examine relative to Australian consumption is the annual rate of growth of the population. If this was principally a result of natural increase, it would be an indicator of the dependency ratio, with higher dependency ratios being associated with higher rates of growth of the population and hence, because of the lower proportion representing the productive work force, with lower per capita consumption levels, other things being equal. In fact, the growth of the Australian population includes a large migrant component, and this confuses the issue somewhat, since a high proportion of migrants are of productive ages and so should contribute to a larger national income. Nevertheless, because of the delay in fully absorbing migrants into the work force, it is likely that in years of high migrant intake, consumption on a per capita basis will be lower than it otherwise would have been. Thus one might extend the model of equation (10.4.8) to include the current rate of growth of the population, $P_t$.

$$C_t = \alpha_0 + \alpha_1 Y_t + \alpha_2 Y_{t-1} + \alpha_3 Y_{t-2} + \alpha_4 I_t + \alpha_5 P_t + \alpha_6 L_{t-2} + \alpha_7 C_{t-1} + \epsilon_t. \tag{10.5.1}$$

This model was estimated and appears in Display 10.5.1.

Considering the estimates in Display 10.5.1 relative to those in Display 10.4.6, it is apparent that the additional variable has had little effect on the other partial regression estimates. The coefficient $\hat{\alpha}_5$ is of correct sign but not significantly nonzero at the 5 percent level on a one-tail test. When the estimation period was extended to 1982-83, the t-statistics for $\hat{\alpha}_5$ dropped to 1.25, whereas that for $\hat{\alpha}_6$ increased to 1.95, thus throwing further doubt of the relevance of the additional variable. Chow's F-statistic for the additional periods takes the value $F_{3,24} = 1.92$; thus it does not support a hypothesis of a changed regression structure in recent years, even though the most influential observations relate to the early 1950s. Overall, however, one would prefer the results of Display 10.4.6 as being

DISPLAY 10.5.1

```
-- REGR C2 7 C4 C13 C14 C6 C7 C11 C12 C15

NOTE - THE FOLLOWING VARIABLES ARE HIGHLY
CORRELATED WITH OTHER PREDICTOR VARIABLES
C12

   32 CASES USED
   11 CASES CONTAINED MISSING VALUES

THE REGRESSION EQUATION IS
Y =    .0636 + 0.318 X1 + 0.264 X2
     - 0.139 X3 -   1.25 X4 -   2.65 X5
     + .0531 X6 + 0.496 X7

                                          ST. DEV.       T-RATIO =
      COLUMN      COEFFICIENT             OF COEF.       COEF/S.D.
      --              0.06357             0.08060            0.79
X1    C4              0.31829             0.04140            7.69
X2    C13             0.26359             0.09436            2.79
X3    C14            -0.13928             0.04425           -3.15
X4    C6             -1.2469               0.2678           -4.66
X5    C7              -2.649                1.694           -1.56
X6    C11             0.05308             0.03257            1.63
X7    C12              0.4958              0.1331            3.73

THE ST. DEV. OF Y ABOUT REGRESSION LINE IS
S = 0.03141
WITH (   32- 8) =  24 DEGREES OF FREEDOM

R-SQUARED = 99.8 PERCENT
R-SQUARED = 99.8 PERCENT, ADJUSTED FOR D.F.

ANALYSIS OF VARIANCE

 DUE TO        DF          SS        MS=SS/DF
REGRESSION      7    15.29348         2.18478
RESIDUAL       24     0.02368         0.00099
TOTAL          31    15.31716

             X1          Y     PRED. Y     ST.DEV.
ROW          C4         C2       VALUE     PRED. Y      RESIDUAL      ST.RES.
  9        3.20    2.71678     2.72604     0.02730      -0.00926      -0.60 X
 11        3.81    2.90885     2.94459     0.02256      -0.03574      -1.63 X
 12        3.22    2.81616     2.80991     0.02674       0.00625       0.38 X
 13        3.22    2.68550     2.69619     0.02623      -0.01069      -0.62 X
 20        3.50    3.12399     3.05010     0.01232       0.07389       2.56R

R DENOTES AN OBS. WITH A LARGE ST. RES.
X DENOTES AN OBS. WHOSE X VALUE GIVES IT LARGE INFLUENCE.

DURBIN-WATSON STATISTIC = 1.66
```

Reprinted with permission of Minitab, Inc.

likely to relate to the most appropriate model under both the permanent income and life cycle hypotheses. It is also reasonably appropriate under the absolute income hypothesis, in preference to (or as an extension of) the model presented in Display 10.2.4. Only the model presented in Display 10.3.6, which was preferred under the relative income hypothesis, is distinctly different in form, so an investigation of the compatibility of the results in Displays 10.3.6 and 10.4.6 will form the next section of this study.

## 10.6 RECONCILIATION OF RESULTS

Considering equations (10.4.8) and (10.3.6), one possible approach to the reconciliation of the results discussed above would be to write down a comprehensive model such as

$$C_t = \alpha_0 + \alpha_1 Y_t^{**} + \alpha_2 Y_{t-1}^{**} + \alpha_3 Y_{t-2}^{**} + \alpha_4 Z_{1t} + \alpha_5 Z_{1t-1} + \alpha_6 Z_{1t-2} + \alpha_7 I_t + \alpha_8 I_{t-1} + \alpha_9 I_{t-2} + \alpha_{10} L_t + \alpha_{11} L_{t-1} + \alpha_{12} L_{t-2} + \alpha_{13} C_{t-1} + \epsilon_t, \qquad (10.6.1)$$

and then, after estimation, drop nonsignificant variables until a stable result is achieved. Possible multicollinearity affecting the significance of particular variables is always a potential problem for such an approach since the model is specified in original variables rather than first differences. However, this approach was adopted and following estimation it was found that $\alpha_6$, $\alpha_8$, $\alpha_9$, and $\alpha_{10}$ were nonsignificant. When the corresponding variables had been dropped, the model was reestimated, the results being presented in Display 10.6.1.

It is apparent from the results of Display 10.6.1 that this model represents a significant improvement on the previous best results of Display 10.4.6, as the residual variance has decreased by over 20 percent, while the Durbin-Watson statistic is fully consistent with the hypothesis of no first-order autocorrelation in the disturbances. One feature of the results, however, deserves attention. This is the apparent almost equality in magnitude (apart from sign) of $\alpha_2$ and $\alpha_3$ and of $\alpha_{11}$ and $\alpha_{12}$, which is consistent with a possible model in first differences of the corresponding variables given the standard deviations of the coefficients. Thus, in a different notation, the new model would be specified as

$$C_t = \gamma_0 + \gamma_1 Y_t^{**} + \gamma_2 \Delta Y_{t-1}^{**} + \gamma_3 Z_{1t} + \gamma_4 Z_{1t-1} + \gamma_5 I_t + \gamma_6 \Delta L_{t-1} + \gamma_7 C_{t-1} + \epsilon_t, \qquad (10.6.2)$$

with current consumption being functionally related to $Y_t^{**}$ and its most recent previous change; to $Z_{1t}$ and $Z_{1t-1}$, the current and lagged increases

DISPLAY 10.6.1

```
-- REGR C2 9 C16-C20 C7 C14 C15 C12 C24


NOTE - THE FOLLOWING VARIABLES ARE HIGHLY
CORRELATED WITH OTHER PREDICTOR VARIABLES
C12

   32 CASES USED
   11 CASES CONTAINED MISSING VALUES

THE REGRESSION EQUATION IS
Y = - 0.127 + 0.489 X1 + 0.255 X2
    - 0.233 X3 + 0.199 X4 + 0.311 X5
    -  1.50 X6 - 0.289 X7 + 0.269 X8
    + 0.486 X 9

                                           ST. DEV.     T-RATIO =
       COLUMN      COEFFICIENT             OF COEF.     COEF/S.D.
       --            -0.12748              0.03996         -3.19
X1     C16            0.48890              0.07202          6.79
X2     C17             0.2553               0.1199          2.13
X3     C18           -0.23280              0.07105         -3.28
X4     C19            0.19865              0.05617          3.54
X5     C20            0.31099              0.08869          3.51
X6     C7             -1.5001               0.2439         -6.15
X7     C14            -0.2886               0.1285         -2.25
X8     C15            0.26929              0.09610          2.80
X9     C12             0.4857               0.1347          3.61

THE ST. DEV. OF Y ABOUT REGRESSION LINE IS
S = 0.02871
WITH (  32-10) =  22 DEGREES OF FREEDOM

R-SQUARED = 99.9 PERCENT
R-SQUARED = 99.8 PERCENT, ADJUSTED FOR D.F.

ANALYSIS OF VARIANCE

 DUE TO      DF          SS        MS=SS/DF
REGRESSION    9     15.29903        1.69989
RESIDUAL     22      0.01813        0.00082
TOTAL        31     15.31716

            X1          Y     PRED. Y    ST.DEV.
ROW        C16         C2       VALUE    PRED. Y     RESIDUAL      ST.RES.
 12       3.22    2.81616    2.79916    0.02658     0.01700       1.57 X
 20       3.50    3.12399    3.06290    0.01281     0.06109       2.38R
 35       5.31    4.48452    4.47191    0.02574     0.01261       0.99 X

R DENOTES AN OBS. WITH A LARGE ST. RES.
X DENOTES AN OBS. WHOSE X VALUE GIVES IT LARGE INFLUENCE.

DURBIN-WATSON STATISTIC = 1.99
```

(if any) in per capita income; to the inflation rate $I_t$, to the lagged change in per capita liquid assets; and to lagged consumption.

This model based on the relative income hypothesis seems to be consistent with an adaptive learning process being utilized by consumers as they attempt to process all the new information that continually floods in upon them (Severn, 1983). The model was estimated and is presented in Display 10.6.2. It is clear that the hypotheses $\alpha_2 = -\alpha_3$, $\alpha_{11} = -\alpha_{12}$ are accepted by the data since there is a small improvement in the residual variance estimate. All of the parameters are estimated fairly precisely, and the Durbin-Watson statistic supports the hypothesis of no first-order autocorrelation in the disturbances. The long-run marginal propensity to consume out of $Y_t^{**}$, given by $\hat{\gamma}_1/(1-\hat{\gamma}_7)$, is estimated at 0.9812, while the long-run propensity to consume out of $Z_1$, given by $(\hat{\gamma}_3 + \hat{\gamma}_4)/(1 - \hat{\gamma}_7)$, is estimated at 0.9631. The long-run propensity to consume from $\Delta Y_{t-1}^{**}$, given by $\hat{\gamma}_2/(1 - \hat{\gamma}_7)$, is much lower, being estimated at 0.4705. This is consistent with the original proposal by Duesenberry relating to the behavior of consumers in response to a change in their established income level. The sign of $\gamma_6$ gives cause for some concern since one might expect an increased level of consumption to follow an increase in the level of liquid wealth, but a negative sign could be consistent with components of the life cycle hypothesis (Bladen-Hovell and Richards, 1983; Hendry and Ungern-Sternberg, 1981; Davidson and Hendry, 1981).

Considering this final model, which has an $\bar{R}^2 = 0.9984$, while it seems to be based principally on the relative income hypothesis, there are some features inconsistent with this hypothesis. For example, the intercept is significantly negative rather than nonsignificant. This is probably related to the variables in the model other than relative income, so too much should not be made of this point. The inflation variable seems to have increased in importance relative to the model of Display 10.3.6, with a point estimate of the elasticity of consumption with respect to inflation now being $\epsilon = -0.06$ at the sample means of the variables. This implies as a point estimate, a decline in consumption of \$30.35 per capita at 1979-80 prices per 1 percent increase in the inflation rate. If one interprets the concept of permanent income in a more flexible fashion than does the standard adaptive expectations hypothesis, all of the variables $Y_t^{**}$, $\Delta Y_{t-1}^{**}$, $Z_{1t}$, $Z_{1t-1}$, and $\Delta L_{t-1}$ could be considered as contributors to the formation of a consumer's current understanding of permanent income. (The time delays involved would, however, be inconsistent with the rational expectations hypothesis.)

Overall, then, the model (10.6.2), which is accepted at this stage as a tentatively best model that can be subjected to further testing, would seem not to allow one to discriminate clearly among the alternative hypotheses but instead, to incorporate elements of several of the initially outlined hypotheses. Given that the Chow F-test for structural change in recent years takes a value of $F_{3,24} = 2.34$, it may well be that this model will soon be shown to be inadequate. It has, however, been considered against models in

DISPLAY 10.6.2

```
-- REGR C2 7 C16 C21 C19 C20 C6 C15 C12 C23

   32 CASES USED
   11 CASES CONTAINED MISSING VALUES

THE REGRESSION EQUATION IS
Y = - 0.119 + 0.487 X1 + 0.233 X2
    + 0.186 X3 + 0.292 X4 -  1.51 X5
    - 0.260 X6 + 0.504 X7

                                             ST. DEV.     T-RATIO =
         COLUMN      COEFFICIENT             OF COEF.     COEF/S.D.
         --            -0.11945              0.03251        -3.67
X1       C16            0.48693              0.06191         7.87
X2       C21            0.23348              0.06505         3.59
X3       C19            0.18616              0.04856         3.83
X4       C20            0.29177              0.06574         4.44
X5       C6            -1.5060                0.1694        -8.89
X6       C15           -0.26036              0.07315        -3.56
X7       C12            0.50374              0.07099         7.10

THE ST. DEV. OF Y ABOUT REGRESSION LINE IS
S = 0.02768
WITH (  32- 8) =  24 DEGREES OF FREEDOM

R-SQUARED = 99.9 PERCENT
R-SQUARED = 99.8 PERCENT, ADJUSTED FOR D.F.

ANALYSIS OF VARIANCE

 DUE TO      DF          SS       MS=SS/DF
REGRESSION    7     15.29877       2.18554
RESIDUAL     24      0.01839       0.00077
TOTAL        31     15.31716

FURTHER ANALYSIS OF VARIANCE
SS EXPLAINED BY EACH VARIABLE WHEN ENTERED IN THE ORDER GIVEN

 DUE TO      DF          SS
REGRESSION    7     15.29877
C16           1     15.16639
C21           1      0.00154
C19           1      0.01169
C20           1      0.00278
C6            1      0.07166
C15           1      0.00613
C12           1      0.03858

            X1          Y     PRED. Y     ST.DEV.
ROW        C16         C2       VALUE     PRED. Y     RESIDUAL      ST.RES.
 11       3.38    2.90885     2.92769     0.02016     -0.01884      -0.99 X
 12       3.22    2.81616     2.79319     0.02346      0.02297       1.56 X
 20       3.50    3.12399     3.06217     0.00931      0.06182       2.37R
 35       5.31    4.48452     4.47462     0.02411      0.00990       0.73 X

R DENOTES AN OBS. WITH A LARGE ST. RES.
X DENOTES AN OBS. WHOSE X VALUE GIVES IT LARGE INFLUENCE.

DURBIN-WATSON STATISTIC = 2.05
```

Reprinted with permission of Minitab, Inc.

first differences, models in logarithms, and models in first differences of logarithms, in research not reported here, and found to be more acceptable than any of those alternatives.

Turning to Hendry's chi-square test of forecasts (see Davidson et al., 1978, 674), this is based on the asymptotic approximation

$$\frac{\Sigma_{i=1}^{F} (C_{T+i} - \hat{C}_{T+i})^2}{\hat{\sigma}^2} \sim \chi^2(F),$$

where F is the number of periods over which forecasts are produced, $\hat{\sigma}^2$ the residual variance estimated for the periods used to fit the regression model, and $(C_{T+i} - \hat{C}_{T+i})$ the difference between the actual realized value at time T + i and the point forecast made for T + i using the actual future values of the regressor variables. In this case estimation has been carried out over the periods 1948-49 to 1979-80, with forecasts being generated for the following financial years. The chi-square value obtained was $\chi^2(3) = 9.97$, a value that would be exceeded in only 2 percent of cases if the model was correct, so this suggests that our final model is still inadequate even though the standard error is only 0.6 percent of the 1979-80 consumption level.

Considering possible reasons for model inadequacy, one that has been largely ignored in this chapter is the simultaneous existence of the identity which involves household savings, $S_t$, which is

$$Y_t = C_t + S_t,$$

along with the consumption function currently being estimated. As discussed in Chapter 11, this has the effect of introducing bias and inconsistency into OLS estimates of the parameters of the consumption relationship. Knowledge of this problem, which affects all the estimates in this chapter, should make for caution in accepting particular point estimates of coefficients. For example, the derived long-run MPC for model (10.6.2) seems to be rather too large. However, given the fit obtained to the empirical data, it is likely that the effect of simultaneous equations bias on the estimated coefficients should be quite small, so that ignoring the simultaneous equations problem in this instance may be acceptable.

Further, although the lagged dependent variable is used as a regressor in equation (10.6.2), which introduces an element of small-sample bias into parameter estimates, the potential problem of correlation between this variable and the disturbance (discussed in detail in Chapter 11) does not seem to have been encountered. Overall, therefore, equation (10.6.2) seems to be a reasonable base from which to begin a future study of the Australian consumption function. As we conclude this first part of the book, it is our hope that the "thinking out loud" process presented in this chapter will be of assistance to you in your future attempts to come to grips with the challenge presented by the task of measuring economic relationships. Good luck!

# Part II
# MORE ADVANCED TOPICS

# 11

# Stochastic Regressors, Instrumental Variables, and Simultaneous Equations

## 11.1 STOCHASTIC REGRESSORS[1]

In our early discussion of the regression model[2]

$$Y_t = \beta_0 + \beta_1 X_{1t} + \beta_2 X_{2t} + \cdots + \beta_K X_{Kt} + u_t$$

it was assumed that the disturbance $u_t$ satisfied the three conditions

(a) $E(u_t) = 0$;

(b) $E(u_t^2)$ is constant; and

(c) $E(u_t u_s) = 0$ when $t \neq s$.

In addition, it was assumed that the regressors were fixed numbers and not random variables. By this we mean that if it were possible to repeat the experiment, while the values of the dependent variable $Y_t$ would be different (because of the influence of the random variable $u_t$), the values of the regressors would remain the same.

Under these conditions we could establish that ordinary least squares produced estimators that were linear and unbiased and had the smallest variances of any such estimators. The next stage in our study was to relax the assumptions about the disturbance $u_t$ and to discover how OLS could be modified to produce good estimates when either (b) or (c) did not apply. This enabled us to deal with the practical problems of heteroscedasticity and autocorrelation.

The assumption about fixed regressor variables may be quite realistic in the context of a controlled experiment in which the designer sets the values of the variables (e.g., how much superphosphate to apply to a wheat crop). However, in the uncontrolled situation in which the econometrician usually has to work, it is much more likely that the regressor variables as well as the dependent variable will be stochastic variables. We will outline just three situations that will ensure the stochastic nature of the regressors.

## Errors in Measurement

Suppose for simplicity that the values that a regressor $X_t$ takes are fixed, but that there are errors of measurement, so that what is observed is $\tilde{X}_t$, say, where

$$\tilde{X}_t = X_t + \epsilon_t$$

and $\epsilon_t$ is a measurement error. Such errors would almost always be random, implying that $\tilde{X}_t$ must then also be a random variable.

## Lagged Dependent Variables

In our study of the consumption function, we have seen that a model of the form

$$C_t = \beta_0 + \beta_1 C_{t-1} + \beta_2 Y_t + u_t$$

is sometimes used. Now if $C_t$ is a stochastic variable (as it must be), $C_{t-1}$ is also stochastic. We thus have at least one stochastic regressor in this model.

## Simultaneous Causality (Feedback)

It has always been assumed that the X's cause the dependent variable Y, but that Y does not cause the X's. Let us now suppose that Y does cause or influence some of the regressors. In the context of a simple regression model,

$$Y_t = \beta_0 + \beta_1 X_t + u_t$$

expresses the influence that $X_t$ exerts on $Y_t$. If $Y_t$ also influences $X_t$, there must also be a relationship, perhaps of the form

$$X_t = \gamma_0 + \gamma_1 Y_t + v_t.$$

Using the first equation to eliminate $Y_t$ from the second equation, we have

$$X_t = \frac{\gamma_0 + \gamma_1 \beta_0}{1 - \gamma_1 \beta_1} + \frac{\gamma_1 u_t + v_t}{1 - \gamma_1 \beta_1}.$$

Being a linear combination of random variables, the term $(\gamma_1 u_t + v_t)/(1 - \gamma_1 \beta_1)$ must be a random variable, and hence $X_t$ is stochastic. In general, whenever there is any simultaneous feedback from dependent variable to regressor variable(s), the regressor(s) must be stochastic.

We will provide one more (familiar) example of this phenomenon. The simple Keynesian consumption function is

$$C_t = \beta_0 + \beta_1 Y_t + u_t.$$

We also know that there is the accounting identity

$$Y_t = C_t + A_t,$$

where $A_t$ is autonomous expenditure, taking investment, government expenditure, exports, and imports into account. Thus while $Y_t$ influences $C_t$, $C_t$ also influences $Y_t$.

Eliminating $C_t$, we have

$$Y_t = (\beta_0 + \beta_1 Y_t + u_t) + A_t$$

or

$$Y_t = \frac{\beta_0}{1 - \beta_1} + \frac{1}{1 - \beta_1} A_t + \frac{u_t}{1 - \beta_1}.$$

The term $u_t/(1 - \beta_1)$ on the right-hand side guarantees that $Y_t$ must be stochastic.

Having established that fixed regressors are likely to be the exception rather than the rule, let us return to the derivation of the properties of OLS and see where the assumption of fixed regressors was used. As usual, we will look only at the simple regression model.

First, it was shown [see equations (2.3.5) and (2.3.6)] that

$$\hat{\beta}_1 = \sum_{t=1}^{N} b_t Y_t, \tag{11.1.1}$$

where

$$b_t = \frac{x_t}{\sum_{i=1}^{N} x_i^2}.$$

From this expression for $\hat{\beta}_1$ we inferred that it was a linear estimator, that is, a linear combination of the observations on the dependent variable $Y_t$. However, (11.1.1) is a linear combination of the $Y_t$ only if the $b_t$ are fixed numbers. If the X's are fixed, this will be true, whereas if they are stochastic, it will not.

Second, it has been shown that under the usual assumptions on $u_t$, $\hat{\beta}_1$ is unbiased. This was proved as follows: Algebraic manipulations enabled us to express $\hat{\beta}_1$ in the form

$$\hat{\beta}_1 = \beta_1 + \frac{\Sigma xu}{\Sigma x^2}. \tag{11.1.2}$$

Then, taking expectations

$$E(\hat{\beta}_1) = \beta_1 + E\left(\frac{\Sigma xu}{\Sigma x^2}\right)$$

$$= \beta_1 + \frac{\Sigma x E(u)}{\Sigma x^2} = \beta_1$$

as $E(u) = 0$. Thus, $E(\beta_1) = \beta_1$ proving that $\hat{\beta}_1$ is an unbiased estimator. The crucial step in this derivation is

$$E\left(\frac{\Sigma xu}{\Sigma x^2}\right) = \frac{\Sigma x E(u)}{\Sigma x^2}.$$

This equality is true only if the x's are fixed numbers. Thus, when the regressors are stochastic, it is no longer necessarily true that OLS is unbiased.

Clearly, in taking our analysis one step closer to the "real world," by recognizing the stochastic nature of most regressors in the uncontrolled environment, we must reassess the properties of our fundamental tool, ordinary least squares. To do this, it is necessary to introduce a few additional statistical concepts.

## 11.2 SOME ASYMPTOTIC DEFINITIONS AND RESULTS

A consequence of the fact that OLS is no longer a linear estimator when the regressors are stochastic is that it is usually no longer possible to find the distribution of $\hat{\beta}_1$. However, by using theorems analogous to the central limit theorem, it is usually possible to find the approximate distribution when the sample size is large. The study of statistical behavior when samples are large (or, strictly, infinite) is called asymptotic theory. Although a thorough study of this branch of statistics is completely beyond the scope of this book, an appreciation of a few basic concepts will stand us in good stead.

### Consistency

The most important large-sample property of an estimator is consistency. We will illustrate this concept using a (hopefully) familiar example from elementary statistics.

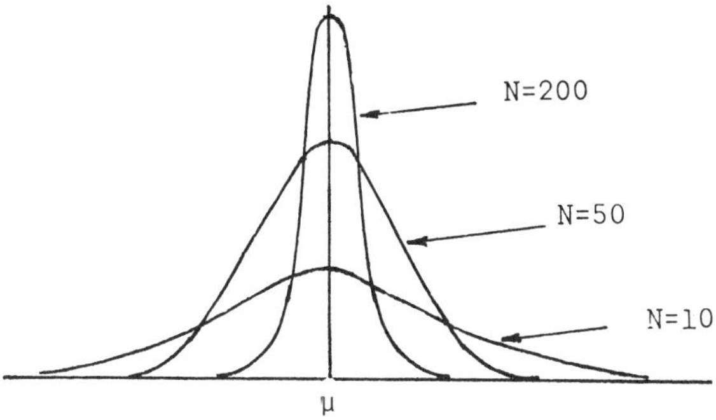


Figure 11.2.1. Distribution of $\bar{X}$ for different values of N.

We know that the estimator of the population mean $\mu$ which is usually used is the sample mean $\bar{X}$. Also, if the population is normal,

$$\bar{X} \sim N(\mu, \sigma^2/N).$$

Thus the mean of the distribution of $\bar{X}$ is always $\mu$, but the variance decreases as the sample size N increases. That is, as N increases the distribution of $\bar{X}$ "collapses" about the population mean $\mu$, as illustrated in Figure 11.2.1. This means that the difference between $\bar{X}$ and $\mu$ can almost certainly be made as small as we like (e.g., $|\bar{X} - \mu| < 10^{-3}$ or $|\bar{X} - \mu| < 10^{-6}$) by making the sample size large enough.

Whenever an estimator has the property that its distribution collapses about a parameter as the sample size increases, it is said to be a consistent estimator of that parameter. In the example above, the estimator $\bar{X}$ is an unbiased estimator of $\mu$ $[E(\bar{X}) = \mu]$. However, this will not always be the case, and biased estimators are often consistent. A sufficient (but not necessary) condition for an estimator $\hat{\alpha}$, based on a sample of size N, to be a consistent estimator of the parameter $\alpha$ is that

$$\text{(i)} \quad \lim_{N \to \infty} E(\hat{\alpha}) = \alpha; \text{ and}$$

$$\text{(ii)} \quad \lim_{N \to \infty} \text{var}(\hat{\alpha}) = 0. \qquad (11.2.1)$$

That is, if there is any bias, it diminishes to zero with increasing sample size, and the variance also decreases to zero.

## Probability Limit (plim)

Let us denote by $X^{(N)}$ a random variable whose value is determined by a sample of size N. For example, $X^{(N)}$ could be a sample mean, a sample variance, a sample correlation coefficient, and a regression estimate $\hat{\beta}$—to give just a few common examples. If the distribution of $X^{(N)}$ collapses around a fixed number $\alpha$ as $N \rightarrow \infty$, we say that "the probability limit of $X^{(N)}$ is $\alpha$" and write[3]

$$\text{plim } X^{(N)} = \alpha. \tag{11.2.2}$$

Clearly, if an estimator $\hat{\alpha}$ is a consistent estimator of a parameter $\alpha$, then plim $\hat{\alpha} = \alpha$.

There are some very simple rules for finding probability limits of functions of random variables. We state them without proof.

(i) If C is a constant,

$$\text{plim } C = C.$$

(ii) If $X_1$ and $X_2$ are two random variables,

$$\text{plim}(X_1 X_2) = (\text{plim } X_1)(\text{plim } X_2).$$

(iii) $\text{plim}(X_1 + X_2) = \text{plim}(X_1) + \text{plim}(X_2)$.

(iv) If $g(X)$ is a continuous function of X, then

$$\text{plim}[g(X)] = g[\text{plim}(X)].$$

This last result is very important. It tells us, for example, that if plim $X = \alpha$, then

$$\text{plim } X^2 = \alpha^2,$$

$$\text{plim}(1/X) = 1/\alpha,$$

and so on.

## Asymptotic Efficiency

So far we have used the term efficiency in connection with unbiased estimators. If $\hat{\beta}$ and $\tilde{\beta}$ are two unbiased estimators, we say that $\hat{\beta}$ is more efficient than $\tilde{\beta}$ if

$$\text{var}(\hat{\beta}) < \text{var}(\tilde{\beta}).$$

Furthermore, we would say that $\hat{\beta}$ is efficient (relative to some class of estimators) if $\text{var}(\hat{\beta})$ is smaller than that of any other estimator belonging to the class. For example, the OLS estimator (under the usual assumptions) is efficient among the class of linear unbiased estimators. When we are considering large-sample properties, a question arises, as to whether it is possible to talk about asymptotic efficiency of consistent estimators, given that the variances are almost always zero (for samples of infinite size). We will approach this question by once again using an example.

We have seen that $\bar{X}$ is a consistent estimator of $\mu$, with distribution $N(\mu, \sigma^2/N)$. The <u>median</u> $\bar{X}_M$, as N becomes very large, has a distribution which approximates to $N(\mu, \pi\sigma^2/2N)$. Thus, as $N \to \infty$, the variances of both estimators approach zero. If, instead of $\bar{X}$ and $\bar{X}_M$ we consider $\sqrt{N}(\bar{X} - \mu)$ and $\sqrt{N}(\bar{X}_M - \mu)$, we have that

$$\sqrt{N}(\bar{X} - \mu) \sim N(0, \sigma^2)$$

$$\sqrt{N}(\bar{X}_M - \mu) \sim N(0, \pi\sigma^2/2)$$

and $\sigma^2 < \pi\sigma^2/2$. On this basis we would say that asymptotically $\bar{X}$ is more efficient than $\bar{X}_M$.

We generalize this to any pair of consistent estimators $\hat{\beta}$ and $\tilde{\beta}$. We say that $\hat{\beta}$ is asymptotically more efficient than $\tilde{\beta}$ if

$$\text{var}[\sqrt{N}(\hat{\beta} - \beta)] < \text{var}[\sqrt{N}(\tilde{\beta} - \beta)]. \qquad (11.2.3)$$

## 11.3 PROPERTIES OF OLS WHEN THE REGRESSORS ARE STOCHASTIC

It will be seen that the properties of OLS when the regressors are stochastic depends crucially on the relationship between the regressors and the disturbance. We consider three different categories of this relationship.

### The Regressors Are Independent of the Disturbance

Suppose that we have the model[4]

$$Y_t = \beta_0 + \beta_1 X_t + u_t,$$

where the observations on X are u are <u>independent</u>. It is important to realize that we are assuming that <u>every</u> observation on X is independent of <u>every</u> observation on u. That is, $X_t$ and $u_s$ are independent, regardless of the values of t and s. Before we consider the implications of this assumption, we will pause to see when it is likely to be realistic.

Let us return to the simple Keynesian consumption function with the accounting identity.

$$C_t = \beta_0 + \beta_1 Y_t + u_t,$$

$$Y_t = C_t + A_t.$$

We will assume that $A_t$ is stochastic but autonomous. By this we mean that it is not influenced in any way by the levels of consumption ($C_t$) or income ($Y_t$). By implication from the first of these equations, $A_t$ must also be independent of $u_t$.

Suppose now that we eliminate $Y_t$ from the consumption equation. Thus

$$C_t = \beta_0 + \beta_1(C_t + A_t) + u_t.$$

That is,

$$C_t = \frac{\beta_0}{1 - \beta_1} + \frac{\beta_1}{1 - \beta_1}A_t + \frac{u_t}{1 - \beta_1}.$$

This equation is an example of a regression model in which the regressor (A) is stochastic but independent of the disturbance $[u(1 - \beta_1)]$. It is also clear from our argument that every value of A must be independent of every value of u. In general, we can say that provided that the causation goes from regressor to dependent variable but not vice versa, the assumption of independence is likely to be realistic.

We will now examine the expectation of the OLS estimator $\hat{\beta}$ under the independence assumption. We have [see equation (11.1.2)]

$$\hat{\beta}_1 = \beta_1 + \frac{\Sigma xu}{\Sigma x^2}.$$

Therefore,

$$E(\hat{\beta}_1) = \beta_1 + E\left(\frac{\Sigma xu}{\Sigma x^2}\right). \tag{11.3.1}$$

Looking in detail at the last expression, we have

$$E\left(\frac{\Sigma xu}{\Sigma x^2}\right) = E\left(\frac{x_1u_1}{\Sigma x^2} + \frac{x_2u_2}{\Sigma x^2} + \cdots + \frac{x_Nu_N}{\Sigma x^2}\right)$$

$$= E\left(\frac{x_1u_1}{\Sigma x^2}\right) + E\left(\frac{x_2u_2}{\Sigma x^2}\right) + \cdots + E\left(\frac{x_Nu_N}{\Sigma x^2}\right).$$

The first of these terms is

$$E\left(\frac{x_1u_1}{\Sigma x^2}\right) = E\left(\frac{x_1}{x_1^2 + x_2^2 + \cdots + x_N^2}u_1\right).$$

Now if A and B are any two independent random variables

$$E(AB) = E(A)E(B).$$

As $u_1$ is independent of all the x's, it must be independent of the random variable $x_1/(x_1^2 + x_2^2 + \cdots + x_N^2)$. Thus

$$E\left(\frac{x_1}{x_1^2 + x_2^2 + \cdots + x_N^2} u_1\right) = E\left(\frac{x_1}{x_1^2 + x_2^2 + \cdots + x_N^2}\right) E(u_1) = 0,$$

because $E(u_1) = 0$.

Applying the same reasoning to all the other terms, we obtain

$$E\left(\frac{\Sigma xu}{\Sigma x^2}\right) = \Sigma E\left(\frac{x}{\Sigma x^2}\right) E(u) = 0,$$

and it follows that

$$E(\hat{\beta}_1) = \beta_1. \tag{11.3.2}$$

Thus, when X is stochastic but independent of u, OLS is unbiased. Because $\hat{\beta}_1$ is no longer a linear estimator, it is difficult, if not impossible, to determine the distribution of $\hat{\beta}_1$ for finite samples. It is important, therefore, to examine the asymptotic properties. We have

$$\hat{\beta}_1 = \beta_1 + \frac{\Sigma xu}{\Sigma x^2} = \beta_1 + \frac{\Sigma xu/N}{\Sigma x^2/N}.$$

Therefore,

$$\text{plim } \hat{\beta}_1 = \beta_1 + \frac{\text{plim}(\Sigma xu/N)}{\text{plim}(\Sigma x^2/N)}. \tag{11.3.3}$$

For $\hat{\beta}_1$ to be a consistent estimator of $\beta_1$ (i.e., plim $\hat{\beta}_1 = \beta_1$), we are going to need

(i) $\text{plim}(\Sigma xu/N) = 0$,

(ii) $\text{plim}(\Sigma x^2/N) \neq 0$.

Condition (ii) involves only the regressor variable and means that there is some variability in the observations on the regressor. We will assume that this is always true. We can indicate why condition (i) is fulfilled when x and u are independent, as follows:

Independence implies zero covariance. Thus

$$\begin{aligned} 0 = \text{cov}(x, u) &= E[(x - E(x))(u - E(u))] \\ &= E[(x - E(x))u] \quad \text{as } E(u) = 0 \\ &= E(xu) - E(x)E(u) = E(xu), \quad \text{again because } E(u) = 0. \end{aligned}$$

Thus, independence between x and u implies that $E(xu) = 0$. But $\Sigma xu/N$ is simply the sample analog of $E(xu)$. Thus, as $N \to \infty$, we would expect the sample value to approach the population value. That is,[5]

$$\text{plim}\left(\frac{\Sigma xu}{N}\right) = E(xu) = 0.$$

It follows then that

$$\text{plim } \hat{\beta}_1 = \beta_1; \tag{11.3.4}$$

that is, $\hat{\beta}_1$ is a consistent estimator of $\beta_1$. Furthermore, it can be shown that the usual OLS expressions for $\hat{\sigma}^2$ and $S^2_{\hat{\beta}_1}$ are unbiased and consistent estimators for $\sigma^2$ and $\text{var}(\hat{\beta})$. There is one further point of interest: If the disturbance $u_t$ is assumed to be normally distributed, OLS is asymptotically efficient.

We summarize the results of this section as follows: If the regressors are stochastic, but independent of the disturbance, OLS has the following properties:

(i) The estimators $\hat{\beta}_1$, $\hat{\sigma}^2$, and $S^2_{\hat{\beta}_1}$ are unbiased in finite samples.

(ii) These estimators are consistent.

(iii) OLS estimators are asymptotically efficient if the disturbances are normally distributed.

## Regressors and Disturbance Contemporaneously Independent

In the preceding section we assumed that all values of the regressors were independent of all values of the disturbance. We now relax this asumption somewhat and assume that only observations belonging to the same time period are independent. Consider the following model:

$$Y_t = \beta_0 + \beta_1 X_t + u_t,$$

where

$$X_t = Y_{t-1}.$$

That is, the regressor is a lagged dependent variable.

Clearly, $Y_t$ depends on $u_t$ and therefore $Y_{t-1}$ (= $X_t$) depends on $u_{t-1}$. Furtheremore, $X_t$ (= $Y_{t-1}$) depends on $Y_{t-2}$ and hence on $u_{t-2}$, and so on. Thus the current value of the regressor depends on all past values of the disturbance. Provided that there is no serial correlation in the disturbances, it is reasonable to assume that $X_t$ is independent of $u_t$. However, regressor and disturbance are only contemporaneously independent. This will be true whenever the regressors include lagged dependent variables and the disturbance is not affected by autocorrelation.

Given a model of this type, we may well ask: What is the expectation of $\hat{\beta}_1$? From (10.1.2),

$$E(\hat{\beta}_1) = \beta_1 + E\left(\frac{\Sigma xu}{\Sigma x^2}\right).$$

The first term is again

$$E\left(\frac{x_1}{x_1^2 + x_2^2 + \cdots + x_N^2} u_1\right),$$

but now while $u_1$ is independent of $x_1$ (and $x_1^2$), it is not independent of $x_2^2, \ldots, x_N^2$. Thus

$$E\left(\frac{x_1}{x_1^2 + x_2^2 + \cdots + x_N^2} u_1\right) \neq E\left(\frac{x_1}{x_1^2 + x_2^2 + \cdots + x_N^2}\right) E(u_1),$$

and we are therefore unable to establish that $\hat{\beta}_1$ is unbiased. In fact, it can be shown that in general $\hat{\beta}_1$ is a biased estimator of $\beta_1$.

However, even though $\hat{\beta}_1$ is biased in small samples, it may possess desirable asymptotic properties. Referring back to the consistency proof in the preceding section, we can see that the consistency or otherwise of $\hat{\beta}_1$ depends crucially on whether $\text{plim}(\Sigma xu/N) = 0$. Now

$$\text{plim}\left(\frac{\Sigma xu}{N}\right) = \text{plim}\left(\frac{x_1u_1 + x_2u_2 + \cdots + x_Nu_N}{N}\right)$$

and thus involves only contemporaneous covariances (i.e., $x_1$ with $u_1$, $x_2$ with $u_2$, etc.). The fact that x and u are contemporaneously independent is sufficient to ensure that $\text{plim}(\Sigma xu/N) = 0$[6] and hence

$$\text{plim}\ \hat{\beta}_1 = \beta_1 .$$

The OLS estimator is therefore consistent, even though it is biased in small samples. It can also be shown that $\hat{\sigma}^2$ and $S^2_{\hat{\beta}_1}$ are consistent estimators of $\sigma^2$ and $\text{var}(\hat{\beta}_1)$ and that if $u_t$ is normal, OLS is asymptotically efficient.

We have considered the properties of OLS under the assumption that the regressors and disturbance are contemporaneously independent. As the only case of practical interest is when lagged dependent variables occur among the regressors and when there is no autocorrelation in the disturbances, we will summarize the results in these terms as follows: When the regressors include lagged dependent variables and the disturbance is not autocorrelated, OLS has the following properties:

(i) OLS estimators of $\beta$, $\sigma^2$, and $\text{var}(\beta)$ are biased in small samples.
(ii) These estimators are consistent.
(iii) OLS estimators are asymptotically efficient if the disturbance is distributed normally.

Regressors and Disturbance Contemporaneously Correlated

Contemporaneous correlation between regressors and disturbance arises usually from two situations.

(i) The regressors contain lagged dependent variables and there is autocorrelation in the disturbance. We will illustrate this using the simple example of the preceding section, namely

$$Y_t = \beta_0 + \beta_1 Y_{t-1} + u_t.$$

We have seen that $Y_{t-1}$ depends on $u_{t-1}, u_{t-2}, \ldots$. If there is autocorrelation, $u_t$ will also depend on at least some of the past disturbances. This will give rise to correlation between $Y_{t-1}$ and $u_t$.

(ii) Whenever there is feedback, that is, not only do the regressors cause the dependent variable, but it also simultaneously influences the regressors—then this correlation occurs. Consider again the consumption function

$$C_t = \beta_0 + \beta_1 Y_t + u_t$$

with the accounting identity

$$Y_t = C_t + A_t.$$

Eliminating $C_t$, we obtain

$$Y_t = \frac{\beta_0}{1 - \beta_1} + \frac{1}{1 - \beta_1} A_t + \frac{u_t}{1 - \beta_1}.$$

Clearly, $Y_t$ is influenced by (and hence, correlated with) $u_t$.

A further example will reinforce this point. Consider the supply and demand for an agricultural commodity.

$$\text{Supply:} \quad Q_t = \beta_0 + \beta_1 P_t + \beta_2 R_t + u_t$$

$$\text{Demand:} \quad Q_t = \gamma_0 + \gamma_1 P_t + \gamma_2 Y_t + v_t,$$

where $P_t$ and $Q_t$ are price and quantity, $R_t$ is rainfall, and $Y_t$ is income. We know that price and quantity are determined simultaneously, and this will imply that $P_t$ is correlated with both the disturbances $u_t$ and $v_t$. We can prove this as follows. Eliminating $Q_t$,

$$P_t = \frac{\beta_0 - \gamma_0}{\gamma_1 - \beta_1} + \frac{\beta_2}{\gamma_1 - \beta_1} R_t - \frac{\gamma_2}{\gamma_1 - \beta_1} Y_t + \frac{u_t - v_t}{\gamma_1 - \beta_1}.$$

Thus $P_t$ is influenced by the disturbance $(u_t - v_t)/(\gamma_1 - \beta_1)$ and is therefore correlated with both $u_t$ and $v_t$.

Our analysis of the previous sections enables us to infer the properties of OLS. First, it is still the case that

$$E\left(\frac{x_1}{x_1^2 + x_2^2 + \cdots + x_N^2}\right) u_1 \neq 0,$$

and it therefore follows that $\hat{\beta}_1$ is biased. Second, recalling that $\text{plim}(\Sigma xu/N)$ is essentially the population contemporaneous covariance between regressor and disturbance, it follows that $\text{plim}(\Sigma xu/N) \neq 0$ and hence

$$\text{plim}\ \hat{\beta} \neq \beta. \qquad (11.3.5)$$

Thus in this situation, OLS is not only biased but also inconsistent. This means that no matter how large our sample, there is no reason to believe that OLS can give us good estimates. Obviously, when regressors and disturbance are correlated, another method of estimation must be found. The inconsistency of OLS is illustrated in Figure 11.3.1 for the simple regression model. Assume that $X_t$ and $u_t$ are positively correlated. Then large positive values of $u_t$ will tend to be associated with large $X_t$ and large negative values with small $X_t$. Clearly $\hat{\beta}_1$ will be positively biased, and increasing the sample size will not eliminate this bias.

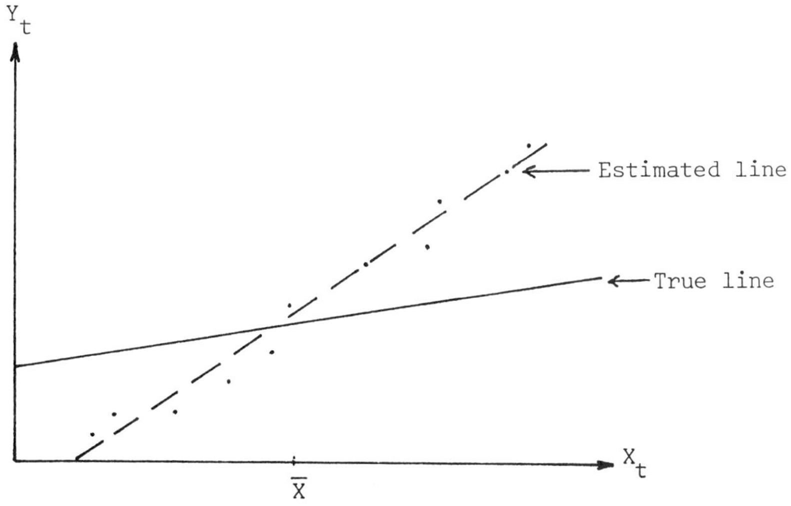


Figure 11.3.1. Bias of OLS even when the sample is large.

Table 11.3.1

| | | Properties of OLS | |
|---|---|---|---|
| Case | Usual occurrence | Finite sample | Asymptotic[a] |
| I. Regressors fixed in repeated samples | Controlled experiments | BLUE | Consistent efficient |
| II. All observations on regressors independent of all observations on disturbance | Regressors stochastic but autonomously determined | Unbiased | Consistent efficient |
| III. Regressors and disturbance contemporaneously uncorrelated | Regressors include lagged dependent variable; disturbance not autocorrelated | Biased | Consistent efficient |
| IV. Regressors and disturbance contemporaneously correlated | Regressors include lagged dependent variable and disturbance is autocorrelated or Simultaneous feedback from dependent variable to regressors | Biased | Inconsistent |

[a]The efficiency results depend on the disturbance being normally distributed.

We summarize as follows: When the regressors and disturbance are contemporaneously correlated, OLS is biased in small samples and inconsistent. We now draw together the major conclusions on the properties of OLS estimators in Table 11.3.1.

## 11.4 THE INSTRUMENTAL VARIABLE TECHNIQUE

We have seen that whether or not OLS is a consistent estimator depends on how the regressor is related to the disturbance. In particular, if

$\text{plim}(\Sigma xu/N) = 0$, OLS is consistent; otherwise, it is not. The instrumental variable (IV) technique is a method of constructing a consistent estimator when $\text{plim}(\Sigma xu/N) \neq 0$. We will approach the development of the IV estimator in an intuitive way using a simple regression model. Let us suppose that we can find a variable $Z_t$ (denote $Z_t - \bar{Z}$ by $z_t$) which is (a) correlated with $X_t$ ($\text{plim}\ \Sigma zx/N \neq 0$); and (b) uncorrelated with $u_t$ ($\text{plim}\ \Sigma zu/N = 0$).

If (a) holds, the information in $X_t$ will to a certain extent be reflected in $Z_t$. If (b) holds, $Z_t$ will avoid the fundamental problem of $X_t$ being correlated with $u_t$. In deviation form,

$$y_t = \beta_1 x_t + u_t.$$

Also, if $\tilde{\beta}_1$ is some estimate of $\beta_1$,

$$y_t = \tilde{\beta}_1 x_t + e_t, \tag{11.4.1}$$

where $e_t$ is the residual (not the disturbance).

Let us now choose $\tilde{\beta}_1$ so that the residuals reflect the desired characteristic of the population, namely $\text{plim}(\Sigma zu/N) = 0$. Thus we will choose $\tilde{\beta}_1$ in such a way that $\Sigma ze/N = 0$. From (11.4.1)

$$\begin{aligned} \Sigma ze &= \Sigma(zy - \tilde{\beta}_1 zx) \\ &= \Sigma zy - \tilde{\beta}_1 \Sigma zx. \end{aligned}$$

Therefore, if

$$\tilde{\beta}_1 = \frac{\Sigma zy}{\Sigma zx}, \tag{11.4.2}$$

then $\Sigma ze = 0$. We define the instrumental variable estimator of $\beta_1$ by (11.4.2). This estimator will exist provided that $\Sigma zx \neq 0$, that is, provided Z and X are correlated. It can easily be shown that the condition $\Sigma e = 0$ implies that

$$\tilde{\beta}_0 = \bar{Y} - \tilde{\beta}_1 \bar{X}. \tag{11.4.3}$$

We will now examine the properties of the IV estimator. From (11.4.2),

$$\tilde{\beta}_1 = \frac{\Sigma z(\beta_1 x + u)}{\Sigma zx} = \beta_1 \frac{\Sigma zx}{\Sigma zx} + \frac{\Sigma zu}{\Sigma zx}.$$

Therefore,

$$\tilde{\beta}_1 = \beta_1 + \frac{\Sigma zu/N}{\Sigma zx/N}.$$

Taking probability limits, using the conditions $\text{plim}(\Sigma zu/N) = 0$ and $\text{plim}(\Sigma zx/N) \neq 0$, we obtain

$$\text{plim}\ \tilde{\beta}_1 = \beta_1,$$

which establishes the consistency of the IV estimator.

It can be shown that

$$\text{var}(\tilde{\beta}_1) = \frac{\sigma^2 \Sigma z^2}{(\Sigma zx)^2} \tag{11.4.4}$$

and

$$\text{var}(\tilde{\beta}_0) = \sigma^2 \left[\frac{1}{N} + \frac{\bar{X}^2 \Sigma z^2}{(\Sigma zx)^2}\right]. \tag{11.4.5}$$

It is worthwhile setting down the expressions for instrumental variable estimators in the case when there are two regressors $X_1$ and $X_2$.

$$\tilde{\beta}_1 = \frac{(\Sigma z_1 y)(\Sigma z_2 x_2) - (\Sigma z_2 y)(\Sigma z_1 x_2)}{(\Sigma z_1 x_1)(\Sigma z_2 x_2) - (\Sigma z_2 x_1)(\Sigma z_1 x_2)}, \tag{11.4.6}$$

$$\tilde{\beta}_2 = \frac{(\Sigma z_2 y)(\Sigma z_1 x_1) - (\Sigma z_1 y)(\Sigma z_2 x_1)}{(\Sigma z_1 x_1)(\Sigma z_2 x_2) - (\Sigma z_2 x_1)(\Sigma z_1 x_2)}, \tag{11.4.7}$$

$$\tilde{\beta}_0 = \bar{Y} - \tilde{\beta}_1 \bar{X}_1 - \tilde{\beta}_2 X_2. \tag{11.4.8}$$

The expressions for the variances of the $\tilde{\beta}$'s are somewhat complicated in this case, and readers are referred to Koutsoyiannis (1977, 272).

The following points on instrumental variable estimation are worth noting:

(a) It may be that only some of the regressors are correlated with the disturbance. For example, suppose that the model is

$$Y_t = \beta_0 + \beta_1 X_{1t} + \beta_2 X_{2t} + u_t,$$

and while $X_{1t}$ is correlated with $u_t$, $X_{2t}$ is not. In such a case we would use two instruments $Z_{1t}$, $Z_{2t}$ where $Z_{2t} = X_{2t}$. Clearly, $X_{2t}$ satisfies the following two conditions for an instrument:

(i) It is uncorrelated with $u_t$.
(ii) It is correlated with $X_{2t}$—in this case perfectly correlated!

(b) Any variables that satisfy the two conditions can be used as instruments and will provide consistent estimators. Of course, they will not all be equally efficient. The problem in practice is to find suitable instruments. If a variable is correlated with a regressor, which is itself correlated with the disturbance, the variable in question would also usually be correlated with the disturbance.

(c) For the simple regression model, we have seen that

$$\text{var}(\tilde{\beta}_1) = \sigma^2 \frac{\Sigma z^2}{(\Sigma zx)^2}.$$

Now the sample correlation between $Z_t$ and $X_t$ is given by

$$r_{xz} = \frac{\Sigma xz}{(\Sigma x^2 \Sigma z^2)^{\frac{1}{2}}}.$$

Thus

$$\mathrm{var}(\tilde{\beta}_1) = \frac{\sigma^2}{\Sigma x^2} \frac{1}{r^2_{xz}} \cdot \qquad (11.4.9)$$

Comparing this equation with equation (2.4.2), derived as the variance of OLS when the regressors are fixed, we see the similarities and difference. The conditions on $\sigma^2$ and $X_t$ that give rise to good OLS estimation carry through to IV estimation. In addition, it is important that the correlation between $X_t$ and $Z_t$ be as large (in magnitude) as possible to minimize the variance of the estimator.

## 11.5 TWO-STAGE LEAST SQUARES AS INSTRUMENTAL VARIABLE ESTIMATION

As we have already seen, one of the reasons for contemporaneous correlation between regressors and disturbance (and hence the inconsistency of OLS) is that there is feedback from dependent variable to regressors. In this case it is essential that single equations be seen as belonging to a system of equations. This in turn demands methods of estimation which are appropriate to systems. Although a study of the topic of systems methods of estimation is outside the scope of this book, we will briefly look at one such method, known as two-stage least squares (2SLS). This will provide us with an example of the use of instrumental variables. Before we proceed, we mention two concepts that are associated with the study of systems.

(a) Variables are always divided into two classes: those which are simultaneously determined by the system (known as endogenous variables) and those which are not (known as predetermined variables). The latter class contains variables determined autonomously (exogenous variables) and lagged values of endogenous variables.

(b) Prior to estimation it is necessary to check whether the observations on the variables contain the information needed to estimate the parameters. We are not here talking about whether the estimation will be good or poor, but rather whether estimation can in principle be carried out at all. This aspect of a system is known as the problem of identification. If an equation within a system is not identified, then even if thousands of observations are available, the parameters cannot be estimated. Conversely, if an equation is identified, it is possible, in principle, to estimate its parameters. Any econometrics book that considers systems of equations discusses the identification problem. A very simple introduction to this problem is found in Pindyck and Rubinfeld (1981, 186-190).

As usual, we will introduce concepts through a simple example, the two-equation system consisting of the consumption function and the accounting identity. We state without any further discussion that the consumption equation is identified, and hence the parameters can be estimated.

Our system is

$$C_t = \beta_0 + \beta_1 Y_t + u_t, \tag{11.5.1}$$

$$Y_t = C_t + A_t. \tag{11.5.2}$$

The endogenous variables are $C_t$ and $Y_t$ and the predetermined variable is $A_t$. If we eliminate $C_t$, we obtain the reduced-form equation for $Y_t$,[7]

$$Y_t = \pi_0 + \pi_1 A_t + v_t, \tag{11.5.3}$$

where

$$\pi_0 = \frac{\beta_0}{1 - \beta_1}, \qquad \pi_1 = \frac{1}{1 - \beta_1}, \qquad v_t = \frac{u_t}{1 - \beta_1}.$$

It has already been shown that the fact that $A_t$ is autonomously determined guarantees that $A_t$ and $v_t$ are uncorrelated, and hence OLS will yield consistent estimators of the reduced-form parameters $\pi_0$ and $\pi_1$. In fact, it is almost always true that reduced-form parameters can be consistently estimated by OLS.[8]

From the estimated reduced-form equation we obtain a prediction of $Y_t$, given by

$$\hat{Y}_t = \hat{\pi}_0 + \hat{\pi}_1 A_t, \tag{11.5.4}$$

and as usual

$$Y_t = \hat{Y}_t + e_t, \tag{11.5.5}$$

where $e_t$ is the residual (not disturbance) in the OLS estimation of the reduced form. Obtaining $\hat{Y}_t$ is the first stage of the estimation.

We may now rewrite the consumption equation (11.5.1) in the form

$$C_t = \beta_0 + \beta_1(\hat{Y}_t + e_t) + u_t$$

or

$$C_t = \beta_0 + \beta_1 \hat{Y}_t + \epsilon_t, \tag{11.5.6}$$

where $\epsilon_t = u_t + \beta_1 e_t$. Notice carefully the essential difference between (11.5.1) and (11.5.6). In the first equation the regressor is $Y_t$ and the disturbance $u_t$, and these are correlated. In the second equation the regressor is $\hat{Y}_t$ and the disturbance is $\epsilon_t$, and these are not correlated.[9] We may thus regress $C_t$ on $\hat{Y}_t$ using OLS and obtain consistent estimators of $\beta_0$ and $\beta_1$. This regression is the second stage.

We now generalize from our simple example and describe the steps of 2SLS as follows:

(a) The system of equations is "solved," so that each endogenous variable is expressed in terms of the predetermined variables. These equations are the reduced-form equations.
(b) OLS is performed on each reduced-form equation and predictions of the endogenous variables are obtained. This is the first stage.
(c) The endogenous variables on the right-hand sides of each of the original equations are replaced by their corresponding predictions. OLS is used on these equations to estimate the original parameters. This is the second stage, and these estimators are consistent.

Let us now return to our example. The 2SLS estimate of $\beta_1$ (denoted by $\tilde{\beta}_1^*$) is obtained by OLS on equation (11.5.6). It is therefore given by

$$\tilde{\beta}_1^* = \frac{\Sigma c\hat{y}}{\Sigma \hat{y}^2}. \tag{11.5.7}$$

We have also shown [see equation (c) in note 9] that

$$\text{plim}\left(\frac{\Sigma \hat{Y}u}{N}\right) = 0.$$

Assuming then that $\hat{Y}_t$ is correlated with $Y_t$ (a not unreasonable assumption!), it follows from our discussion on instrumental variables that $\hat{Y}_t$ is a suitable instrument for $Y_t$. An IV estimator is then given by

$$\tilde{\beta}_1 = \frac{\Sigma c\hat{y}}{\Sigma y\hat{y}}. \tag{11.5.8}$$

This must be a consistent estimator. We have thus constructed two apparently different consistent estimators of $\beta_1$. However,

$$\Sigma y\hat{y} = \Sigma(\hat{y} + e)\hat{y} = \Sigma\hat{y}^2 + \Sigma e\hat{y} = \Sigma\hat{y}^2 \quad (\text{as } \Sigma e\hat{y} = 0).$$

Thus $\Sigma y\hat{y} = \Sigma\hat{y}^2$, and it follows that

$$\tilde{\beta}_1^* = \tilde{\beta}_1. \tag{11.5.9}$$

This illustrates what is a quite general principle—namely, 2SLS is a particular case of IV estimation. We have therefore shown one way in which the apparently contradictory requirements of IV estimation (find a variable that is uncorrelated with u, yet at the same time highly correlated with X) can be satisfied.

The 2SLS estimation of the system (11.5.1) and (11.5.2) can easily be accomplished using MINITAB®. We will use the data of Table 2.10.1. Let us assume that $C_t$ is stored in column C1 and $Y_t$ in column C2. The first step

DISPLAY 11.5.1

```
MTB > LET C3=C2-C1
MTB > REGR C2 1 C3-C5

The regression equation is
C2 = 1.96 + 2.50 C3

Predictor        Coef         Stdev      t-ratio
Constant       1.9567        0.4204         4.65
C3            2.50287       0.04458        56.14

s = 1.007        R-sq = 99.0%     R-sq(adj) = 99.0%

Analysis of Variance

SOURCE         DF          SS          MS
Regression      1      3194.9      3194.9
Error          32        32.4         1.0
Total          33      3227.4

Unusual Observations
Obs.      C3        C2        Fit  Stdev.Fit   Residual    St.Resid
 32     13.2    37.560     35.045      0.269      2.515       2.59R

R denotes an obs. with a large st. resid.

MTB > REGR C1 1 C5

The regression equation is
C1 = 0.782 + 0.600 C5

Predictor        Coef         Stdev      t-ratio
Constant       0.7818        0.4524         1.73
C5            0.60046       0.01781        33.71
1

s = 1.007        R-sq = 97.3%     R-sq(adj) = 97.2%

Analysis of Variance

SOURCE         DF          SS          MS
Regression      1      1151.9      1151.9
Error          32        32.4         1.0
Total          33      1184.4
```

Reprinted with permission of Minitab, Inc.

is to use the accounting identity to create the observations on $A_t$. As $A_t = Y_t - C_t$, this is done by

LET C3 = C2 - C1

($A_t$ is now in column C3.) We now regress $Y_t$ on $A_t$, saving the predictions $\hat{Y}_t$,

```
REGR C2 1 C3-C5
```

($\hat{Y}_t$ is stored in C5.) Finally, $C_t$ is regressed on $\hat{Y}_t$:

```
REGR C1 1 C5
```

Results of the 2SLS regression are shown in Display 11.5.1.

The estimated model is

$$\hat{C}_t = \underset{(0.452)}{0.782} + \underset{(0.018)}{0.600}Y_t; \qquad R^2 = 0.973.$$

This result may be compared with the simple regression on page 00.

NOTES

1. The term <u>stochastic variable</u> is used synonymously with <u>random variable</u>.

2. As the problems discussed in this chapter usually occur with time series data, we will use the index t rather than i.

3. From now on, we will drop the superscript N.

4. We are using a simple model for ease of exposition. All conclusions we reach generalize immediately to the case of K regressors.

5. For a rigorous proof, the interested reader is referred to Goldberger (1964, 269-270).

6. Again, a rigorous justification of this result can be found in Goldberger (1964, 274).

7. A reduced-form equation is one in which an endogenous variable is expressed in terms of the predetermined variables alone, together with a disturbance. There are as many reduced-form equations as there are endogenous variables.

8. The exception to this rule occurs when the predetermined variables include lagged endogenous variables and there is autocorrelation in the disturbances. We will ignore this case and assume that OLS always gives consistent estimators.

9. To establish the consistency of 2SLS formally we need only show that

$$\text{plim}\left(\frac{\Sigma \hat{Y}\epsilon}{N}\right) = 0, \tag{a}$$

and from the definition of $\epsilon_t$ this will be achieved if we can show that $\text{plim}(\Sigma \hat{Y}u/N) = \text{plim}(\Sigma \hat{Y}e/N) = 0$. It is a condition of OLS estimation that $\Sigma e = 0$ and $\Sigma eA = 0$ [see (2.3.11) and (2.3.12)]; thus

$$\frac{\Sigma \hat{Y}e}{N} = \frac{\Sigma(\hat{\pi}_0 + \hat{\pi}_1 A)e}{N} = \frac{\hat{\pi}_0 \Sigma e}{N} + \frac{\hat{\pi}_1 \Sigma eA}{N} = 0$$

Therefore,

$$\text{plim}\left(\frac{\Sigma\,\hat{Y}e}{N}\right) = 0. \tag{b}$$

Again,

$$\text{plim}\left(\frac{\Sigma\,\hat{Y}u}{N}\right) = \text{plim}\left(\frac{\Sigma\,(\hat{\pi}_0 + \hat{\pi}_1 A)u}{N}\right)$$

$$= \text{plim}\ \hat{\pi}_0\ \text{plim}\left(\frac{\Sigma\,u}{N}\right) + \text{plim}\ \hat{\pi}_1\ \text{plim}\left(\frac{\Sigma\,Au}{N}\right).$$

It is easily shown that $\text{plim}(\Sigma\,u/N) \equiv \text{plim}(\bar{u}) = 0$ (show this as an exercise), and we have already seen that as A is exogenous, $\text{plim}(\Sigma\,Au/N) = 0$. Thus

$$\text{plim}\left(\frac{\Sigma\,\hat{Y}u}{N}\right) = 0. \tag{c}$$

Equations (b) and (c) show that (a) is true and hence two-stage least squares provides consistent estimators.

# 12

# Maximum Likelihood Estimation

## 12.1 INTRODUCTION

In all our analysis so far, the fundamental tool has been OLS. When the conditions under which OLS is BLUE are not met (e.g., heteroscedasticity and autocorrelation), the problem has been transformed so that OLS becomes BLUE.

There is available, however, a much wider principle of estimation, which was first enunciated by Sir Ronald Fisher in 1921. This principle, known as maximum likelihood (ML) estimation, is necessary in some of the work that follows. It does not undermine any of the earlier material because, as we will see, when ML estimation is applied to the regression model with normally distributed disturbances, the resulting estimators are exactly the same as OLS.

In this chapter our aim is to give the reader a basic appreciation of the maximum likelihood approach to estimation as preparation for the work on binary choice models and specification testing that follows. Section 12.3, which gives some examples of ML estimation, is perhaps too technical for some readers and may be omitted.

## 12.2 MAXIMUM LIKELIHOOD ESTIMATION

As with many statistical methods, maximum likelihood estimation is firmly based in intuition. We will introduce the concepts of the likelihood function and the method of maximum likelihood by way of some simple examples, the first being taken from Mood, Graybill, and Boes (1974).

Consider an urn that contains black and white balls and suppose it is known that the ratio of black to white balls is either 3:1 or 1:3. That is, the probability p of drawing a black ball is either 3/4 or 1/4. To estimate p, 3 balls are drawn (with replacement) and it is found that all the balls are white. Remembering that p is either 1/4 or 3/4, on the basis of this sample evidence almost everyone would intuitively opt for $p = 1/4$. They have, in fact, without realizing it, used the method of maximum likelihood to estimate p.

Let us look at the situation using elementary statistics. If X denotes the number of successes (i.e., black balls) in three trials, X is a binomial random variable with probability function given by

$$P(X = x) = \binom{3}{x} p^x (1 - p)^{3-x}.$$

Thus, if $p = 1/4$,

$$P(X = 0) = \binom{3}{0}\left(\frac{1}{4}\right)^0 \left(\frac{3}{4}\right)^3 = 0.42.$$

Similarly, if $p = 3/4$,

$$P(X = 0) = \binom{3}{0}\left(\frac{3}{4}\right)^0 \left(\frac{1}{4}\right)^3 = 0.02.$$

Thus the probability of obtaining the sample result we actually did get $(X = 0)$ is 0.42 if $p = 1/4$, but only 0.02 if $p = 3/4$. The sample result we actually got is a highly unlikely event if $p = 3/4$, and much more likely if $p = 1/4$. It stands to reason that we should pick the value of p that maximizes the probability of obtaining the sample result we actually did get. Hence we estimate p to be 1/4.

In this simple example, we are maximizing the probability over only two values of p. In most situations, we do not have the information that enables us to restrict the value of p to just two possible values. Usually, p could have any value in the range (0, 1). However, the principle used to estimate p remains the same—we choose as our estimate the value of p that maximizes the probability of obtaining the sample result we actually did get. This probability is called the likelihood of the sample or (when viewed as a function of p) the likelihood function.

Let us change our example to suppose that p can take any value in the range (0, 1) and when a sample size 3 is drawn, one ball is black. Then

$$P(X = 1) = \binom{3}{1} p^1 (1 - p)^{3-1} = 3p(1 - p)^2.$$

The graph of $P(X = 1)$ as a function of p is shown in Figure 12.2.1. The estimate of p, denoted by $\hat{p}$, corresponds to the maximum value of $P(X = 1)$. Using differential calculus, or from the graph, we find $\hat{p} = 1/3$.

We are now in a position to define formally what is known as the likelihood function. There are actually two definitions, depending on whether we are sampling discrete or continuous random variables. In the case of a discrete random variable, the likelihood function is defined to be the probability of getting the sample we actually did get.[1] When sampling from a continuous

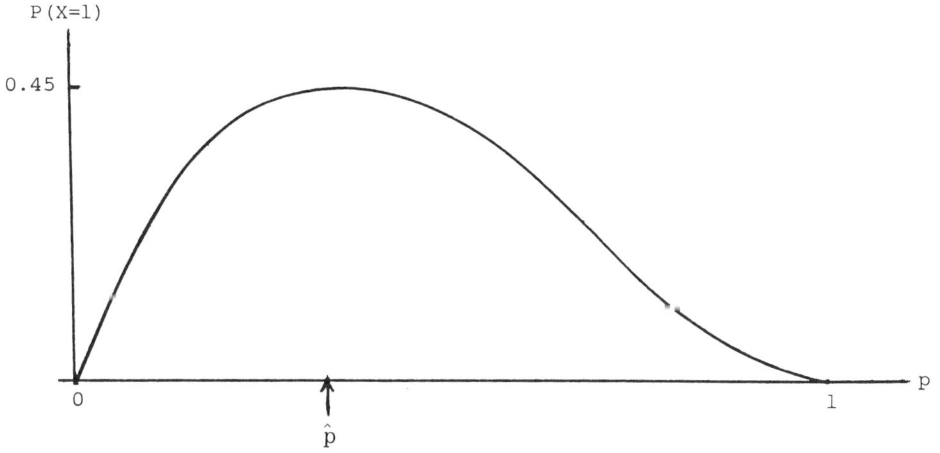


Figure 12.2.1. $P(X = 1) = 3p(1 - p)^2$.

random variable, the likelihood function is proportional to the above probability.

DEFINITION 12.2.1 Suppose that $x_1, x_2, \ldots, x_N$ is the value of a random sample from a discrete random variable having probability function $p(x; \theta)$, where $\theta$ is a vector of parameters. Then the likelihood function $L(\theta)$ is given by

$$L(\theta) = p(x_1; \theta)p(x_2; \theta) \cdots p(x_N; \theta). \tag{12.2.1}$$

DEFINITION 12.2.2 Suppose that $x_1, x_2, \ldots, x_N$ is the value of a random sample from a continuous random variable having density function $f(x; \theta)$. The likelihood function $L(\theta)$ is given by

$$L(\theta) = f(x_1; \theta)f(x_2; \theta) \cdots f(x_N; \theta). \tag{12.2.2}$$

In both cases, the maximum likelihood estimator $\hat{\theta}$ of $\theta$ is defined as the value of $\theta$ that maximizes the likelihood function.

In our introductory examples, the only parameter of the probability function was p. Thus in these cases $\theta = p$. However, in many applications there is more than one parameter. For example, if we are sampling from a normal population, with mean $\mu$ and variance $\sigma^2$, then $\theta$ is the vector $(\mu, \sigma^2)'$. Again, if we are considering a sample $Y_1, Y_2, \ldots, Y_N$ drawn from the regression model, $\theta$ would be the vector containing the $K + 2$ parameters $\beta_0, \beta_1, \ldots, \beta_K, \sigma^2$.

## 12.3 SOME IMPORTANT EXAMPLES

### Estimating Proportions

We have already seen how a proportion can be estimated by maximum likelihood in a very specific example. We will now generalize that example. When sampling for proportions we are actually drawing a sample from a Bernouilli random variable. That is, the random variable X can take only the values 1 and 0 with probabilities p and q (= 1 - p), respectively. Thus the probability function is given by

$$p(x;\theta) = p^x q^{1-x},$$

where $\theta = p$. Our task is to estimate the parameter p. From (12.2.1), the likelihood function L(p) is given by

$$L(p) = \left(p^{x_1} q^{1-x_1}\right)\left(p^{x_2} q^{1-x_2}\right)\cdots\left(p^{x_N} q^{1-x_N}\right)$$

$$= \left(p^{x_1} p^{x_2} \cdots p^{x_N}\right)\left(q^{1-x_1} q^{1-x_2} \cdots q^{1-x_N}\right).$$

That is,

$$L(p) = p^{\Sigma x_i} q^{n-\Sigma x_i} = p^{\Sigma x_i}(1 - p)^{n-\Sigma x_i}.$$

To find the maximum likelihood estimator, we use differentiation to find the value of p that maximizes L(p). Now

$$\frac{dL(p)}{dp} = (\Sigma x_i)p^{\Sigma x_i - 1}(1 - p)^{n-\Sigma x_i} - p^{\Sigma x_i}(n - \Sigma x_i)(1 - p)^{n-\Sigma x_i - 1}.$$

The maximum occurs when dL(p)/dp = 0; that is, when

$$(\Sigma x_i)p^{\Sigma x_i - 1}(1 - p)^{n-\Sigma x_i} = p^{\Sigma x_i}(n - \Sigma x_i)(1 - p)^{n-\Sigma x_i - 1};$$

that is, when

$$(\Sigma x_i)(1 - p) = p(n - \Sigma x_i)$$

or

$$p = \frac{\Sigma x_i}{n}.$$

The maximum likelihood estimator $\hat{p}$ is thus defined by $\Sigma x_i/n$, which is simply the sample proportion. Therefore, the estimator of the unknown proportion p that is always used (namely, the sample proportion) is actually the maximum likelihood estimator. Notice in our simple example that $\hat{p} = 1/3$ = sample proportion.

## Estimating the Parameters of a Normal Distribution

Here

$$f(x;\theta) = \frac{1}{\sqrt{2\pi\sigma^2}} \exp\left[-\frac{1}{2\sigma^2}(x - \mu)^2\right].$$

Thus, by (12.2.1),

$$L(\mu, \sigma^2) = \frac{1}{(2\pi\sigma^2)^{N/2}} \exp\left[-\frac{1}{2\sigma^2}\Sigma(x_i - \mu)^2\right].$$

To find the maximum likelihood estimator $\hat{\mu}$ and $\hat{\sigma}^2$, partial differentiation is employed. We will omit the algebraic details and quote the result:

$$\hat{\mu} = \bar{x}$$

$$\hat{\sigma}^2 = \frac{1}{n}\Sigma(x_i - \bar{x})^2$$

Note that the maximum likelihood estimator of $\mu$ is the usual sample mean, but that the estimator of $\sigma^2$ differs slightly from the sample variance.

## Estimating the Regression Model

We will confine our attention to the simple regression model

$$Y_i = \beta_0 + \beta_1 X_i + u_i,$$

where

$$u_i \sim N(0, \sigma^2).$$

The sample values $u_1, u_2, \ldots, u_N$ of the disturbance are, of course, not observed. They give rise, by (12.2.1), to the likelihood function

$$\frac{1}{(2\pi\sigma^2)^{N/2}} \exp\left(-\frac{1}{2\sigma^2}\Sigma u_i^2\right).$$

Because $u_i = Y_i - \beta_0 - \beta_1 X_i$, we can express the likelihood in terms of the observations $Y_1, Y_2, \ldots, Y_N$ and the unknown parameters $\beta_0$, $\beta_1$, and $\sigma^2$. Thus

$$L(\beta_0, \beta_1, \sigma^2) = \frac{1}{(2\pi\sigma^2)^{N/2}} \exp\left[-\frac{1}{2\sigma^2} \Sigma (Y_i - \beta_0 - \beta_1 X_i)^2\right].$$

To find the maximum likelihood estimators $\beta_0$, $\beta_1$, and $\sigma^2$ we differentiate partially with respect to these parameters and solve the equations that result from setting the differentials equal to zero. Again, omitting the algebra, we obtain

$$N\hat{\beta}_0 + \hat{\beta}_1 \Sigma X_i = \Sigma Y_i$$

$$\hat{\beta}_0 \Sigma X_i + \hat{\beta}_1 \Sigma X_i^2 = \Sigma X_i Y_i.$$

The equations are the same as (2.2.4) and (2.2.5), showing that the maximum likelihood estimators are identical to OLS estimators. The maximum likelihood estimator for $\sigma^2$ is found to be

$$\hat{\sigma}^2 = \frac{1}{N} \Sigma (Y_i - \hat{\beta}_0 - \hat{\beta}_1 X_i)^2$$

$$= \frac{1}{N} \Sigma e_i^2,$$

which is different from the unbiased estimator given in (2.5.5).

## 12.4 PROPERTIES OF MAXIMUM LIKELIHOOD ESTIMATORS

The reason that maximum likelihood estimation has such a central role in statistics is that these estimators possess some very attractive asymptotic properties.

### Consistency

Maximum likelihood estimators are consistent estimators. The meaning and implications of consistency have been discussed in Section 11.2.

### Asymptotic Normality

The asymptotic distribution of all maximum likelihood estimators is normal, regardless of the distribution from which the sample was drawn. This means that when the principle of maximum likelihood is used to obtain estimators, inference can always be based on the normal distribution, provided that the sample is large. This property is analogous to the central limit theorem result on the asymptotic distribution of the sample mean $\bar{x}$.

### Asymptotic Efficiency

The definition of asymptotic efficiency has been given in Section 11.2. Very informally we may say that if a consistent estimator is asymptotically efficient, it is the best estimator that can be used. ML estimators have this property. In Table 11.3.1 we noted (without explanation) that OLS estimators are asymptotically efficient in three common situations, provided that the disturbances are normally distributed. The reason is now apparent—OLS estimators are also ML estimators when the disturbances are normal.

### Invariance

Let us suppose that we are required to estimate a parameter $\phi$, given by $\phi = \ln \mu$, where $\mu$ is the mean of a normal population. An intuitively obvious estimator would be $\hat{\phi} = \ln \hat{\mu} = \ln \bar{X}$. However, this raises the question as to whether $\hat{\phi}$ is the maximum likelihood estimator for $\phi$ just because $\bar{X}$ is for $\mu$. If $\hat{\phi}$ <u>is</u> the ML estimator for $\phi$, it will be consistent, asymptotically efficient, and so on. However, if it is <u>not</u> the ML estimator for $\phi$, we may have no idea what its properties are. The property of invariance guarantees that $\hat{\phi}$ is the ML estimator of $\phi$.

Formally, the invariance property may be defined as follows. Let $\phi$ be some function of a parameter $\theta$, whose ML estimator is $\hat{\theta}$. That is,

$$\phi = f(\theta).$$

Then, if we define an estimator $\hat{\phi}$ by

$$\hat{\phi} = f(\hat{\theta}),$$

where the function $f(\cdot)$ is the same in both of the equations above, $\hat{\phi}$ is the ML estimator of $\phi$.

EXAMPLE We have seen that for the linear regression model with normally distributed disturbances, the ML estimator of $\sigma^2$ is

$$\hat{\sigma}^2 = \frac{1}{N} \Sigma e_i^2.$$

The invariance property ensures that the estimator $(\Sigma e_i^2/N)^{\frac{1}{2}}$ is a consistent and asymptotically efficient estimator of the parameter $\sigma$.

This brief introduction to the concepts of the likelihood function and maximum likelihood estimation will be sufficient to enable the reader to understand what follows. It may not necessarily enable him or her to derive ML estimators in specific problems.

## NOTE

1. The likelihood function is thus slightly different from the probability of a sample result which we have used in the example. However, the difference is only in a constant of proportionality.

# 13

# Estimating Proportions: Binary Choice Models

## 13.1 INTRODUCTION

Before we consider these models, it will be helpful to have another quick look at the classical regression model introduced in Chapter 2, from a slightly different perspective. Suppose that we are drawing a random sample $Y_1, Y_2, \ldots, Y_N$ from a normal distribution, with mean $\mu$ and variance $\sigma^2$. That is,

$$Y_i \sim N(\mu, \sigma^2), \qquad i = 1, 2, \ldots, N.$$

It follows immediately that

$$Y_i - \mu \sim N(0, \sigma^2).$$

If we define the random variable $u_i$ as

$$u_i = Y_i - \mu,$$

we obtain

$$Y_i = \mu + u_i, \tag{13.1.1}$$

where $u_i \sim N(0, \sigma^2)$. Thus

$$E(u_i) = 0, \tag{13.1.2}$$

$$E(u_i^2) = \sigma^2 \tag{13.1.3}$$

(since $\sigma^2 = E(u_i^2) - [E(u_i)]^2$), and, because the $Y_i$ form a random sample,

$$E(u_i u_j) = 0 \quad (j \neq i). \tag{13.1.4}$$

The observations of the random sample can therefore be described by the model (13.1.1) together with the three conditions (13.1.2), (13.1.3), and (13.1.4), which are identical to those of (2.1.2), (2.1.3), and (2.1.4).

If we wanted to estimate the mean $\mu$, obviously we would do this by finding the sample mean $\bar{Y}$. That is,

$$\hat{\mu} = \bar{Y}.$$

Often, however, we are able to carry out a much deeper and more useful analysis by assuming that the mean $\mu$ is determined by known factors $X_1, X_2, \ldots, X_K$. We would then write (13.1.1) as

$$Y_i = f(X_{1i}, X_{2i}, \ldots, X_{Ki}) + u_i \qquad (13.1.5)$$

and we have arrived at the regression model.

If we add the assumption that $f(\cdot)$ is a linear function, we obtain the linear regression model

$$Y_i = \beta_0 + \beta_1 X_{1i} + \cdots + \beta_K X_{Ki} + u_i,$$

which has been the basis of our study to this point.

Regression analysis therefore may be regarded as an attempt to estimate the population mean, when it is believed that the mean is determined by some identifiable factors. Furthermore, a successful regression analysis gives us a very good idea of <u>how</u> the mean is affected by changes in these factors.

We now turn our attention to what may well be the most common problem in all statistics—estimating a proportion. Hardly a day goes by without the major daily newspapers reporting the results of an opinion poll. Almost invariably they estimate proportions—the proportion of the population that supports a particular political party, or is happy with the way the economy is performing, or views a particular television program, and so on. The proportion is simply estimated by the sample proportion.

When we are sampling to estimate a proportion p, the random sample $Y_1, Y_2, \ldots, Y_N$ can be defined by

$$\begin{aligned} Y_i &= 1 \quad \text{if success} \\ &= 0 \quad \text{if failure,} \end{aligned}$$

where "success" means that the person questioned does support the political party or is happy with the economy, and so on. Thus a sample consists of a collection of ones and zeros. For example, with $N = 10$, we may have the sample (1,0,0,0,1,0,0,1,1,0). With this type of sample, the sample proportion is exactly the same as the sample mean $\bar{Y}$. In our example, $\Sigma Y = 4$ and therefore, $\bar{Y} = 4/10 = 0.4$ = sample proportion. Sampling to estimate a proportion can therefore be seen as a special case of estimating a population mean.

For the purposes of this discussion, let us suppose that we are interested in estimating the proportion of a state that vote Democratic. Following our discussion of the regression model, it now seems natural to expect that we may be able to perform a more useful analysis if we can identify factors that influence the proportion of Democratic voters. For example, it is probably reasonable to assume that voter intentions are influenced by education and income level. Therefore, if we define

$X_1$ = years of education, and

$X_2$ = annual income,

we would postulate that

$$p = f(X_1, X_2).$$

We will now discuss four methods that have been used to extend the estimation of proportions in much the same way that regression analysis enables us to generalize estimation of a population mean. An excellent survey is given by Ameniya (1981).

## 13.2 THE LINEAR PROBABILITY MODEL

The most obvious approach to finding how a probability (or proportion) p depends on a set of factors $X_1, X_2, \ldots, X_K$ is to use OLS to estimate the regression model

$$Y_i = \beta_0 + \beta_1 X_{1i} + \cdots + \beta_K X_{Ki} + u_i,$$

where $Y_i = 0$ or 1. However, such an approach gives rise to the following problems:

(i) The disturbance $u_i$ is <u>heteroscedastic</u>. It can be shown (see Goldberger, 1964, 249) that

$$\text{var}(u_i) = E(Y_i)[1 - E(Y_i)].$$

This can be overcome in the usual way by dividing throughout by the estimate $\hat{\sigma}_i$ of $\{E(Y_i)[1 - E(Y_i)]\}^{\frac{1}{2}}$, namely

$$\hat{\sigma}_i = [(\hat{\beta}_0 + \hat{\beta}_1 X_{1i} + \cdots + \hat{\beta}_K X_{Ki})(1 - \hat{\beta}_0 - \cdots - \hat{\beta}_K X_{Ki})]^{\frac{1}{2}}.$$

(ii) The disturbance $u_i$ is not normal. We have that

$$u_i = Y_i - (\beta_0 + \beta_1 X_{1i} + \cdots + \beta_K X_{Ki}).$$

Because $Y_i$ is either 0 or 1, it follows that <u>for given values of</u> $X_{1i}, \ldots, X_{Ki}$, $u_i$ can only take <u>two</u> values. Therefore, it cannot possibly be a normal random variable. This implies that the inference procedures we have used with the regression model are not valid.

(iii) Probabilities will be predicted by

$$\hat{p}_i = \hat{\beta}_0 + \hat{\beta}_1 X_{1i} + \cdots + \hat{\beta}_K X_{Ki}.$$

However, there is no guarantee that all $\hat{p}_i$ will lie in the range (0, 1).

For these three reasons, the linear probability model is not likely to be a successful approach to estimating probabilities as functions of $X_1$, ..., $X_K$.

## 13.3 ESTIMATION WHEN REPEATED OBSERVATIONS ARE AVAILABLE

Let us suppose that there are multiple observations of Y for every possible combination of values of $X_1, \ldots, X_K$. These possible combinations are called cells. Our task then is to estimate the probability associated with each cell. We will denote the number of observations available on the ith cell by $N_i$, and the total number of cells by T.

Clearly, we could estimate the probability associated with each cell by the sample proportions $r_i = n_i/N_i$, where $n_i$ is the number of successes in the ith cell. This is equivalent to assuming that the population is characterized by T unrelated parameters $p_1, p_2, \ldots, p_T$. However, if we assume that the $p_i$ are linearly related to the K variables $X_1, \ldots, X_K$, that is,

$$p_i = \beta_0 + \beta_1 X_{1i} + \cdots + \beta_K X_{Ki}, \tag{13.3.1}$$

then the population is characterized by K + 1 parameters $\beta_0, \beta_1, \ldots, \beta_K$. If K + 1 is much less than T, there will be considerable gains in efficiency by estimating the $\beta_i$ rather than the $p_i$.

If we define the random variable $u_i$ by

$$u_i = r_i - p_i, \tag{13.3.2}$$

then, by the properties of sample proportions, it follows that

$$E(u_i) = 0, \quad \text{var}(u_i) = \frac{p_i(1 - p_i)}{N_i}. \tag{13.3.3}$$

Combining (13.3.1) and (13.3.2), we have the regression model

$$r_i = \beta_0 + \beta_1 X_{1i} + \cdots + \beta_K X_{Ki} + u_i, \tag{13.3.4}$$

where the $u_i$ satisfy (13.3.3). The properties of this model are:

(i) The disturbance $u_i$ is heteroscedastic. This can be accounted for by dividing the ith observation by $[r_i(1 - r_i)/N_i]^{\frac{1}{2}}$.
(ii) As with the linear probability model, $u_i$ is not normal. In fact, it is related to the binomial random variable. However, if $N_i$ is large, that is, greater than 30, then by the central limit theorem, $u_i$ will be approximately normally distributed.
(iii) Once again, there is no guarantee that $\hat{p}_i$, given by

$$\hat{p}_i = \hat{\beta}_0 + \hat{\beta}_1 X_{1i} + \cdots + \hat{\beta}_K X_{Ki},$$

will lie in the range (0, 1). However, if all the $N_i$ are large, this requirement would usually be fulfilled.

Thus provided that correction is made for heteroscedasticity, (13.3.4) may be estimated by OLS and with large samples the usual inferences will be valid. It is the requirement that $N_i$ is large, for all cells, that gives rise to the practical shortcoming of this model. If most of the X variables measure categories and K is small, we may reasonably expect to get a number of observations in each cell, provided that the sample is large. However, if K is large, giving rise to many cells, and if the X's are continuous variables, there will inevitably be many cells with only a single observation. We now turn our attention to two models that guarantee $0 \le \hat{p}_i \le 1$ and do not require more than a single observation per cell.

## 13.4 THE PROBIT MODEL

We will introduce the probit model in the context of a specific example. Let us suppose that a marketing firm is trying to determine the proportion of the community that will buy a video recorder (VCR) in the coming year. That is, viewed from the point of view of a particular individual, the firm is trying to estimate the probability that he or she will purchase a VCR.

It would seem very reasonable to assume that this probability would depend on (at least) two factors, namely whether the individual already owns a VCR and the person's income level. Let

$$X_{1i} = 1 \quad \text{if ith individual owns a VCR,}$$
$$= 0 \quad \text{otherwise,}$$

and $X_{2i}$ is the income level of the ith individual. The variable $X_1$ is an ordinary dummy variable. We will define an index $I_i'$, given by

$$I_i' = \gamma_0 + \gamma_1 X_{1i} + \gamma_2 X_{2i}, \tag{13.4.1}$$

which is a single number summarizing the values of $X_1$ and $X_2$ for the ith individual. For the moment, we will assume that the values of parameters $\gamma_0$, $\gamma_1$, and $\gamma_2$ are known.

If we now consider all the individuals who have the same index value $I_i'$, it is fairly clear that different decisions will be made on whether or not to purchase a video. These differences will come about because of the many other factors which will also determine what decision is taken. For example, one person who does not already own a VCR may decide to buy. Another, on exactly the same income, also without a recorder, may not purchase because it is more important to him to save for an overseas holiday. We will attempt to take account of these individual differences by defining a threshold value $I_{ij}^*$ for each individual j whose index value is $I_i'$. A decision will be made to buy a VCR if the index value exceeds the threshold value. Otherwise, an individual will not purchase the recorder. Those who have a high resistance to buying, for whatever reasons, would be characterized by a high threshold value, and vice versa. Taken over the whole community, we can regard the threshold values I* as values of a random variable, reflecting individual values, tastes, and priorities.

It is well known that personal characteristics such as height, weight, or IQ invariably have a normal distribution. The reason for this is related to the central limit theorem and arises from the fact that such characteristics are the end product of innumerable hereditary and environmental influences. For the same reason it is very reasonable to assume that threshold values I* will be normally distributed.

Thus we will assume that

$$I^* \sim N(\mu, \sigma^2),$$

where $\mu$ and $\sigma^2$ have known values. Then

$$\begin{aligned}
p_i &\equiv P[\text{purchasing} \mid \text{index value } I_i'] \\
&= P[I^* \leq I_i'] \\
&= P\left[\frac{I^* - \mu}{\sigma} \leq \frac{I_i' - \mu}{\sigma}\right] \\
&= P\left[Z \leq \frac{I_i' - \mu}{\sigma}\right],
\end{aligned}$$

where Z is the standardized normal random variable. Therefore, from (13.4.1),

$$\begin{aligned}
p_i &= P\left[Z \leq \left(\frac{\gamma_0 - \mu}{\sigma} + \frac{\gamma_1}{\sigma} X_{1i} + \frac{\gamma_2}{\sigma} X_{2i}\right)\right] \\
&= P[Z \leq (\beta_0 + \beta_1 X_{1i} + \beta_2 X_{2i})],
\end{aligned}$$

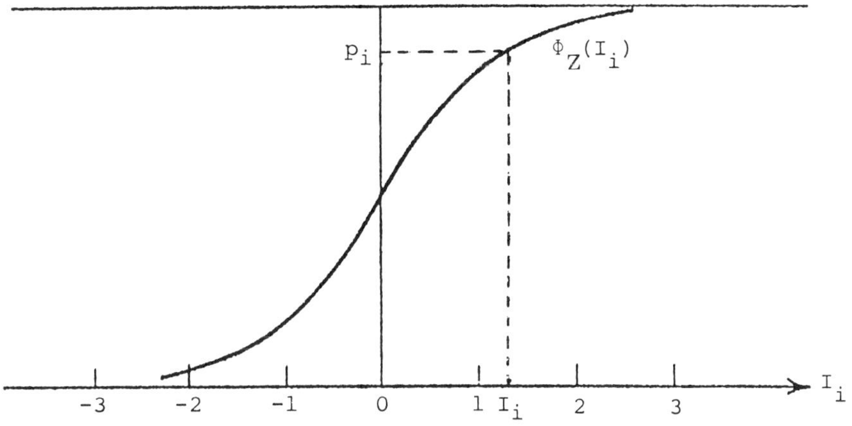


Figure 13.4.1. Using an index value $I_i$ to compute the corresponding probability $p_i$.

where $\beta_0 = (\gamma_0 - \mu)/\sigma$, $\beta_1 = \gamma_1/\sigma$, and $\beta_2 = \gamma_2/\sigma$. We will now redefine the index in terms of the $\beta$ values and denote it by $I_i$ to obtain

$$p_i = P(Z \le I_i) \equiv \Phi_Z(I_i), \tag{13.4.2}$$

where

$$I_i = \beta_0 + \beta_1 X_{1i} + \beta_2 X_{2i}, \tag{13.4.3}$$

and $\Phi_Z(I_i)$ is the cumulative distribution function of the standardized normal variable, evaluated at the point $I_i$. Equation (13.4.2), which is the key to the probit model, is shown graphically in Figure 13.4.1.

To show how equation (13.4.2) would be used, let us assign hypothetical values to $\beta_0$, $\beta_1$, and $\beta_2$. For example, we will take

$$\beta_0 = -7, \quad \beta_1 = -5.4, \quad \text{and} \quad \beta_2 = 0.2.$$

Table 13.4.1

| Individual, i | Owner/nonowner, $X_1$ | Income level, $X_2$ (thousands of dollars) |
|---|---|---|
| 1 | 0 | 30 |
| 2 | 0 | 42 |
| 3 | 1 | 50 |

Table 13.4.2

| Individual, i | Owner/nonowner, $X_1$ | Income level, $X_2$ | Index value, I | Probability of purchase, $p_i$ |
|---|---|---|---|---|
| 1 | 0 | 30 | -1 | 0.16 |
| 2 | 0 | 42 | 1.4 | 0.92 |
| 3 | 1 | 50 | -2.4 | 0.01 |

Consider three individuals characterized as owner/nonowner and by income level, measured in thousands of dollars as shown in Table 13.4.1. Then, for individual 1,

$$I_1 = -7 - 5.4 \times 0 + 0.2 \times 30 = -1.$$

To find the probability that this person will purchase a VCR, we find from equation (13.4.2) and standard normal tables that

$$P(\text{purchase} | I = -1) = P(Z \leq -1) = 0.16.$$

Proceeding in this way, we summarize our analysis for the three individuals in Table 13.4.2.

Thus, given the values of the $\beta$'s, we are able to determine the probability of purchasing a video recorder once we have information on whether a person already owns one and what his or her income is. Individual 1 does not own a video but is on a relatively low salary. For such an individual the probability of purchase is low (0.16). Individual 2, on the other hand, also a nonowner but on a substantially higher income, is likely to buy a recorder (probability is 0.92). Finally, individual 3, on a still higher income level, is already an owner. Such an individual is very unlikely to buy another recorder (probability is only 0.01).

## Marginal Changes in p

In the linear regression model, the $\beta$ coefficients are interpreted as the marginal effect that a unit change in the regressor variable has on the conditional mean (see Section 4.1). Such information is often a very important outcome of a regression analysis. In the same way, we should investigate how a change in the variable $X_j$ (say) affects the probability of success p.

From (13.4.2),

$$\frac{\partial p_i}{\partial X_{ji}} = \frac{\partial}{\partial X_{ji}} \Phi_Z(I_i) = \frac{d\Phi_Z(I_i)}{dI_i} \frac{\partial I_i}{\partial X_{ji}}$$

$$= f_Z(I_i)\beta_j, \tag{13.4.4}$$

where

$$f_Z(I_i) = \frac{d\Phi_Z(I_i)}{dI_i} = (2\pi)^{\frac{1}{2}} \exp\left(-\frac{1}{2}I_i^2\right) \tag{13.4.5}$$

is the probability density function of the standardized normal random variable, shown in Figure 13.4.2.

We can immediately make three deductions from the result (13.4.4):

(i) The sign of $\partial p_i/\partial X_{ji}$ is determined by the sign of $\beta_j$, because $f_Z(I_i)$ is always positive. Thus if $\beta_j$ is positive, an increase in $X_{ji}$ will result in $p_i$ increasing, and vice versa. The sign of $\beta_j$ plays the same role as it does in the regression model.

(ii) The magnitude of $\partial p_i/\partial X_{ji}$ is directly proportional to the magnitude of $\beta_j$.

(iii) The function $f_Z(I_i)$ is approximately zero when $|I_i| > 3$ and has its maximum value when $I_i = 0$ (see Figure 13.4.2). Therefore, for a given value of $\beta_j$, changes in $X_{ji}$ will have very little effect when $I_i$ is large in magnitude. The probability $p_i$ will be most sensitive to changes in $X_j$ when $I_i$ is close to zero.

EXAMPLE Let us examine the effect of a \$1000 increase in income on the probability of purchase of a VCR. Using (13.4.4) and (13.4.5) with $\beta_2 = 0.2$, the results are summarized in Table 13.4.3. As a further illustration, let

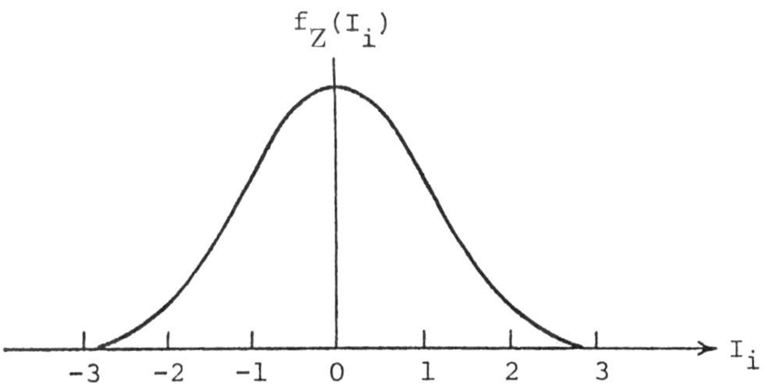


Figure 13.4.2. Density function of the standardized normal random variable.

Table 13.4.3

| Individual, i | Index value, $I_i$ | $f_Z(I_i)$ | $\Delta p_i$ |
|---|---|---|---|
| 1 | -1 | 0.242 | 0.048 |
| 2 | 1.4 | 0.150 | 0.030 |
| 3 | -2.4 | 0.022 | 0.004 |

us consider a fourth individual who does not own a VCR and has an annual income of \$36,000. Here $X_{14} = 0$ and $X_{24} = 36$. Thus

$$I_4 = -7 - 5.4 \times 0 + 0.2 \times 36 = 0.2,$$
$$p_4 = \Phi_Z(I_4) = \Phi_Z(0.2) = 0.58,$$
$$f_Z(I_4) = 0.391,$$

and

$$\Delta p_4 = 0.2 \times 0.391 = 0.08.$$

Thus an increase of \$1000 annual income for this individual will increase the probability of purchasing a VCR by almost 0.1, which is far higher than the increase in the probabilities for the other three individuals. This is due to the small index value $I_4 = 0.2$.

## Estimating the Probit Model

Clearly, the whole analysis we have done so far has depended on knowing the values of $\beta_0$, $\beta_1$, and $\beta_2$. In practice, these values would not be known. We now turn our attention to the problem of estimating the $\beta$'s from sample observations. Let us suppose that the probability p of a success is a function of K factors $X_1, X_2, \ldots, X_K$. This leads to the formulation of an index $I_i$ given by

$$I_i = \beta_0 + \beta_1 X_{1i} + \cdots + \beta_K X_{Ki}. \tag{13.4.6}$$

The method of maximum likelihood will be used to estimate the unknown parameters $\beta_0, \beta_1, \ldots, \beta_K$.

The dependent variable $Y_i$ can take only two values—1 or 0, depending on whether the response is a success or failure. Let us denote the index values of the N individuals in the sample by $I_1, I_2, \ldots, I_N$ and assume, purely for convenience, that the data are ordered so that all the successes (say, m of them) come first, followed by the N - m failures.

Then by equation (13.4.2),

$$P[Y = 1 | I_i] \equiv P[\text{success} | I_i] = \Phi_Z(I_i)$$

and, by implication,

$$P[Y = 0 | I_j] = P[\text{failure} | I_j] = 1 - \Phi_Z(I_j).$$

We know from Chapter 12 that the likelihood function is defined as the probability of obtaining the sample that actually was obtained. For a random sample, this is simply the product of individual probabilities. Thus, here, given our convention on how the sample is ordered,

$$L(y) = \Phi_Z(I_1)\,\Phi_Z(I_2) \cdots \Phi_Z(I_M)[1 - \Phi_Z(I_{m+1})] \cdots [1 - \Phi_Z(I_N)].$$

That is,

$$\ln L(y) = \sum_{i=1}^{m} \ln \Phi_Z(I_i) + \sum_{i=m+1}^{N} \ln[1 - \Phi_Z(I_i)]. \tag{13.4.7}$$

Now $I_i$, defined in equation (13.4.6), is a function of $\beta_0, \beta_1, \ldots, \beta_K$, and it follows that ln L(y) is also a function of these unknown $\beta$'s.

Maximum-likelihood estimation is carried out, finding the particular values $\hat{\beta}_0, \hat{\beta}_1, \ldots, \hat{\beta}_K$ that yield the maximum value for ln L(y). This will be performed using numerical techniques carried out with a computer package such as SHAZAM and the details need not concern us.

## 13.5 THE LOGIT MODEL

The principles underlying the logit model are identical to those of the probit model. In fact, the logit model may be regarded as a variation of the probit model, having some mathematical properties that make it slightly easier to use. Because it is so similar to the probit model, we will not treat it in as much detail.

We recall that the essentials of the probit model are given by equation (13.4.2), namely,

$$p_i = P[Y = 1 | I_i] = \Phi_Z(I_i).$$

Thus when the index value $I_i$ is known, implying that the $\beta$ coefficients have been estimated, we can estimate the probabilities of success and failure.

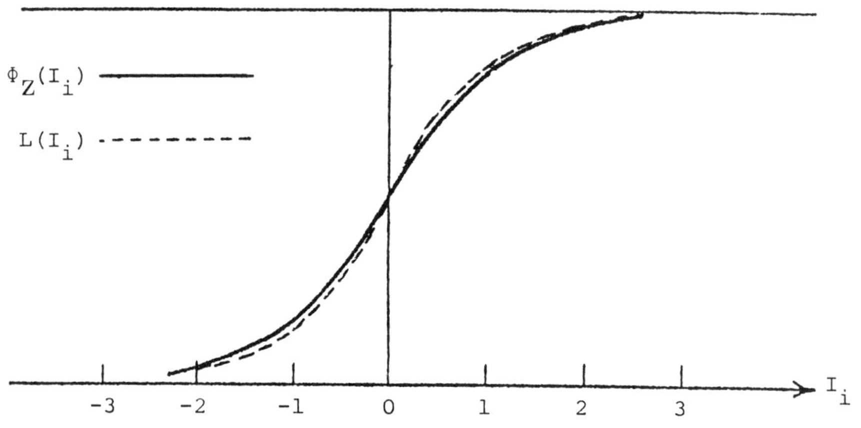


Figure 13.5.1. Graphs of the probit and logit functions $\Phi_Z(I_i)$ and $L(I_i)$.

This is achieved by evaluating $\Phi_Z(I_i)$ by consulting tables of the normal distribution. The function $\Phi_Z(I_i)$ cannot be evaluated by a mathematical formula.

Consider now another function L(x), known as the logistic function, and defined by

$$L(x) = \left[1 + \exp\left(-\frac{\pi x}{\sqrt{3}}\right)\right]^{-1} \qquad (13.5.1)$$

This function is very similar to the normal cumulative distribution function $\Phi_Z(x)$, as shown in Figure 13.5.1. Thus if we were to replace $\Phi_Z(I_i)$ in (13.4.2) by the logistic function $L(I_i)$—hence the term logit model—we would anticipate that our estimated probabilities would be very similar.

EXAMPLE Let us return to our previous example of finding the probabilities of purchasing a video recorder, evaluated for three different people. We assumed $\beta_0 = -7$, $\beta_1 = -5.4$, and $\beta_2 = 0.2$ and for the three individuals found index values

$$I_1 = -1, \quad I_2 = 1, \quad \text{and} \quad I_3 = -2.4.$$

From equation (13.5.1), $L(I_i)$ is now easily evaluated using a hand calculator. A comparison of the probabilities of purchase using the probit and logit models is given in Table 13.5.1. For all practical purposes, the logit and probit probabilities are the same. As a matter of convenience, the $\beta$ coefficients are usually multiplied by the factor $\pi/\sqrt{3}$, and we then have

$$p_i = L(I_i) = [1 + \exp(-I_i)]^{-1}. \qquad (13.5.2)$$

Table 13.5.1

| Individual, $i$ | Index value, $I_i$ | Probit probabilities | Logit probabilities |
|---|---|---|---|
| 1 | -1 | 0.16 | 0.14 |
| 2 | 1 | 0.92 | 0.86 |
| 3 | -2.4 | 0.01 | 0.01 |

This equation is the logit equivalent of the basic probit model (13.4.2).

In our example, this is equivalent to defining

$$I_i = \beta_0 + \beta_1 X_{1i} + \beta_2 X_{2i},$$

where now the $\beta$'s are redefined as

$$\beta_0 = \frac{-7\pi}{\sqrt{3}} = -12.7,$$

$$\beta_1 = \frac{-5.4\pi}{\sqrt{3}} = 9.79,$$

$$\beta_2 = \frac{0.2\pi}{\sqrt{3}} = 0.363.$$

Marginal Changes in p

In a similar way to the probit model we can analyze how marginal changes in the regressors induce changes in $p_i$. From equation (13.5.2),

$$\frac{\partial p_i}{\partial X_{ji}} = \frac{\partial L(I_i)}{\partial X_{ji}} = \frac{dL(I_i)}{dI_i} \frac{\partial I_i}{\partial X_{ji}}$$

$$= \frac{\exp(-I_i)\beta_j}{[1 + \exp(-I_i)]^2}$$

$$= (p_i^{-1} - 1)p_i^2 \beta_j.$$

Thus

$$\frac{\partial p_i}{\partial X_{ji}} = p_i(1 - p_i)\beta_j. \tag{13.5.3}$$

Table 13.5.2

| Individual, i | Probability of purchase, $p_i$ | $\Delta p_i$ |
|---|---|---|
| 1 | 0.14 | 0.04 |
| 2 | 0.86 | 0.04 |
| 3 | 0.01 | 0.004 |
| 4 | 0.59 | 0.09 |

This equation can be interpreted in a similar way to the equivalent probit result (13.4.4). The sign of $\beta_j$ determines whether increases in $X_{ji}$ lead to corresponding increases or decreases in $p_i$. Furthermore, the function $p_i(1 - p_i)$ is very small when $p_i \approx 0$ or $p_i \approx 1$, and has its maximum value of 0.25 when $p_i = 0.5$. Thus changes in $X_{ji}$ will have very little effect on $p_i$ when it is close to the extreme values of 0 and 1, whereas probabilities will be most sensitive to changes in $X_j$ when $p_i$ is close to 0.5 and $\beta_j$ is large.

EXAMPLE In the case of marginal changes in VCR purchase, based on the probit model, we introduced a fourth individual for whom $X_{14} = 0$ and $X_{24} = 36$. Using the logit formulation for $I_4$, we have

$$I_4 = -12.7 - 9.79 \times 0 + 0.363 \times 36 = 0.368.$$

Hence, by (13.5.2),

$$p_4 = [1 + \exp(-0.368)]^{-1} = 0.59.$$

We can now use the probabilities for the four individuals (see Table 13.5.1), together with equation (13.5.3), in which $\beta_2 = 0.363$, to calculate the changes in $p_i$ that will occur from an increase in income of \$1000. These results are given in Table 13.5.2. Comparing Tables 13.4.3 and 13.5.2, we can again see that the probit and logit models yield very similar results.

We close our discussion of the logit model by noting that estimation can be carried out using a standard package such as SHAZAM. As far as the user is concerned, the only difference with probit model estimation is that a different command is accessed.

## 13.6 ASSESSING PROBIT AND LOGIT MODELS

In the linear regression model, a standard test which is always carried out is that of the hypothesis

$$\beta_1 = \beta_2 = \cdots = \beta_K = 0$$

(see Section 6.4). This test is to determine whether the regressors are relevant. Alternatively, it is a test to determine whether the mean of the Y's is a constant or a function of the variables $X_1, X_2, \ldots, X_K$.

In the context of binary choice models, it is also important to assess whether the regressor variables $X_1, X_2, \ldots, X_K$ are relevant. Such a test is based on the likelihood ratio, which is dealt with in more detail in Section 14.4. Briefly, the maximum value of the likelihood function from the probit (or logit) model is compared with the maximum value, which is obtained assuming that the probability of success does not depend on any regressor variables. If the likelihood value from the probit model is much larger, there is support for including the regressors in the model.

More formally, we wish to test the hypothesis

$$H_0: \beta_1 = \beta_2 = \cdots = \beta_K$$

for the model

$$I_i = \beta_0 + \beta_1 X_{1i} + \cdots + \beta_K X_{Ki}.$$

Let us denote the maximum value of the likelihood function under the probit (or logit) model by $\tilde{L}_1$, and the maximum value under $H_0$ (i.e., p is constant) by $\tilde{L}_0$. Then if $H_0$ is true,

$$-2(\tilde{L}_0 - \tilde{L}_1) \sim \chi^2_K, \qquad (13.6.1)$$

and $H_0$ is rejected if this statistic exceeds the critical value for the $\chi^2_K$ distribution. Just as the F-statistic for the routine test of Section 6.4 is printed by most regression packages, the $\chi^2$-statistic (perhaps called "likelihood ratio test") is part of the output for most packages that perform probit and logit analyses.

## 13.7 EMPIRICAL EXAMPLE

A current issue in accounting is that of "segment profit disclosure" by companies. As companies become more complex, their operations are structured in identifiable segments. Investors would prefer to make decisions on the basis of disaggregated financial statements, setting out the performance of the individual segments. In Australia, there is no legal requirement to present disaggregated statements, but many companies voluntarily disclose segment profit. Kelly (1986) examined a cross section of 132 Australian companies and related their decision on segment disclosure to several corporate attributes. In particular, Kelly used probit analysis to estimate disclosure probabilities as a function of the following variables:

(a) Size: defined as the logarithm of market capitalization (measured in thousands of dollars)
(b) Leverage: defined as the ratio of long-term debt to total assets
(c) Return on investment (ROI): defined as the ratio of net profit to total assets
(d) Industry membership:[1] categorized as
  (i) Mining and metals
  (ii) Building contractors and suppliers
  (iii) Manufacturers and retailers
  (iv) Diversified industrial and resources
  (v) Other

A sample of the observations is given in Table 13.7.1. A SHAZAM program to create the required dummy variables and perform a probit analysis on these data is shown below, with the results given in Display 13.7.1.

```
FILE 33 PROB.DTA
READ (33)DEP INDUSTRY SIZE ROI LEVERAGE
IF(INDUSTRY.EQ.2)BUILDING=1
IF(INDUSTRY.EQ.3)MANRETL=1
IF(INDUSTRY.EQ.4)INDRES=1
IF(INDUSTRY.EQ.5)OTHERS=1
GENR SIZESQ=SIZE*SIZE
PROBIT DEP BUILDING MANRETL INDRES OTHERS SIZE ROI LEVERAGE
```

Table 13.7.1

| Company | Size | Leverage | ROI | Industry membership |
|---|---|---|---|---|
| 1 | 8.327 | 0.2792 | 0.0610 | 4 |
| 2 | 7.728 | 0.2636 | 0.0042 | 1 |
| 3 | 7.156 | 0.5350 | 0.0090 | 1 |
| ⋮ | ⋮ | ⋮ | ⋮ | ⋮ |
| 131 | 4.114 | 0.0609 | 0.0146 | 5 |
| 132 | 4.108 | 0.0145 | 0.0664 | 3 |

DISPLAY 13.7.1

| VARIABLE NAME | ESTIMATED COEFFICIENT | ASYMPTOTIC STANDARD ERROR | T-RATIO | ELASTICITY AT MEANS | WEIGHTED AGGREGATE ELASTICITY |
|---|---|---|---|---|---|
| BUILDING | 1.2862 | 0.51390 | 2.5029 | 0.24496 | 0.22350 |
| MANRETL | 0.72393 | 0.44630 | 1.6221 | 0.33702 | 0.20940 |
| INDRES | 1.6864 | 0.52967 | 3.1838 | 0.24979 | 0.22730 |
| OTHERS | 1.0086 | 0.46536 | 2.1673 | 0.28812 | 0.22911 |
| SIZE | 0.47216 | 0.15878 | 2.9737 | 3.4356 | 2.5292 |
| ROI | -16.461 | 5.9649 | -2.7596 | -1.3247 | -0.77196 |
| LEVERAGE | -0.58542 | 1.0838 | -0.54017 | -0.13993 | -0.11185 |
| CONSTANT | -3.0507 | 0.93654 | -3.2574 | -4.2606 | -3.0515 |

LOG-LIKELIHOOD(0) = -75.307 LOG LIKELIHOOD FUNCTION = -60.645
LIKELIHOOD RATIO TEST = 29.3244 WITH 7 D.F.



These results imply, for example, that for a company belonging to the manufacturers and retailers group, the index $I_i$ is estimated as

$$I_i = -2.33 + 0.472\ SIZE_i - 16.5\ ROI_i - 0.585\ LEVERAGE_i. \quad (13.7.1)$$

Consider now a hypothetical firm from this group having a market capitalization of \$510 million, a long-term debt of 25 percent of total assets, and a return on investment of 3 percent. That is, for this firm

$$SIZE = \ln(510) = 6.23,$$

$$ROI = 0.03,$$

$$LEVERAGE = 0.25.$$

From (13.7.1),

$$I_i = -2.33 + 0.472 \times 6.23 - 16.5 \times 0.03 - 0.585 \times 0.25$$
$$= -0.031$$

Thus, denoting by $p_i$ the probability that this firm will voluntarily disclose segment performance, we have by (13.4.2),

$$p_i = \Phi_Z(-0.031) = 0.49.$$

There is therefore approximately a 0.5 probability that such a firm would disclose segment performance.

By contrast, a company having exactly the same corporate attributes, but belonging to the diversified industrial and resources group (here $\hat{\beta}_0 = -3.05 + 1.68 = -1.37$), would have an index value

$$I_i = 0.93,$$

yielding

$$p_i = \Phi_Z(0.93) = 0.82.$$

Such a firm is very likely to disclose segment performance.

## NOTE

1. The industry membership effect was captured by four dummy variables, taking the mining and metals group as base.

# 14

# Specification Testing of the Regression Model

## 14.1 INTRODUCTION

Right from the beginning of our study we have stressed that the properties of the estimation techniques depend on correct specification of the model. Thus if a relevant variable is omitted, estimators are biased. If the disturbances are autocorrelated or heteroscedastic, there is a loss of efficiency and, perhaps more important, standard errors may be biased, leading to incorrect inference. If in the case of stochastic regressors, the disturbance is correlated with the regressors, simultaneity bias results, and so on. All of this implies that estimation is only the first part of a regression analysis. The careful econometrician should invest considerable effort in checking specification to detect inadequacies in his and her models. In this regard, it is worth quoting Pagan (1984) at length:

> Modelling was never meant to be easy. The 1970s taught many econometricians the worth of this aphorism but, strangely, there has grown up a conception that the principal lesson to be learned related to the utility of large-scale models. What should have come from this experience, however, was an appreciation of the weakness of the research strategy in evidence during the late sixties and early seventies. Four steps almost completely describe it: a model is postulated, data gathered, a regression run, some t-statistics or simulation performance provided and another "empical regularity" was forged. What was wrong with this approach? Depending upon the example there could be many specific factors, but three deficiencies are outstanding. First, it failed to be critical. Almost no attempt was made to discover weaknesses in the model selected, or at least to present them to an observer. Once a model agreeing with a priori concepts of "signs" and "significance" was obtained, research terminated. Second, it failed to be informative. Few indices of model adequacy were given and very little discussion on the sensitivity of results was provided. Finally, in very few cases was an

attempt made to reconcile the "new" formulation with earlier research. In many instances, apart from obligatory references at the beginning of the article, a reader could well have gained the impression that there had been no previous research on the topic.

In this chapter we gather together some of the more important techniques available for checking model specification and integrate these with diagnostic procedures we have already encountered. There is an extensive literature on this subject and our discussion is confined to methods which, by and large, are (i) easy to apply with a standard regression package; and (ii) have wide applicability.

## 14.2 ASSESSING MODEL ADEQUACY

Models are constructed to explain, or account for, the variation in the dependent variable. By including a disturbance, we are recognizing that this cannot be perfectly achieved. Nevertheless, we are hoping to take account of the systematic variation in the dependent variable. If the residuals have systematic characteristics, clearly our model is inadequate. Many tests of model adequacy (often called "misspecification tests") are attempts to discover whether the residuals are random or have some structure. Familiar examples are the Durbin-Watson test for autocorrelation, the Goldfeld-Quandt and Breusch-Pagan tests for heteroscedasticity, and the lack-of-fit test. In each of these examples, the characteristics of the residuals are being examined—for time dependency in the case of the Durbin-Watson test, and the nature of their variability in the other three tests. A valuable addition to tests based on OLS residuals are those which use what are known as "recursive residuals." This is a little beyond the scope of this book, but the interested reader is referred to an excellent discussion in Harvey (1983, 148-154). In this section we introduce three further tests of misspecification which are widely applicable.

### The RESET Test (Ramsey, 1969; Ramsey and Schmidt, 1976)

The RESET (Regression Specification Error Test)[1] test is a general procedure designed to detect specification errors due to omitted variables, incorrect functional form, heteroscedasticity, and simultaneity bias. Because it is very easy to apply, it is a useful tool in checking model adequacy.

Let us suppose that our maintained model is

$$Y_i = \beta_0 + \sum_{j=1}^{K} \beta_j X_{ji} + u_i.$$

The RESET test consists of estimating the model by OLS and obtaining the predicted values $\hat{Y}_i$. The model is then augmented by the variables $\hat{Y}^2$, $\hat{Y}^3$, ..., $\hat{Y}_i^r$. That is, we now estimate

$$Y_i = \beta_0 + \sum_{j=1}^{K} \beta_j X_{ji} + \sum_{j=2}^{r} \gamma_j \hat{Y}_i^j + u_i \tag{14.2.1}$$

and test the hypothesis $\gamma_2 = \cdots = \gamma_r = 0$ using an F-test.

If the hypothesis is rejected, model inadequacy is indicated. Computer simulation experiments have suggested that choosing $r = 4$ will often give good results. Note that the variable $\hat{Y}$ is not included. This is because $\hat{Y}$ is perfectly collinear with $X_1$, ..., $X_K$ and OLS would break down.

## Tests of Model Stability

When time-series data are used, it is quite likely that the true model will change through time. It is important, therefore, to test a model's adequacy with respect to its stability over time. There are two procedures that are commonly used to check model stability.

(a) The Chow test. This test was discussed in Section 7.3 and illustrated in Section 9.2.

(b) A test of postsample predictions. Suppose, as usual, that we have N sample observations. If only N - r observations are used for estimation, the predictions of $y_{N-r+1}$, ..., $y_N$ can be compared with the actual observations. If the model is stable, the prediction errors should be small. It can be shown that, asymptotically,

$$\frac{\hat{Y}_{N-r+j} - Y_{N-r+j}}{S} \sim N(0, 1^2) \tag{14.2.2}$$

for $j = 1, 2, \ldots, r$, where S is the standard error of the prediction errors.

The actual computation of the prediction errors and their standard errors can be accomplished using a minor modification of the technique given in Section 7.7, due to Salkever (1976). Let us define r dummy variables $D_1$, $D_2$, ..., $D_r$ so that

$$\begin{aligned} D_{it} &= 1 \quad \text{when } t = N - r + i\text{; and} \\ &= 0 \quad \text{otherwise.} \end{aligned}$$

Then if we run the model

$$Y_t = \beta_0 + \sum_{j=1}^{K} \beta_j X_{jt} + \sum_{j=1}^{r} \gamma_j D_{jt} + u_t, \tag{14.2.3}$$

the OLS estimates can be shown to have the following meaning:

(i)' $\hat{\beta}_0, \hat{\beta}_1, \ldots, \hat{\beta}_K$ are the estimates of the $\beta$'s using the first N - r observations.

(ii)' $\hat{\gamma}_1, \ldots, \hat{\gamma}_r$ are the negative of the prediction errors for observations $Y_{N-r+1}, \ldots, Y_N$.

(iii)' The standard errors of the $\hat{\gamma}$'s are those of the prediction errors.

Thus, to judge whether the prediction errors are small (i.e., the model is stable) we simply compare the t-ratios of the $\hat{\gamma}$'s with the appropriate critical value from the standard normal distribution (e.g., 1.96 for a test of size 0.05).

The modification we have made to the technique given in Section 7.7 is that the last r observations on the dependent variable for the regression here are the actual observations $Y_{N-r+1}, \ldots, Y_N$, whereas in Section 7.6 they are the mean of the first N - r observations. As can be seen from equation (7.7.1), this enables the prediction errors to be computed directly as the $\hat{\gamma}$'s. The price we pay is that the $R^2$ value arising from the regression is not the same as we would obtain if we simply regressed Y on $X_1, \ldots, X_K$ using the first N - r observations.

When the model has either autocorrelated disturbances and/or lagged dependent variables, the method of producing prediction errors and corresponding standard errors has to be modified. The details may be found in Pagan and Nicholls (1984).

ILLUSTRATION In Section 9.3 the relative income hypothesis was applied to the data of Table 9.2.1 and the model (9.3.5) estimated. In this illustration we estimate the model using data from 1948-49 to 1979-80, leaving the last three observations to compare with postsample predictions. A MINITAB® program to accomplish this is given below.

```
NAME C1='C'  C2='Y**'  C3='Z1'  C4='I'
SET C50
32(0)  1  2  3
INDICATOR  C50  C51  C5-C7
NAME  C5='D1'  C6='D2'  C7='D3'
BRIEF  1
REGR  C1  6  C2-C7
STOP
```

The relevant results are shown in Display 14.2.1. We see that the postsample prediction errors are -0.097, -0.127, and -0.102, having t-ratios of -1.48, -1.92, and -1.51, respectively. As none of these t-values is significantly different from zero at the 5 percent level, we have no reason to doubt the stability of this model over the period covered by the data.

DISPLAY 14.2.1

```
MTB > REGR C1 6 C2-C7

The regression equation is
C = 0.0151 + 0.898 Y** + 0.417 Z1 - 1.16 I + 0.0970 D1 + 0.127 D2 + 0.102 D3
```

| Predictor | Coef | Stdev | t-ratio |
|---|---|---|---|
| Constant | 0.01511 | 0.05820 | 0.26 |
| Y** | 0.89810 | 0.01526 | 58.85 |
| Z1 | 0.41658 | 0.07890 | 5.28 |
| I | -1.1627 | 0.2310 | -5.03 |
| D1 | 0.09705 | 0.06577 | 1.48 |
| D2 | 0.12689 | [illegible] | [illegible] |
| D3 | 0.10179 | 0.06754 | 1.51 |

```
s = 0.06192        R-sq = 99.5%        R-sq(adj) = 99.4%
```

Reprinted with permission of Minitab, Inc.

A fundamental characteristic of misspecification tests is that if a model is shown to be inadequate, the tests of themselves give no indication of how the model should be modified. For example, if in the illustration above, the prediction errors were found to be significantly large, we would have concluded that the relative income hypothesis, at least in the form of equation (9.3.5), was inadequate as a model for these data. However, the test would not give any indication of how the model should be modified. This contrasts with <u>specification</u> tests, which we will discuss shortly.

## 14.3 DISCRIMINATING BETWEEN MODELS

A common strategy in model building is to estimate a number of models and then choose the best model on some goodness-of-fit criterion. We have dealt in considerable detail with the uses and abuses of $R^2$ and $\bar{R}^2$ as means of discriminating between competing models (see Section 5.2). The test between linear and log-linear models (Section 6.5) is also based on goodness of fit. In essence, a comparison is made between residual sums of squares, once units of measurement have been chosen to make such a comparison valid. We now present two further criteria which have certain advantages over $R^2$ and $\bar{R}^2$ as a means of discrimination.

### The Prediction Criterion

As Judge et al. (1985, 863) remark, "the most uncomfortable aspect of both $R^2$ and adjusted $R^2$ measures is that they do not include a consideration of the losses associated with choosing an incorrect model." Amemiya (1980) has advanced the prediction criterion (PC), which minimizes the mean-square prediction error, which he takes as a measure of the cost of choosing the incorrect model. The PC measure of goodness of fit is given by

$$PC = \hat{\sigma}^2\left(1 + \frac{K + 1}{N}\right). \tag{14.3.1}$$

Amemiya has shown that minimizing PC is equivalent to maximizing a modified $R^2$, denoted by $\tilde{R}^2$, where

$$\tilde{R}^2 = 1 - (1 - R^2)\,\frac{N + K + 1}{N - K - 1}. \tag{14.3.2}$$

If we compare this with equation (5.2.2), which defines $\bar{R}^2$, it can be seen that $\tilde{R}^2$ imposes a greater penalty to increasing the number of regressors than does $\bar{R}^2$. We recall that $R^2$ can be increased simply by adding regressor variables, even if they are completely irrelvant. This leads to the introduction of $\bar{R}^2$, which balances increase in $R^2$ against reduction in degrees of freedom. However, as pointed out at the end of Section 6.3, $\bar{R}^2$ will increase provided that $F > 1$. Thus $\bar{R}^2$ will increase sometimes even when added variables, as a group, do not appear to be relevant when judged by an F-test.

Use of $\tilde{R}^2$, derived by minimizing a cost function, takes the process of adjustment of $R^2$ further, particularly when sample sizes are not large. Amemiya has shown that when variables are added, $\tilde{R}^2$ will increase if, and only if, $F > 2N/(N + K + 1)$. For values of $K/N < 0.1$, as would be usual, the ratio $2N/(N + K + 1)$ is fairly close to 2. Thus as a "rule of thumb" we can say that $R^2$ increases if $F > 1$, whereas $\tilde{R}^2$ increases only when $F > 2$. There is therefore less likelihood of an inconsistency arising between goodness of fit and F-tests when $\tilde{R}^2$ is used.

### The Akaike Information Criterion (AIC)

The Akaike information criterion (AIC) is based on the maximum value $\tilde{L}$ of the log of the likelihood function, making allowance for the number of parameters K.[2] It is defined by

$$AIC = -\frac{2}{N}\,\tilde{L} + \frac{2K}{N}. \tag{14.3.3}$$

For <u>linear</u> models, AIC is very similar to PC (Amemiya, 1980). It does, however, have the advantage of being a natural measure of goodness of fit when models are nonlinear and are estimated by maximum likelihood. It is routinely produced as part of the output in a number of regression packages.

## 14.4 TESTS OF SPECIFICATION I: NESTED MODELS

When a model is a special case of a more general model, it is said to be nested within the general model. It follows that when we are testing restrictions on a model, the restricted model is nested within the original model. All the hypothesis tests that we have dealt with so far, using t- or F-tests, are in fact tests of nested models. Characteristic of all these tests has been the following:

(i) Both restricted and unrestricted models are linear in the parameters $\beta_0, \beta_1, \ldots, \beta_K$.
(ii) The disturbance $u_t$ obeys the usual assumptions (2.1.2)-(2.1.4) and is normally distributed.

There are, however, many practical problems in which one or both of (i) and (ii) are not fulfilled. In such situations standard t- and F-tests cannot be used and other testing procedures must be devised. The three testing principles that we discuss—the likelihood ratio, and Wald and Lagrange multiplier principles—provide tests based on the likelihood of a sample.

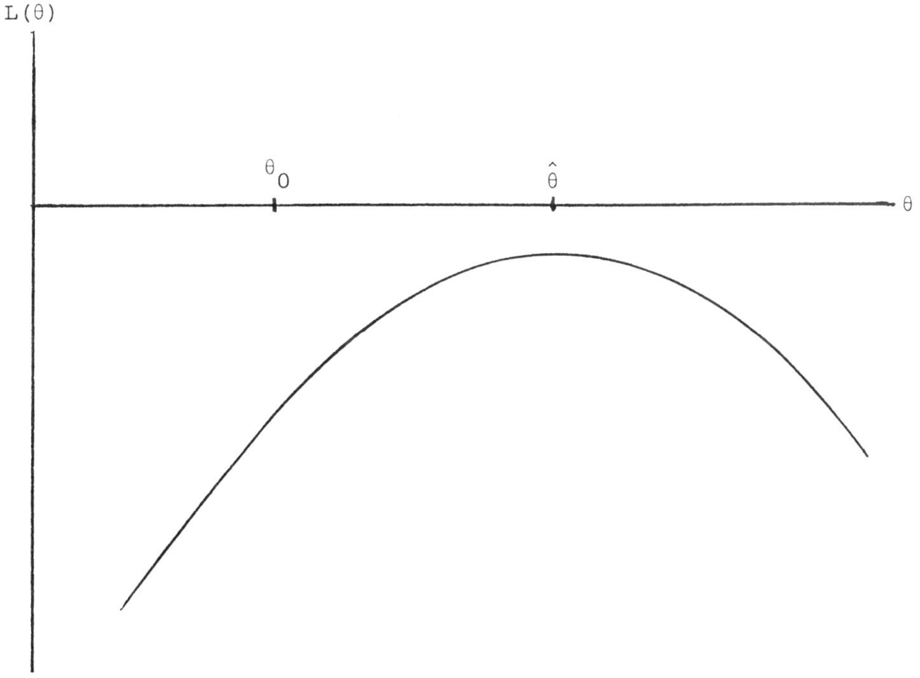


Figure 14.4.1. Typical log-likelihood function.

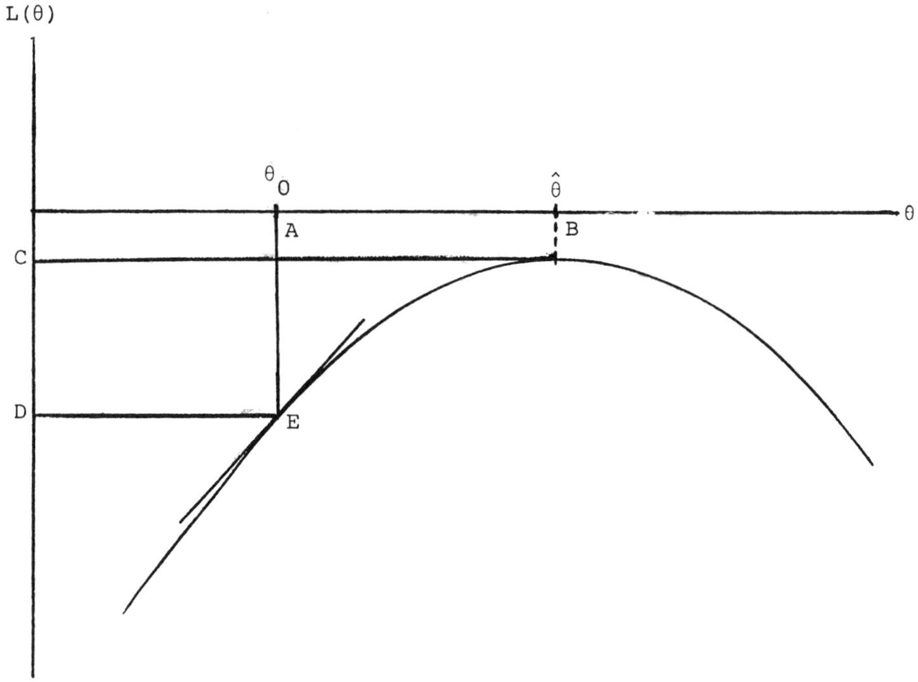


Figure 14.4.2. Log-likelihood function, showing different ways of measuring the distance between $\hat{\theta}$ and $\theta_0$.

Buse (1982) has given a simple graphical exposition which shows the relationship between these principles for the special case of a null hypothesis that involves only a single parameter, $\theta$. Consider the logarithm of the likelihood function $L(\theta)$, shown in Figure 14.4.1. The value of $\theta$ specified by the null hypothesis $\theta = \theta_0$ is shown, together with the maximum-likelihood estimate $\hat{\theta}$. It stands to reason that we would normally accept the null hypothesis if $\hat{\theta}$ is close to $\theta_0$. Conversely, if $\hat{\theta}$ is nowhere near $\theta_0$, we would take this to be evidence that the null is false. There are actually three different ways in which we can assess whether or not $\hat{\theta}$ and $\theta_0$ are close. Consider Figure 14.4.2.

(a) The direct approach is to compare $\hat{\theta}$ with $\theta_0$ or, equivalently, compare $\hat{\theta} - \theta_0$ with zero. If $\hat{\theta} - \theta_0$ is close to zero, $H_0$ is accepted; otherwise, it is rejected. This is the Wald principle. Application requires testing whether $\hat{\theta} - \theta_0$ is significantly different from zero, and this involves being able to find the standard error of $\hat{\theta}$.

(b) Another approach is to compare the two points C and D. If $\hat{\theta}$ is close to $\theta_0$, then C will be close to D. Now the distance CD is simply $L(\hat{\theta}) - L(\theta_0)$. Comparing $L(\hat{\theta})$ with $L(\theta_0)$ is the likelihood-ratio principle.

(c) In Figure 14.4.2 we have depicted the slope of the likelihood function when $\theta = \theta_0$ (i.e., at the point E). How, by definition, the slope of the likelihood function is zero when $\theta = \hat{\theta}$. Clearly, if $\hat{\theta}$ and $\theta_0$ are close, the slope at $\theta_0$ must be nearly zero. Thus the final way of assessing whether $H_0$ should be rejected is to see whether $(dL/d\theta)_{\theta=\theta_0}$ is significantly different from zero. This principle is known as the Lagrange multiplier principle.

Without a thorough competence in likelihood methods, the Wald test may be difficult to construct. However, simple rules often enable likelihood ratio and Lagrange multiplier tests to be formulated more easily. For this reason we concentrate on these two testing principles.

## Likelihood-Ratio Tests

We have seen that in the simple case of a model that has only a single parameter $\theta$, the likelihood-ratio (LR) principle results in comparing $L(\hat{\theta})$ to $L(\theta_0)$. In more general situations it is convenient to think of an unrestricted model and a model restricted by a null hypothesis $H_0$. We are familiar with this approach in constructing F-tests. We will define

$\tilde{L}_1$ = maximum value of the log-likelihood function for the unrestricted model; and

$\tilde{L}_0$ = maximum value of the log-likelihood function for the restricted model.

The basic result that enables us to formulate tests is the following:

THEOREM 14.4.1 If $H_0$ is true, then

$$-2(\tilde{L}_0 - \tilde{L}_1) \sim \chi^2_M, \tag{14.4.1}$$

where M is the number of constraints imposed by $H_0$.

This result is actually an asymptotic result, but will usually give reliable results provided that the sample size is not too small. The practicality of the LR test comes about because most computer packages (e.g., SHAZAM) routinely compute the maximum value of the log-likelihood function. It is then a simple matter to use (14.4.1) to test the validity of the restriction.

EXAMPLE The constant elasticity of substitution (CES) production function takes the form

$$\ln q_t = \alpha + (\beta - 1) \ln L_t - \beta\gamma^{-1} \ln(1 - \delta + \delta k_t^{-\gamma}) + u_t, \qquad (14.4.2)$$

where $q_t = Q_t/L_t$, $k_t = K_t/L_t$. As usual $Q_t$, $K_t$, and $L_t$ are quantity, capital, and labor, respectively. The advantage of the CES production function over the Cobb-Douglas is that it allows an elasticity of substitution other than unity.

It can be shown that when $\gamma = 0$, (14.4.2) becomes

$$\ln Q_t = \alpha + \beta(1 - \delta) \ln L_t + \beta\delta \ln K_t + u_t,$$

DISPLAY 14.4.1

*Final Results of CES estimation:*

```
FINAL STATISTICS :

TIME =  0.000 SEC.          ITER. NO.      18           FUNCT. EVALUATIONS     33
LOG-LIKELIHOOD FUNCTION =  25.04643
COEFFICIENTS
   1.671955            0.6354147           -5.781470            0.7673487
GRADIENT
   0.2850415E-08       0.1369749E-07        0.3634127E-12       0.3238566E-09

MAXIMUM LIKELIHOOD ESTIMATE OF SIGMA-SQUARED = 0.20758E-02

           COEFFICIENT   ST. ERROR        T-RATIO

ALPHA         1.6720       0.85848         1.9476
BETA          0.63541      0.18709         3.3963
GAM          -5.7815      21.099          -0.27401
DEL           0.76735      0.41864         1.8330
```

*Final Results of CD estimation:*

```
FINAL STATISTICS :

TIME = 0.000 SEC.           ITER. NO.      8            FUNCT. EVALUATIONS     27
LOG-LIKELIHOOD FUNCTION = 24.98640
COEFFICIENTS
   1.539567            0.2048794           0.4600489
GRADIENT
  -0.5974786E-06     -0.2929923E-05     -0.3395874E-05

MAXIMUM LIKELIHOOD ESTIMATE OF SIGMA-SQUARES = 0.20925E-02


BETA0         1.5396       0.82622         1.8634
BETA1         0.20488      0.23561         0.86957
BETA2         0.46005      0.12068         3.8122
```

which is the Cobb-Douglas function with returns to scale equal to $\beta$. Thus the Cobb-Douglas function is nested within the CES function. We will now use the data of Table 4.7.1 to test the hypothesis $\gamma = 0$. The CES production function is nonlinear, and we use nonlinear estimation in SHAZAM. The program is shown below, and the results in Display 14.4.1.

```
FILE 33 CD.DTA
READ(33) K L Q
GENR QL=Q/L
GENR KL=K/L
GENR LNQL=LOG(QL)
GENR LNL=LOG(L)
GENR LNQ=LOG(Q)
NL 1 / NCOEF=4
EQ LNQL=ALPHA+(BETA-1)*LNL-(BETA/GAM)*LOG(1-DEL+DEL*(KL)**(-GAM))
COEF ALPHA 2 BETA 0.7 GAM 0. DEL 3
END
NL 1 / NCOEF=3
EQ LNQ=BETA0+BETA1*LNL+BETA2*LNK
END
STOP
```

The unconstrained maximum of the log-likelihood has the value

$$\tilde{L}_i = 25.046.$$

The constrained function is linear and may be estimated by OLS. However, we estimated this also by nonlinear least squares to obtain

$$\tilde{L}_0 = 24.986.$$

Thus the likelihood-ratio statistic (LR) is

$$-2(24.986 - 25.046) = 0.12$$

with $M = 1$. As the critical 5 percent value for $\chi_1^2$ is 3.84, we cannot reject the null hypothesis $\gamma = 0$. The Cobb-Douglas production function therefore seems adequate to describe these data.

## Lagrange Multiplier Tests

We have seen that the Lagrange multiplier (LM) principle is to test the significance of the slope of the log-likelihood at the value of $\theta$ specified by the null hypothesis. As with the Wald test, application of the principle to specific problems may be difficult. However, Breusch and Pagan (1980) have shown that in quite a wide class of models, application of the LM principle involves only two OLS regressions. We will confine our attention to this class of models and summarize the Breusch and Pagan results, without making any attempt to provide the link between principle and operation.

Let us suppose that the model may be written in the form

$$Y_t = g(X_t; \theta) + u_t = g_t + u_t,$$

where

(i) $u_t$ is normal and has the usual properties;
(ii) $X_t$ stands for regressor variables that are independent of $u_t$; and
(iii) $\theta = (\theta_1, \theta_2, \ldots, \theta_K)$ represents the parameters of the model.

Generally, $g(X_t; \theta) \equiv g_t$ would be a nonlinear function of $\theta_1, \theta_2, \ldots, \theta_K$. If $g_t$ is a linear function and the null hypothesis is linear, it may be tested by the conventional t- or F-tests. Suppose that the restricted model has been estimated, giving parameter estimates $\tilde{\theta}_1, \tilde{\theta}_2, \ldots, \tilde{\theta}_K$ and residuals $\tilde{e}_t$ ($t = 1, 2, \ldots, N$). We will define K variables

$$\frac{\partial \tilde{g}_t}{\partial \theta_i} \quad (i = 1, 2, \ldots, K)$$

where "~" indicates that the derivatives of $g_t$ are evaluated at $\tilde{\theta}_1, \tilde{\theta}_2, \ldots, \tilde{\theta}_K$. The basic result is now stated:

THEOREM 14.4.2 (Breusch and Pagan, 1980) If $H_0$ is true, then provided that N is large,

$$NR^2 \sim \chi^2_M, \tag{14.4.3}$$

where

(i) N is the sample size and M the number of constraints implied by $H_0$;
(ii) $R^2$ is the $R^2$ obtained when $\tilde{e}_t$ is regressed on the K regressors $\partial \tilde{g}_t / \partial \theta_i$ ($i = 1, 2, \ldots, K$).

The null hypothesis is rejected if $NR^2$ is greater than the critical value of $\chi^2_M$.

We will illustrate the construction of LM tests by first considering tests of linear models, then citing two examples given by Breusch and Pagan, and finally by testing the Cobb-Douglas production function against the CES function.

EXAMPLE 1: Testing a Linear Hypothesis. Consider the model

$$Y_t = \beta_0 + \beta_1 X_{1t} + \cdots + \beta_K X_{Kt} + u_t$$

and $H_0: \beta_1 = \beta_2 = 0$. Here $M = 2$,

$$\frac{\partial \tilde{g}_t}{\partial \beta} = \ln L_t + \tilde{\delta} \ln k_t - \frac{\tilde{\delta}(1 - \tilde{\delta})}{2} \tilde{\gamma}(\ln k_t)^2$$

$$= \ln L_t + \tilde{\delta} \ln k_t \quad (\text{as } \tilde{\gamma} = 0),$$

$$\frac{\partial \tilde{g}_t}{\partial \delta} = \tilde{\beta} \ln k_t \quad (\text{as } \tilde{\gamma} = 0),$$

$$\frac{\partial \tilde{g}_t}{\partial \gamma} = \frac{-\tilde{\beta}\tilde{\delta}(1 - \tilde{\delta})}{2} (\ln k_t)^2.$$

The second stage regression is of $\tilde{e}_t$ against $\ln L_t + \tilde{\delta} \ln k_t$, $\tilde{\beta} \ln k_t$, and $-[\tilde{\beta}\tilde{\delta}(1 - \tilde{\delta})/2](\ln k_t)^2$ with an intercept. However, as in the preceding example, this model may be reparametrized without changing $R^2$. We are therefore led to the second regression of $\tilde{e}_t$ against $\ln L_t$, $\ln k_t$, $(\ln k_t)^2$ with an intercept. $NR^2$ from this second regression is compared with $\chi_1^2$.

A MINITAB® program to perform the two regressions, using observations on K, L, and Q from Table 4.6.1 (stored in 'CD.DTA') is shown below. Relevant results are given in Display 14.4.2.

```
READ 'CD.DTA' C1-C3
#K, L, Q IN C1, C2, C3
LET C11 = LOGE(C1)
LET C12 = LOGE(C2)
LET C13 = LOGE(C3)
NAME C11 'LN(K)' C12 'LN(L)' C13 'LN(Q)'
LET C14 = C11-C12
LET C15 = C14*C14
NAME C14 'LN(K/L)' C15 'LNK/L**2'
REGR C13 2 C11 C12;
RESID C20.
NAME C20 'RESID'
REGR C20 3 C12 C14 C15
STOP
```

From the results,[3]

$$NR^2 = 15\left(\frac{ESS}{TSS}\right) = \frac{15 \times .000273}{0.031387} = 0.13.$$

As the critical value for $\chi_1^2$ at the 5 percent level is 3.84, we do not reject $H_0 : \gamma = 0$.

Each of the four examples cited above shares a common characteristic:



$$g_t = \beta_0 + \beta_1 X_{1t} + \cdots + \beta_K X_{Kt},$$

and

$$\theta = (\beta_0, \beta_1, \ldots, \beta_K).$$

The restricted model is

$$Y_t = \beta_0 + \beta_3 X_{3t} + \cdots + \beta_K X_{Kt} + u_t.$$

This may be estimated by OLS giving

$$\tilde{\theta} = (\tilde{\beta}_0, 0, 0, \tilde{\beta}_3, \ldots, \tilde{\beta}_K)$$

and the residuals $\tilde{e}_t$ ($t = 1, 2, \ldots, N$). Then

$$\frac{\partial \tilde{g}_t}{\partial \beta_0} = 1, \quad \frac{\partial \tilde{g}_t}{\partial \beta_1} = X_{1t}, \quad \frac{\partial \tilde{g}_t}{\partial \beta_2} = X_{2t}, \quad \text{etc.}$$

Thus, in the second regression, $\tilde{e}_t$ is regressed on $X_{1t}, X_{2t}, \ldots, X_{Kt}$ with an intercept. If $NR^2$ from this regression exceeds the critical value of $\chi_2^2$, $H_0$ is rejected. The simple rule that emerges for linear hypotheses on linear models is that the residuals from the restricted model are regressed on the original variables and $NR^2$ from this regression is compared with $\chi_M^2$.

EXAMPLE 2 Suppose that $Y_t = \beta_0 + \beta_1 X_{1t} + \cdots + \beta_K X_{Kt} + u_t$ and $u_t = \rho u_{t-1} + v_t$. Here $u_t$ follows a first-order autoregressive process, but $v_t$ has the desired properties. We wish to test $H_0 : \rho = 0$. It follows that

$$Y_t = \beta_0(1 - \rho) + \rho Y_{t-1} + \beta_1 X_{1t} - \beta_1 \rho X_{1,t-1} + \cdots + \beta_K X_{Kt} - \beta_K \rho X_{K,t-1} + v_t.$$

Thus

$$Y_t = g_t + v_t,$$

where

$$g_t = \beta_0(1 - \rho) + \rho Y_{t-1} + \sum_{i=1}^{K} \beta_i (X_{it} - \rho X_{i,t-1}).$$

Notice, $g_t$ is nonlinear in the parameter $\rho$. The restricted model (setting $\rho = 0$) is

$$Y_t = \beta_0 + \beta_1 X_{1t} + \cdots + \beta_K X_{Kt} + v_t.$$

This can be estimated by OLS, yielding estimates $\tilde{\beta}_0, \tilde{\beta}_1, \ldots, \tilde{\beta}_K$, and $\tilde{\rho} = 0$. Denote the residuals by $\tilde{e}_t$. Now

$$\frac{\partial \tilde{g}_t}{\partial \beta_0} = 1 - \tilde{\rho} = 1 \quad (\text{as } \tilde{\rho} = 0),$$

$$\frac{\partial \tilde{g}_t}{\partial \beta_i} = X_{it} - \tilde{\rho} X_{i,t-1} = X_{it} \quad (i = 1, 2, \ldots, K),$$

$$\frac{\partial \tilde{g}_t}{\partial \rho} = -\tilde{\beta}_0 + Y_{t-1} - \tilde{\beta}_1 X_{1,t-1} - \cdots - \tilde{\beta}_K X_{K,t-1}$$

$$= Y_{t-1} - [\tilde{\beta}_0 + \tilde{\beta}_1 X_{1,t-1} + \cdots + \tilde{\beta}_K X_{K,t-1}]$$

$$= \tilde{e}_{t-1}.$$

Thus, to test $H_0 : \rho = 0$, we regress $\tilde{e}_t$ on 1, $X_{1t}, \ldots, X_{Kt}, \tilde{e}_{t-1}$. $NR^2$ from this regression is compared with $\chi_1^2$.

EXAMPLE 3 Konstas and Khouja (1969) proposed a model for the demand for money of the form

$$M_t = \gamma Y_t + \beta (R_t - \alpha)^{-1} + u_t,$$

where $M_t$ is money demand, $Y_t$ is income, and $R_t$ is rate of interest. They tested for the liquidity trap by considering $H_0 : \alpha = 0$. Here

$$g_t = \gamma Y_t + \beta (R_t - \alpha)^{-1}.$$

The restricted model is

$$M_t = \gamma Y_t + \beta R_t^{-1} + u_t,$$

which can be fitted by regressing $M_t$ on $Y_t$ and $R_t^{-1}$ without an intercept. The residuals are $\tilde{e}_t$ and parameter estimates $\tilde{\gamma}$, $\tilde{\beta}$, $\tilde{\alpha}$ (= 0). Then

$$\frac{\partial \tilde{g}_t}{\partial \gamma} = Y_t,$$

$$\frac{\partial \tilde{g}_t}{\partial \beta} = (R_t - \tilde{\alpha})^{-1} = R_t^{-1},$$

and

$$\frac{\partial \tilde{g}_t}{\partial \alpha} = \tilde{\beta}(R_t - \tilde{\alpha})^{-2} = \tilde{\beta} R_t^{-2}.$$

Thus, at the second stage, we regress $\tilde{e}_t$ on $Y_t$, $R_t^{-1}$, $\tilde{\beta} R_t^{-2}$ (no intercept).



Our model is

$$\tilde{e}_t = \gamma_1 Y_t + \gamma_2 R_t^{-1} + \gamma_3 (\tilde{\beta} R_t^{-2}) + v_t$$

$$= \gamma_1 Y_t + \gamma_2 R_t^{-1} + \gamma_3^* R_t^{-2} + v_t,$$

where $\gamma_3^* = \gamma_3 \tilde{\beta}$. Making this reparametrization leaves $R^2$ uncha at the second stage we regress $\tilde{e}_t$ on $Y_t$, $R_t^{-1}$ and $R_t^{-2}$ without an and compare $NR^2$ with $\chi_1^2$.

EXAMPLE 4 We will construct an LM test of the Cobb-Douglas function against the CES function, again using the data of Table have seen that the CES production function is

$$\ln q_t = \alpha + (\beta - 1) \ln L_t - \beta \gamma^{-1} \ln (1 - \delta + \delta k_t^{-\gamma}) + u_t$$

and when $\gamma = 0$ it reduces to the Cobb-Douglas function

$$\ln Q_t = \alpha + \beta(1 - \delta) \ln L_t + \beta\delta \ln K_t + u_t.$$

This is the restricted model. If we estimate it in the form

$$\ln \tilde{Q}_t = \tilde{\beta}_0 + \tilde{\beta}_1 \ln L_t + \tilde{\beta}_2 \ln K_t,$$

then

$$\tilde{\alpha} = \tilde{\beta}_0,$$

$$\tilde{\beta} = \tilde{\beta}_1 + \tilde{\beta}_2,$$

$$\tilde{\delta} = \frac{\tilde{\beta}_2}{\tilde{\beta}_1 + \tilde{\beta}_2}$$

$$\tilde{\gamma} = 0.$$

Setting

$$g_t = \alpha + (\beta - 1) \ln L_t - \beta \gamma^{-1} \ln (1 - \delta + \delta k_t^{-\gamma}),$$

we can expand $g_t$ about $\gamma = 0$ by Taylor's expansion to obtain

$$g_t = \alpha + (\beta - 1) \ln L_t + \beta \left[ \delta \ln k_t - \frac{\delta(1 - \delta)}{2} \gamma (\ln k_t)^2 \right] + \text{terms}$$

As derivatives of $g_t$ are evaluated with $\gamma = 0$, the terms in $\gamma^2$ be ignored. Then

$$\frac{\partial \tilde{g}_t}{\partial \alpha} = 1,$$

DISPLAY 14.4.2

```
MTB    REGR C13 2 C11 C12;
SUBC   RESID C20.
```

The regression equation is
LN(Q) = 1.54 + 0.460 LN(K) + 0.205 LN(L)

| Predictor | Coef | Stdev | t-ratio |
|---|---|---|---|
| Constant | 1.5396 | 0.9210 | 1.67 |
| LN(K) | 0.4600 | 0.1321 | 3.48 |
| LN(L) | 0.2049 | 0.2631 | 0.78 |

s = 0.05114     R-sq = 71.3%     R-sq(adj) = 66.5%

```
MTB    REGR C20 3 C12 C14 C15
```

The regression equation is
RESID = 0.16 - 0.036 LN(L) + 0.022 LN(K/L) + 0.38 LNK/L**2

| Predictor | Coef | Stdev | t-ratio |
|---|---|---|---|
| Constant | 0.163 | 1.092 | 0.15 |
| LN(L) | -0.0363 | 0.2385 | -0.15 |
| LN(K/L) | 0.0221 | 0.1547 | 0.14 |
| LNK/L**2 | 0.378 | 1.215 | 0.31 |

s = 0.05318     R-sq = 0.9%     R-sq(adj) = 0.0%

Analysis of Variance

| SOURCE | | | |
|---|---|---|---|
| Regression | 3 | 0.000273 | 0.000091 |
| Error | 11 | 0.031114 | 0.002829 |
| Total | 14 | 0.031387 | |

Reprinted with permission of Minitab, Inc.

namely, that the restricted model is linear and can be estimated by OLS. It is this property that makes LM tests attractive, as recourse to nonlinear methods of estimation can be avoided and we only need access to a standard regression package such as MINITAB.® Thus, even if the general model is highly nonlinear (e.g., the CES production function), an LM test can often be constructed which involves only linear regression.

## 14.5 TESTS OF SPECIFICATION II: NONNESTED MODELS

In the preceding section we discussed tests between models in which the null hypothesis specifies a model which is a restricted version of that of the alternative hypothesis. Such tests become tests of parameter restrictions. In these situations, once a level of significance has been set, there is no ambiguity in the result. Either $H_0$ is rejected—implying acceptance of $H_1$—or $H_0$ is accepted.

We now turn our attention to testing between models where neither is a special case of the other. Such models are known as nonnested or separate models. Tests of separate models differ fundamentally from tests of nested models in that the outcome can be ambiguous. The procedure is that a model is specified as the null hypothesis only temporarily. Once $H_0$ has been tested against $H_1$, the models are reversed with the previous $H_1$ now becoming $H_0$, and vice versa. There are four possible outcomes to such a procedure:

(i) Both models are accepted.
(ii) Both models are rejected.
(iii) Model 1 is accepted and model 2 rejected.
(iv) Model 2 is accepted and model 1 rejected.

Only in (iii) and (iv) do we have an unambiguous statement about which is the preferred model. At this stage it is appropriate to consider briefly the relationship between nonnested tests and the model discrimination (or selection) methods discussed in Section 14.3. MacKinnon (1983) has said:

> It is customary in the econometric literature to treat model selection criteria and nonnested hypothesis tests as being very closely related, perhaps even as rival procedures. That is a very misleading point of view. Model selection criteria are appropriate when one wishes to choose one out of a group of rival models. The objective is to choose the "best" model, explicitly trading off goodness of fit and parsimony in parametrization. Either one assumes in advance that one of the models must be "true," or one simply does not care whether any of the models is satisfactory.
>
> Nonnested hypothesis tests, on the other hand, are tests of model specification, just like tests for serial correlation or omitted variables.

Because nonnested tests are hypothesis tests in the usual sense, probabilities can be assigned to the incorrect rejection of a null hypothesis. No such probabilities are available when using model discrimination criteria.

The basic principle underlying many nonnested tests is due to Cox (1962) and is an extension of the likelihood ratio test discussed in the preceding section. We again denote the maximum values of the log-likelihoods under $H_0$ and $H_1$ by $\tilde{L}_0$ and $\tilde{L}_1$, respectively. The Cox principle is to compare $\tilde{L}_0 - \tilde{L}_1$ with the value we would expect $\tilde{L}_0 - \tilde{L}_1$ to take if $H_0$ were true. Thus if $\tilde{L}_0 - \tilde{L}_1$ is close to what we would expect it to be if $H_0$ were true, we would accept $H_0$. Otherwise, $H_0$ would be rejected.

Two points should be noted at this stage.

(i) If $H_0$ is rejected, this does not imply that $H_1$ is accepted. Rejection of $H_0$ says nothing about the adequacy of $H_1$. This contrasts to the classical procedure followed when testing nested models.

(ii) If the models are in fact nested and not separate, then if $H_0$ is true, the expected (i.e., average) value of $\tilde{L}_0 - \tilde{L}_1$ must be zero. This can be inferred from Figure 14.4.2. The Cox principle then reduces to comparing $\tilde{L}_0 - \tilde{L}_1$ to zero, which is identical to comparing $\tilde{L}_0$ with $\tilde{L}_1$—the likelihood ratio principle. It is in this sense that the Cox principle can be regarded as extending the likelihood ratio principle to separate models.

The theory of testing separate models is a vast and growing area of modern econometrics. In this brief section we discuss just three tests that can be used, and our guiding principle is simplicity. The reader who wishes to pursue the subject further is referred to review articles by MacKinnon (1983), Mizon (1984), and McAleer (1987).

Before considering these tests, we will set up the framework for our discussion. We will assume, for simplicity, that the two competing models are linear and of the form

$$\begin{aligned} H_0: Y_i &= \beta_0 + \sum_{j=1}^{K_0} \beta_j X_{ji} + \sum_{j=1}^{K} \delta_j W_{ji} + u_{0i}; \\ H_1: Y_i &= \gamma_0 + \sum_{j=1}^{K_1} \gamma_j Z_{ji} + \sum_{j=1}^{K} \delta_j W_{ji} + u_{1i}. \end{aligned} \tag{14.5.1}$$

Thus the variables $X_1, \ldots, X_{K_0}$ and $Z_1, \ldots, Z_{K_1}$ are unique to $H_0$ and $H_1$, respectively. On the other hand, the variables $W_1, W_2, \ldots, W_K$ are common to both models. If the models have no variables in common, then W variables are simply omitted. The disturbances $u_{0i}$ and $u_{1i}$ are normal and have the usual properties with variances $\sigma_0^2$ and $\sigma_1^2$, respectively.

It can be shown that the Cox principle can be applied to specific problems by performing tests based on artificial regressions, and this is the approach we will take. The three tests we will consider are (a) the orthodox F-test; (b) the J-test (Davidson and MacKinnon, 1981), and (c) the JA-test (Fisher and McAleer, 1981).

Orthodox F-test

This test is not based on the Cox principle but on the ordinary F-test procedures with which we are familiar. An obvious approach to testing $H_0$ against $H_1$ is to form the comprehensive model

$$Y_i = \beta_0 + \sum_{j=1}^{K_0} \beta_j X_{ji} + \sum_{j=1}^{K_1} \gamma_j Z_{ji} + \sum_{j=1}^{K} \delta_j W_{ji} + u_i$$

and to test the hypothesis $\gamma_1 = \gamma_2 = \cdots = \gamma_{K_1} = 0$. This can be done by a standard F-test and is an example of nested hypothesis testing. The two competing hypotheses have been nested in the comprehensive model, which contains all the regressors from both models. Although this procedure is very straightforward, it has at least two problems:

(i) In practice there is often very high multicollinearity between the X and Z variables; and
(ii) If $K_1$ is quite large, that is, there are many Z variables, then the degrees of freedom will be reduced.

Both of these problems lead to a loss of power. The term power of a test is used by statisticians to describe the ability of a test to reject $H_0$ when it is false. If the probability of this happening is high, the test is said to be powerful. Clearly, we would always like our tests to be powerful. The problem with a test with low power is that $H_0$ will nearly always be accepted, whether it is true or false.

The J-test

This test comes about by applying the Cox principle in a slightly modified form. While the details of its derivation are far beyond the scope of this book, its application is very simple.

Let us denote by $\hat{Y}_i^{(1)}$ the prediction of $Y_i$ that is obtained by running the regression (14.5.1) appropriate to the alternative hypothesis $H_1$. Then the J-test is applied by computing the artificial regression

$$Y_i = \beta_0 + \sum_{j=1}^{K_0} \beta_j X_{ji} + \sum_{j=1}^{K} \delta_j W_{ji} + \alpha \hat{Y}_i^{(1)} + u_i \qquad (14.5.2)$$

and the significance of $\hat{\alpha}$ is tested in the usual way by a t-test. If $\hat{\alpha}$ is significantly different from zero, $H_0$ is rejected. Even though we are not explaining how the Cox principle translates into the artificial regression

(14.5.2), we can still grasp intuitively why it is a reasonable way to test $H_0$ against $H_1$. We have seen in Section 4.1 that $\alpha$, the coefficient of $\hat{Y}_i^{(1)}$, measures the marginal effect of $\hat{Y}^{(1)}$ on the mean of Y once the effects of $X_1, \ldots, X_{K_0}$ and $W_1, \ldots, W_K$ have been removed. Thus if $\hat{\alpha}$ is not significantly different from zero, $H_1$ has nothing to contribute over and above $H_0$. If, on the other hand, $\hat{\alpha}$ is significantly different from zero, then $H_1$ does have something to contribute apart from $H_0$ and we would reject $H_0$ as being alone an adequate description of how the values of Y are generated. Notice that such a rejection cannot imply that $H_1$ is an adequate description. This could only come about if, having reversed the roles of $H_0$ and $H_1$, the (new) $H_0$ is accepted.

The J-test procedure therefore requires two regressions and may be summarized as follows:

(i) Obtain the predictions of $Y_i$ from the alternative hypothesis $H_1$.
(ii) Add these predictions as another variable to the null hypothesis model.
(iii) Test the significance of the coefficient of the predictions variable and reject $H_0$ if it is significantly different from zero.

There are three additional points on the J-test that should be noted at this stage:

(i) This test is an asymptotic test and therefore may not be reliable in small samples. If a test is carried out at (say) the 5 percent level, there is a probability of 0.05 that the null hypothesis will be rejected when it is true. When an asymptotic test is applied to small samples, the probability of rejecting a true null may be considerably different from that implied by the nominal level of significance. A review of computer simulation experiments given by McAleer (1987) shows that when $N = 20$, the actual size of the J-test is often higher than the nominal size, showing that it rejects a true $H_0$ too often.
(ii) Because $\hat{Y}^{(1)}$ is a single variable, only one extra variable is being added to the model of $H_0$ to carry out the test. This contrasts to the F-test, which adds $K_1$ variables. If $K_1$ is much larger than 1, the J-test permits more degrees of freedom in the second estimation and consequently could be expected to result in a more powerful test. A number of studies have confirmed this [see, e.g., Pesaran (1982), Davidson and MacKinnon (1982)]. If $K_1 = 1$, the J-test and F-test are completely equivalent, implying that in this case the J-test is also valid for small samples.
(iii) We have discussed the test in the context of linear $H_0$ and $H_1$. If either (or both) hypotheses involve nonlinear models, the basic J-test procedure remains unchanged, except that estimation will have to be carried out using nonlinear regression. There are many computer packages (e.g., SHAZAM) available that have this facility.

## The JA-test

This test is very similar in concept to the J-test and can be regarded as a modified J-test which has an interesting property. The J-test is conducted by augmenting the null hypothesis model by one extra variable and testing the significance of the coefficient of this variable. The JA-test is the same in this respect, but differs in how the augmenting variable is constructed.

In the J-test, we recall, the augmenting variable $\hat{Y}^{(1)}$ was obtained by finding the predictions of Y obtained by the alternative hypothesis model. In the JA-test the first step is to find the predictions of Y from the null hypothesis, denoted by $\hat{Y}^{(0)}$. These predictions are then treated as the dependent variable for the alternative hypothesis model and the predictions of this variable $\hat{Y}^{(0)}$ from the alternative hypothesis are obtained. We will denote these predictions by $\hat{Y}^{(0,1)}$. This variable is now used to augment the null hypothesis model and the procedure is then identical to the J-test from then on. We may summarize the JA-test procedure as follows:

(i) Obtain the predictions $\hat{Y}_i^{(0)}$ of $Y_i$ from the null hypothesis $H_0$.

(ii) Obtain the predictions of $\hat{Y}_i^{(0)}$ from the alternative hypothesis $H_1$; these predictions we denote by $\hat{Y}_i^{(0,1)}$.

(iii) Augment the null hypothesis model by the single variable $\hat{Y}^{(0,1)}$, and test the significance of its coefficient. $H_0$ is rejected if the coefficient is significantly different from zero.

As with the J-test, if $K_1 = 1$, it is equivalent to the F-test and, by implication, to the J-test.

The JA-test requires one additional estimation to the J-test. The interesting property, due to a result of Milliken and Graybill (1970), is that a t-test of the coefficient of $\hat{Y}^{(0,1)}$ is an exact test. That is, such a test is valid for small samples and the size of the test will be correct. This contrasts with the J-test, which is only asymptotically valid and may have an actual size which is different (often larger) from the nominal size when the sample is small.

It may be thought that this property makes the JA-test far more useful than the J-test. However, a number of studies have shown that the JA-test may be less powerful than the J-test. For this reason it is advisable to use a range of tests when testing nonnested hypotheses in order to gain the advantage of the good points from the tests.

## Empirical Example

In Chapter 10 we have recorded a fairly extensive analysis of the consumption function, using the Australian data shown in Table 10.2.1. Models were estimated based on four well-known economic theories of consumption. In the terms of this chapter, these models are nonnested or separate. We will use the data of Table 10.2.1 to test betweeen two of these theories as an illustration of the techniques we have discussed. The theories are the Keynesian absolute income hypothesis, estimated in the form (10.2.2)

$$C_t = \beta_0 + \beta_1 Y_t + \beta_2 I_t + u_t,$$

and Duesenberry's relative income hypothesis in the empirical form (10.3.5) namely

$$C_t = \beta_0 + \beta_1 (Y^*_t + Z_{2t}) + \beta_2 Z_{1t} + \beta_4 I_t + u_t.$$

The variables $Y^*$, $Z_1$, $Z_2$, and I have been defined in Section 10.3. We will denote by $Y_2$ the variable $Y^* + Z_2$. Thus our model's representation of the relative income hypothesis is

$$C_t = \beta_0 + \beta_1 Y_{2t} + \beta_2 Z_{1t} + \beta_4 I_t + u_t. \qquad (14.5.3)$$

The necessary regressions to carry out the F-, J-, and JA-tests are summarized as follows:

(i) F-test

C on Y, I, $Y_2$, $Z_1$ (comprehensive model)
C on Y, I (absolute income hypothesis)
C on $Y_2$, $Z_1$, I (relative income hypothesis

In the latter two regressions the fitted or predicted values are saved as $\hat{C}^{(a)}$ and $\hat{C}^{(r)}$, respectively, for the J- and JA-tests to follow.

(ii) J-test

C on Y, I, $\hat{C}^{(r)}$ (absolute income hypothesis)
C on $Y_2$, $Z_1$, I, $\hat{C}^{(a)}$ (relative income hypothesis)

The significance of the coefficient of the predicted variables, that is, $\hat{C}^{(r)}$ or $\hat{C}^{(a)}$, is examined using the t ratio.

(iii) JA-test

$\hat{C}^{(a)}$ on $Y_2$, $Z_1$, I

The predicted values of $\hat{C}^{(a)}$ are stored as the variable $\hat{C}^{(a,r)}$.

C on Y, I, $\hat{C}^{(a,r)}$ (absolute income hypothesis)
$\hat{C}^{(r)}$ on Y, I

The predicted values are stored as $\hat{C}^{(r,a)}$.

C on $Y_2$, $Z_1$, $\hat{C}^{(r,a)}$ (relative income hypothesis)

In this case it is the significance of the coefficients of $\hat{C}^{(a,r)}$ and $\hat{C}^{(r,a)}$ that is of interest.

These regressions (all linear) were run using SHAZAM, which was chosen because of its better accuracy than MINITAB® in the presence of high multicollinearity. An appropriate SHAZAM program is shown below. Note that in this program the variables $\hat{C}^{(a)}$, $\hat{C}^{(r)}$, $\hat{C}^{(a,r)}$, and $\hat{C}^{(r,a)}$ are denoted by ABSPRED, RELPRED, JAABS, and JAREL, respectively.

```
FILE 33 CH11C.DTA
READ(33)C Y P Y1 Z1 Z2 I CLAG Y2
*
*    (A)   F-TEST
*
*    UNRESTRICTED (COMPREHENSIVE) MODEL
OLS C Y I  Y2 Z1/RSTAT ANOVA
*    RESTRICTED BY ABSOLUTE INCOME HYPOTHESIS
OLS C Y I /RSTAT ANOVA PREDICT=ABSPRED
*    RESTRICTED BY RELATIVE INCOME HYPOTHESIS
OLS C Y2 Z1 I /RSTAT ANOVA PREDICT=RELPRED
*
--------------------------------------------
*
*    (B)   J-TEST
*
*   HO IS ABSOLUTE INCOME HYPOTHESIS
OLS C Y I  RELPRED
*
*   HO IS RELATIVE INCOME HYPOTHESIS
OLS C Y2 Z1 I  ABSPRED
*
-------------------------------------------
*
*    (C)   JA-TEST
*
* HO IS ABSOLUTE INCOME HYPOTHESIS
OLS ABSPRED Y2 Z1 I /PREDICT=JAABS
OLS C Y I  JAABS
*
*  HO IS RELATIVE INCOME HYPOTHESIS
OLS RELPRED Y I/PREDICT=JAREL
OLS C Y2 Z1 I  JAREL
*
-------------------------------------------
*
STOP
```

The relevant results for the F-test were as follows:

Comprehensive model: $RSS(H_1) = 0.12949$, $N - K - 1 = 30$
Absolute income hypothesis: $RSS(H_0) = 0.29816$, number of restrictions $M = 2$
Relative income hypothesis: $RSS(H_0) = 0.13274$, number of restrictions $M = 1$

Table 14.5.1

| $H_0$ | Test F | J | JA |
|---|---|---|---|
| Absolute income hypothesis | 19.54[a] | 6.22[a] | 0.59 |
| Relative income hypothesis | 0.75 | -0.87 | -0.87 |

[a]Significant at the 1 percent level.

These results enable us to compute F-statistics in the usual way. A summary of the F-statistics, together with the t-statistics from the J- and JA-tests, is given in Table 14.5.1.

The results indicate a clear preference for the relative income hypothesis which is not rejected by any of the tests. On the other hand, the absolute income hypothesis is clearly rejected by the F- and J-tests. The nonsignificant result for the JA-test may possibly arise because of the lack of power of the JA-test, alluded to previously.

Another feature of the tests is illustrated by the second row of Table 14.5.1. The absolute income hypothesis model has only one variable, namely Y, which is not also in the relative income hypothesis model. Accordingly, the three tests are equivalent, and the J-test is valid for small samples. We note that for both J and JA the t-statistic is -0.87 and that $(-0.87)^2 = 0.76$, which (apart from rounding error) is the value of the F-statistic. In such cases, of course, there is no advantage in calculating more than one test.

As a matter of interest, the tests were performed again on the models (10.2.4) and (10.3.6). That is, the variable $C_{t-1}$ was added to each model. The results are shown in Table 14.5.2. Now none of the tests reject either model. This seems to confirm the important role $C_{t-1}$ plays in explaining movements in $C_t$, although the basis for its inclusion in terms of the two economic theories being tested is not at all clear.

Table 14.5.2

| $H_0$ | Test F | J | JA |
|---|---|---|---|
| Absolute income hypothesis | 0.83 | 0.48 | 1.21 |
| Relative income hypothesis | 1.42 | -1.19 | -1.19 |

## 14.6 TESTING EXOGENEITY

In our discussion of stochastic regressors (Chapter 11), we saw that when there is feedback (or mutual causality between $Y_t$ and some or all of the regressors) OLS estimators are not consistent. Such regressors are known as endogenous variables. On the other hand, if the causality is entirely from regressors to dependent variable $Y_t$, OLS estimators are unbiased, consistent, and asymptotically efficient. This is known as the case of exogenous regressors. Clearly, it would be advantageous to have some way of testing whether the regressors are exogenous. We will describe here a test of exogeneity due to Hausman (1978).

Let us suppose that our model (in deviation form) is

$$y_t = \beta_1 x_t + u_t. \tag{14.6.1}$$

Our hypotheses are

$H_0$: $x_t$ is exogenous;
$H_1$: $x_t$ is not exogenous.

That is,

$$H_0: \operatorname{plim} \frac{\Sigma xu}{N} = 0$$

$$H_1: \operatorname{plim} \frac{\Sigma xu}{N} \neq 0.$$

If $x_t$ is not exogenous, then $\beta_1$ is estimated consistently by the method of instrumental variables (IV) (Section 11.4). We will assume that a suitable instrumental variable $z_t$ $(= Z_t - \bar{Z})$ is known. That is, $\operatorname{plim}(\Sigma zu/N) = 0$. If we regress $x_t$ on $z_t$, we obtain

$$\begin{aligned} x_t &= \hat{\gamma} z_t + e_t \\ &= \hat{x}_t + e_t, \end{aligned} \tag{14.6.2}$$

where $\hat{x}_t = \hat{\gamma} z_t$. A crucial result here is that

$$\Sigma \hat{x}_t e_t = \hat{\gamma}\, \Sigma (Z_t - \bar{Z}) e_t = \hat{\gamma}\, \Sigma Z_t e_t - \hat{\gamma}\bar{Z}\, \Sigma e_t = 0$$

by the fundamental properties (2.3.11) and (2.3.12) of OLS residuals. Thus because $\Sigma \hat{x}_t e_t = 0$, the variables $\hat{x}$ and e are orthogonal. Furthermore,

$$\operatorname{plim} \frac{\Sigma \hat{x} u}{N} = \operatorname{plim} \hat{\gamma} \frac{\Sigma zu}{N} = 0, \tag{14.6.3}$$

so $\hat{x}_t$ is also an instrument for $x_t$.

Substituting (14.6.2) into (14.6.1), we have

$$y_t = \beta_1\hat{x}_t + \beta_1^* e_t + u_t, \tag{14.6.4}$$

where $\beta_1^* = \beta_1$. Thus regressing y on $\hat{x}$ and e will yield two estimates for $\beta_1$. Because the regressors are orthogonal, these two estimates are the same as we would obtain by regressing y on $\hat{x}$ and y on e separately.

Consider now the first estimator obtained by regressing y on $\hat{x}$. Here

$$\hat{\beta}_1 = \frac{\Sigma\hat{x}y}{\Sigma\hat{x}^2}.$$

Now

$$\Sigma\hat{x}^2 = \Sigma\hat{x}(x - e) = \Sigma\hat{x}x - \Sigma\hat{x}e$$

$$= \Sigma\hat{x}x \quad \text{as} \quad \Sigma\hat{x}e = 0.$$

Therefore,

$$\hat{\beta}_1 = \frac{\Sigma\hat{x}y}{\Sigma\hat{x}x}, \tag{14.6.5}$$

which is the IV estimator for $\beta_1$ and is therefore always consistent, regardless of whether $H_0$ is true or false.

Turning to the second estimator, obtained by regressing y on e, and denoted by $\hat{\beta}_1^*$, we have

$$\text{plim}\,\frac{\Sigma eu}{N} = \text{plim}\,\frac{\Sigma xu}{N} - \text{plim}\,\frac{\Sigma\hat{x}u}{N}$$

$$= \text{plim}\,\frac{\Sigma xu}{N} \quad \text{by (14.6.3)}.$$

Therefore, $\hat{\beta}_1^*$ is consistent only if $H_0$ is true. Thus, at least in large samples, we could expect $\hat{\beta}_1$ and $\hat{\beta}_1^*$ to be approximately equal if $H_0$ is true, but not otherwise. This becomes the basis of a test of exogeneity. We may rewrite (14.6.4) as follows:

$$y_t = \beta_1\hat{x}_t + (\beta_1 e_t - \beta_1 e_t) + \beta_1^* e_t + u_t$$

$$= \beta_1(\hat{x}_t + e_t) + (\beta_1^* - \beta_1)e_t + u_t.$$

That is,

$$y_t = \beta_1 x_t + \alpha e_t + u_t, \tag{14.6.6}$$

where

$$\alpha = \beta_1^* - \beta_1.$$

We may now use the familiar t- or F-test to test $\alpha = 0$. Rejection of $\alpha = 0$ would imply that x is endogenous. Although, as usual, we have developed the theory on the simple regression model, the results generalize to multiple regression.

Table 14.6.1

| t | Steel sales, Y | GNP, X | Housing construction, Z |
|---|---|---|---|
| 1 | 385 | 443.6 | 20.7 |
| 2 | 390 | 445.6 | 20.2 |
| 3 | 398 | 444.5 | 19.9 |
| 4 | 396 | 450.3 | 20.0 |
| 5 | 368 | 453.4 | 19.8 |
| 6 | 365 | 453.2 | 19.6 |
| 7 | 364 | 455.2 | 20.8 |
| 8 | 374 | 448.2 | 22.8 |
| 9 | 372 | 437.5 | 24.9 |
| 10 | 371 | 439.5 | 25.4 |
| 11 | 387 | 450.7 | 24.7 |
| 12 | 407 | 461.6 | 23.8 |
| 13 | 406 | 468.6 | 23.6 |
| 14 | 432 | 479.9 | 22.1 |
| 15 | 431 | 475.0 | 21.1 |
| 16 | 425 | 480.4 | 20.7 |
| 17 | 445 | 490.2 | 20.8 |
| 18 | 437 | 489.7 | 21.1 |
| 19 | 427 | 487.3 | 21.7 |
| 20 | 417 | 483.7 | 22.6 |
| 21 | 430 | 482.6 | 23.1 |
| 22 | 453 | 497.8 | 23.9 |
| 23 | 470 | 501.5 | 24.2 |
| 24 | 489 | 511.7 | 23.9 |

Source: Koutsoyiannis (1978).

The procedure can be summarized as follows:

(i) Regress each of the suspected endogenous variables on the instrumental variables $Z_1, Z_2, \ldots, Z_r$. Save the residuals.
(ii) Augment the original model with these residuals.
(iii) Use an F-test to test whether the coefficients of the residuals are jointly zero. If only one variable is suspected to be endogenous, the test on the residual coefficient is just the standard t-test.

EXAMPLE A very simple model for the demand for steel is

$$Y_t = \beta_0 + \beta_1 X_t + u_t,$$

where $Y_t$ represents steel sales and $X_t$ is GNP. However, it could be argued that while the level of GNP influences the demand for steel, GNP is itself affected by the demand for steel. If this is the case, GNP is endogenous and the OLS estimator will suffer from simultaneity bias. We will test the exogeneity of GNP.

$H_0$ : GNP(X) is exogenous;
$H_1$ : GNP is endogenous.

It has been suggested that a suitable instrumental variable is housing construction denoted by Z. The justification is that while it is clear that housing construction affects the demand for steel, it is very unlikely that the demand for steel causes housing construction. Data are shown in Table 14.6.1.

To carry out the test we first regress X on Z and save the residuals e. Then Y is regressed on X and e—this is the unrestricted regression. Finally, Y is regressed on X and this is the restricted regression. These regressions are shown in Display 14.6.1.

We have

$$RSS(H_0) = 3712,$$

$$RSS(H_1) = 3582.$$

Therefore,

$$F = \frac{(3712 - 3582)/1}{3582/(24 - 2 - 1)} = 0.762.$$

The 5 percent critical value for $F_{1,21}$ is 4.32. Therefore, $H_0$ is not rejected and X, that is, GNP, may be regarded as an exogenous variable.

Notice that because we are only testing one variable for exogeneity, we could more simply just look at the t-value associated with the coefficient of the residual in the unconstrained regression. From the Display 14.6.1

$$t = -0.87,$$

DISPLAY 14.6.1

```
MTB > REGR C2 1 C3;
SUBC> RESIDUALS C4.

The regression equation is
C2 = 426 + 1.91 C3

Predictor        Coef          Stdev      t-ratio
Constant       425.76          56.10         7.59
C3              1.907          2.525         0.76

s = 22.17        R-sq = 2.5%        R-sq(adj) = 0.0%

Analysis of Variance

SOURCE        DF          SS           MS
Regression     1       280.4        280.4
Error         22     10813.7        491.5
Total         23     11094.1


MTB > REGR C1 2 C2 C4

The regression equation is
C1 = - 591 + 2.14 C2 - 0.688 C4

Predictor        Coef          Stdev      t-ratio
Constant       -590.7          365.0        -1.62
C2             2.1382         0.7800         2.74
C4            -0.6877         0.7900        -0.87

s = 13.06        R-sq = 87.0%       R-sq(adj) = 85.8%

Analysis of Variance

SOURCE        DF          SS           MS
Regression     2       24035        12017
Error         21        3582          171
Total         23       27617

MTB > REGR C1 1 C2

The regression equation is
C1 = - 277 + 1.47 C2

Predictor        Coef          Stdev      t-ratio
Constant      -277.01          57.77        -4.79
C2             1.4679         0.1233        11.90

s = 12.99        R-sq = 86.6%       R-sq(adj) = 85.9%

Analysis of Variance

SOURCE        DF          SS           MS
Regression     1       23905        23905
Error         22        3712          169
Total         23       27617
```

which is not significant. Notice also that (apart from rounding error) $t^2 = F$ (see the end of Section 6.2).

## NOTES

1. Even though it is tautological to use the term RESET test, this is how it is usually referred to in the literature.

2. When dealing with the linear regression model, we have used K to denote the number of $\beta$ parameters, excluding the intercept. As we may be considering nonlinear models, we use K in a more general sense.

3. We use ESS/TSS for $R^2$ to avoid excessive rounding error. In this example, with $R^2 = 0.009$, $NR^2 = 15 \times 0.009 = 0.14$, which is substantially the same result.

# References


Aigner, D. J. (1971), Basic Econometrics, Prentice-Hall, Englewood Cliffs, N.J.

Aigner, D. J. (1974), MSE dominance of least squares with errors of observation, J. Econometrics 2(4), 365-72.

Almon, S. (1965), The distributed lag between capital appropriations and expenditures, Econometrica 33(1), 178-96.

Amemiya, T. (1980), Selection of regressors, Internat. Econom. Rev. 21(2), 331-54.

Amemiya, T. (1981), Qualitative response models: a survey, J. Econom. Lit. 19, 1483-1536.

Anderson, K. (1974), Distributed lags and barley acreage response analysis, Austral. J. Agric. Econom. 18(2), 119-32.

Anscombe, F. J. (1973), Graphs in statistical analysis, Amer. Statist. 27(1), 17-21.

Australian Bureau of Statistics (1981), Australian National Accounts: Concepts, Sources and Methods, Australian Government Publishing Service, Canberra, Australia.

Battese, G. E. (1977), Agricultural economists, response functions and lack-of-fit tests, Rev. Marketing Agric. Econom. 45(3), 85-94.

Bilson, J. F. O. (1980), The rational expectations approach to the consumption function: a multi-country study, European Econom. Rev. 13(2), 273-99.

Bladen-Hovell, R. C. and Richards, G. M. (1983), Inflation and Australian savings behaviour, 1959-1981, Austral. Econom. Papers 22(41), 290-301.

Box, G. E. P. and Cox, D. R. (1964), An analysis of transformations, J. Roy. Statist. Soc. Ser. B 26(2), 211-43.

Breusch, T. S. and Pagan, A. R. (1979), A simple test for heteroscedasticity and random coefficient variation, Econometrica 47(5), 1287-94.

Breusch, T. S. and Pagan, A. R. (1980), The Lagrange multiplier test and its applications to model specification in econometrics, Rev. Econom. Stud. 47, 239-53.

Buse, A. (1982), The likelihood ratio, Wald and Lagrange multiplier tests: an expository note, Amer. Statist. 36(4), 153-57.

Cagan, P. (1956), The monetary dynamics of hyperinflation, in Friedman, M. (ed.), Studies in the Quantity Theory of Money, University of Chicago Press, Chicago.

Chow, G. C. (1960), Tests of equality between sets of coefficients in two linear regressions, Econometrica 28(3), 591-605.

Cox, D. R. (1962), Further results on tests of separate families of hypotheses, J. Roy. Statist. Soc. Ser. B 24(2), 406-24.

Davidson, J. E. H. and Hendry, D. F. (1981), Interpreting econometric evidence: the behaviour of consumers' expenditure in the U.K., European Econom. Rev. 16(1), 177-92.

Davidson, J. E. H., Hendry, D. F., Srba, F. and Yeo, S. (1978), Econometric modelling of the aggregate time-series relationship between consumers' expenditure and income in the United Kingdom, Econom. J. 88(4), 661-92.

Davidson, R. and MacKinnon, J. G. (1981), Several tests for model specification in the presence of alternative hypotheses, Econometrica 49(3), 781-93.

Davidson, R. and MacKinnon, J. G. (1982), Some non-nested hypothesis tests and the relations among them, Rev. Econom. Stud. 49, 551-65.

Davy, G. M. (1975), A consumption function with money illusion: the Australian experience, Econom. Anal. Policy 6(1), 23-41.

Deaton, A. and Muellbauer, J. (1980), Economics and Consumer Behaviour, Cambridge University Press, New York.

Dhrymes, P. J. (1971), Distributed Lags: Problems of Estimation and Formulation, Holden-Day, San Francisco.

Doran, H. E. (1981), Omission of an observation from a regression analysis: a discussion on efficiency loss with applications, J. Econometrics 16(3), 367-74.

Doran, H. E. and Griffiths, W. E. (1978), Inconsistency of the OLS estimator of the partial adjustment—adaptive expectations model, J. Econometrics 7(2), 133-46.

Duesenberry, J. S. (1949), Income Saving and the Theory of Consumer Behaviour, Harvard University Press, Cambridge, Mass.

Durbin, J. (1960), Estimation of parameters in time-series regression models, J. Roy. Statist. Soc. Ser. B 22(1), 139-53.

Durbin, J. (1970), Testing for serial correlation in least-squares regression when some of the regressors are lagged dependent variables, Econometrica 38(3), 410-21.

Durbin, J. and Watson, G. S. (1950), Testing for serial correlation in least-squares regression, Biometrika 37(4), 409-28.

Durbin, J. and Watson, G. S. (1951), Testing for serial correlation in least-squares regression II, Biometrika 38(2), 159-78.

Edwards, J. B. (1969), The relation between the F-test and $\bar{R}^2$, Amer. Statist. 23(5), 28.

Ferber, R. (1973), Consumer economics, a survey, J. Econom. Lit. 11(4), 1303-42.
Fisher, I. (1937), Note of a short-cut method for calculating distributed lags, Bull. Inst. Internat. Statist. 29, 323-28.
Fisher, G. and McAleer, M. (1981), Alternative procedures and associated tests of significance for non-nested hypotheses, J. Econometrics 16(1), 103-19.
Fox, K. A. (1968), Intermediate Economic Statistics, Wiley, New York.
Freebairn, J. W. (1977), Inflation and stability of the household consumption-savings function, Econom. Record 53(142), 198-218.
Friedman, M. (1957), A Theory of the Consumption Function, Princeton University Press, Princeton, N.J.
Goldberger, A. S. (1964), Econometric Theory, Wiley, New York.
Goldfeld, S. M. and Quandt, R. E. (1965), Some tests for homoscedasticity, J. Amer. Statist. Assoc. 60(310), 539-47.
Granger, C. W. J. (1966), The typical spectral shape of an economic variable, Econometrica 34(1), 150-61.
Griliches, Z. and Rao, P. (1969), Small sample properties of several two-stage regression methods in the context of autocorrelated errors, J. Amer. Statist. Assoc. 64(325), 253-72.
Gujarati, D. (1972), The behaviour of unemployment and unfilled vacancies: Great Britain, 1958-1971, Econom. J. 82(1), 195-202.
Gujarati, D. (1978), Basic Econometrics, McGraw-Hill, New York.
Haitovsky, Y. (1969), A note on the maximisation of $\bar{R}^2$, Amer. Statist. 23(1), 20-21.
Haitovsky, Y. (1973), Regression Estimation from Grouped Observations, Hafner Press, New York.
Harvey, A. C. (1983), The Econometric Analysis of Time Series, Philip Allan, Deddington, Oxford, England.
Hausman, J. A. (1978), Specification tests in econometrics, Econometrica 46(6), 1251-72.
Hendry, D. F. and Ungern-Sternberg, T. Von (1981), Liquidity and inflation effects on consumers' expenditure, in Deaton, A. S. (ed.), Essays in the Theory and Measurement of Consumer Behaviour, Cambridge University Press, Cambridge.
Hoerl, A. E. and Kennard, R. W. (1970), Ridge regression: biased estimation of non-orthogonal problems, Technometrics 12(1), 55-67.
Johnston, J. (1972), Econometric Methods, 2nd ed., McGraw-Hill, New York.
Judge, G. G., Hill, R. C., Griffiths, W. E., Lütkepohl, H. and Lee, T.-C. (1982), Introduction to the Theory and Practice of Econometrics, Wiley, New York.
Judge, G. G., Griffiths, W. E., Hill, R. C., Lütkepohl, H. and Lee, T.-C. (1985), The Theory and Practice of Econometrics, 2nd ed., Wiley, New York.

Kelejian, H. H. and Oates, W. E. (1981), Introduction to Econometrics: Principles and Applications, 2nd ed., Harper & Row, New York.
Kelly, G. J. (1986), An examination of voluntary segment profit disclosures by major Australian companies, paper presented to Accounting Association of Australia and New Zealand Annual Conference, Auckland.
Keynes, J. M. (1936), The General Theory of Employment Interest and Money, Macmillan, London.
Kmenta, J. (1971), Elements of Econometrics, Macmillan, New York.
Konstas, P. and Khouja, M. (1969), The Keynesian demand for money function, J. Money Credit Banking 1, 765-77.
Koutsoyiannis, A. (1977), Theory of Econometrics, 2nd ed., Macmillan, London.
Koyck, L. M. (1954), Distributed Lags and Investment Analysis, North-Holland, Amsterdam.
MacKinnon, J. G. (1983), Model specification tests against non-nested alternatives, Econometric Rev. 2(1), 85-110.
Maddala, G. S. (1977), Econometrics, McGraw-Hill, New York.
Maeshiro, A. (1979), On the retention of the first observation in serial correlation adjustment of regression models, Internat. Econom. Rev. 20(1), 149-57.
McAleer, M. (1981), A small sample test for non-nested regression models, Econom. Lett. 7, 335-38.
McAleer, M. (1987), Specification tests for separate models: a survey. In King, M. L. and Giles, D. E. A. (eds.), Specification Analysis in the Linear Model, Routledge & Kegan Paul, London.
McCallum, B. T. (1972), Relative asymptotic bias from errors of omission and measurement, Econometrica 40(4), 757-58.
Milliken, G. A. and Graybill, F. A. (1970), Extensions of the general linear hypothesis model, J. Amer. Statist. Assoc. 65(2), 797-807.
Mizon, G. F. (1984), The encompassing approach in econometrics. In Hendry, D. F. and Wallis, K. F. (eds.), Econometrics and Quantitative Economics, Blackwell Scientific, Oxford.
Modigliani, F. and Ando, A. (1957), Tests of the life cycle hypothesis of savings: comments and suggestions, Bull. Oxford Univ. Inst. Statist. 19(2), 99-124.
Mood, A. M., Graybill, F. A. and Boes, D. C. (1974), Introduction to the Theory of Statistics, 3rd ed., McGraw-Hill, New York.
Nerlove, M. (1958a), Adaptive expectations and the cobweb phenomena, Quart. J. Econom. 73(2), 227-40.
Nerlove, M. (1958b), The Dynamics of Supply: Estimation of Farmers' Response to Price, Johns Hopkins University Press, Baltimore.
Pagan, A. R. (1984), Model evaluation by variable addition, in Hendry, D. F. and Wallis, K. F. (eds.), Econometrics and Quantitative Economics, Blackwell Scientific, Oxford.

Pagan, A. R. and Nicholls, D. F. (1984), Estimating predictions, prediction errors and their standard deviations using constructed variables, J. Econometrics 24(3), 293-310.

Pesaran, M. H. (1982), Comparison of local power of alternative tests of non-nested regression models, Econometrica 50(5), 1287-1305.

Pierce, D. A. (1977), Relationships—and the lack thereof—between economic time series, with special reference to money and interest rates, J. Amer. Statist. Assoc. 72(357), 11-22.

Pindyck, R. S. and Rubinfeld, D. S. (1981), Econometric Models and Economic Forecasts, 2nd ed., McGraw-Hill, New York.

Ramsey, J. B. (1969), Tests for specification errors in classical linear least-squares regression analysis, J. Roy. Statist. Soc. Ser. B 31(2), 350-71.

Ramsey, J. B. and Schmidt, P. (1976), Some further results in the use of OLS and BLUS residuals in specification error tests, J. Amer. Statist. Assoc. 67(1), 180-86.

Rao, P. and Miller, R. L. (1971), Applied Econometrics, Wadsworth, Belmont, Calif.

Rossi, N. and Schiantarelli, F. (1982), Modelling consumers' expenditure: Italy 1965-1977, European Econom. Rev. 17(2), 371-91.

Ryan, T. A., Joiner, B. L. and Ryan, B. F. (1976), Minitab Student Handbook, Duxbury Press, North Scituate, Mass.

Salkever, D. S. (1976), The use of dummy variables to compute predictions, prediction errors, and confidence intervals, J. Econometrics 4(4), 393-97.

Schmidt, P. (1974), A modification of the Almon distributed lag, J. Amer. Statist. Assoc., 69(347), 679-81.

Schmidt, P. and Waud, R. N. (1973), The Almon lag technique and the monetary versus fiscal policy debate, J. Amer. Statist. Assoc. 68(343), 11-19.

Severn, A. K. (1983), Formation of inflation expectations in the U.K.: pitfalls in the use of the error-learning model, European Econom. Rev. 20(2), 349-63.

Smithies, A. (1945), Forecasting postwar demand: I, Econometrica 13(1), 1-14.

Theil, H. (1971), Principles of Econometrics, North-Holland, Amsterdam.

Whan, I. F. (1972), The economic interpretation of grain sorghum response to fertilizer in the Ord River Valley of Western Australia, M.Ec. dissertation, University of New England, Armidale, Australia.

Wickens, M. R. (1972), A note on the use of proxy variables, Econometrica 40(4), 759-61.

Zellner, A., Huang, D. S. and Chau, L. C. (1965), Further analysis of the short run consumption function with emphasis on the role of liquid assets, Econometrica 33(3), 570-81.

# Index